Modern Combat Pistols

Modern Combat Pistols

Maxim Popenker
Edited by Anthony G. Williams

The Crowood Press

First published in 2007 by
The Crowood Press Ltd
Ramsbury, Marlborough
Wiltshire SN8 2HR

www.crowood.com

British Library Cataloguing-in-Publication Data
A catalogue record for this book is available from the British Library.

ISBN-10 186126 894 7
ISBN-13 978 1 86126 894 5

Designed and typeset by Focus Publishing, Sevenoaks, Kent
Printed and bound in Great Britain by The Cromwell Press, Trowbridge

Contents

Acknowledgements

Special thanks
To my wife Elena, for her love and understanding.
To my father, for teaching me about guns.

Images from following companies and persons were used in this book:
Beretta, Italy (www.beretta.com)
Cherry's Fine Guns, USA (www.cherrys.com)
Ceská Zbrojovka – Uhersky Brod (www.czub.cz)
FN Herstal, Belgium (www.fn-herstal.com)
Heckler und Koch, Germany (www.Heckler-Koch.de)
KBI Inc./Charles Daly (www.charlesdaly.com)
Ruger Firearms, USA (www.ruger.com)
Smith & Wesson, USA (www.Smith-Wesson.com)
Sphinx Systems, Switzerland (www.sphinx-systems.ch)
Springfield Armory, USA (www.springfield-armory.com)
Tanfoglio Fratelli, Italy (www.tanfoglio.it)
Taurus USA, USA (www.taurus-usa.com)

Platoon magazine, South Korea

Anthony Williams (www.quarry.nildram.co.uk)
Oleg Volk (www.olegvolk.net)
Bradley W. Taylor (www.berettacollection.com)
Ken Lunde (www.praxagora.com/lunde/

Also, I would thank following people for helping me with this book:
Alexander 'Doc©'
Alejandro Albertengo (Argentina)
Alexandre Beraldi (Brazil)
Anthony Williams (UK)
Pierangelo Tendas (Italy)
Yan Brazhnikov (Israel)
Mateusz Kurmanow (Poland)
Leszek Erenfeicht (Poland)
Oleg Volk (USA)
Steve Pinner (USA)
Dick Frazer (USA)
Lionel Helias (France)
Romeo Streigner (Croatia)
Miroslav Novak (Czech republic)
Bojan Kavedzic (Serbia)
Vladimir Nenadovic (Serbia)
Yrij Danilov (Russia)
Denis Salakhov (Russia)

Part I

INTRODUCTION

The American 9mm Goncz GA-9, which was included in the infamous 'Assault weapons ban', was a large and well-made weapon, but of little practical value for either government or civilian users.

The AA Arms AP9 weapon represented the other end of the 'Assault pistols' line; based on the infamous Intratec TEC-DC9, this was a rather large, crude and only marginally reliable gun, hardly suitable for anything more serious than shooting at empty beer cans.

Chapter One

Modern Combat Pistols: an Introduction

What is a Pistol?

To start with, we shall define the subject of this book. A pistol or, more accurately, a 'semi-automatic pistol' is a hand-held firearm, intended for personal use. It is employed using one or two hands, typically without any other support (such as a shoulder rest, bipod or mount). 'Semi-automatic' means that the pistol utilizes part of the energy, generated by each shot, to complete the reloading cycle (extract and eject the spent cartridge case, load a fresh cartridge and cock the action). Most handguns fire in semi-automatic mode only, that is, they fire only one shot for each trigger pull. However, a few specialized pistols can also fire in full automatic or burst mode, firing for as long as the trigger is being pressed and ammunition is present in magazine, or for a pre-determined number of rounds (usually two or three). Finally, 'combat' in this book means that the pistols being discussed have been developed to be used as service weapons, primarily or secondarily for military, law-enforcement and security applications. Of course, most of these pistols (except for a few specialized weapons) could be, and in many cases actually are or were, used for other purposes, such as sport and recreational shooting and for civilian self-defence. However, to narrow the scope, this book will not discuss weapons which are not intended primarily as service weapons, such as small-bore target pistols. Of the many pocket-sized defensive pistols, we will also discuss only those that were originally intended for use by police or military personnel.

Offensive versus Defensive Handguns

Like almost any other weapon, handguns can be used in both offensive and defensive roles. The key difference between offensive and defensive scenarios is in the primary objectives achieved by using the weapons. In most defensive scenarios, the key objective is to defend one's own life, or the lives of other persons. The fate of the adversary is generally irrelevant, as long as the primary objective is achieved. In other words, in the defensive scenarios our goal is to not allow the adversary to achieve *his* objectives. If we can succeed in this without actually shooting (and killing) the adversary (either scaring him off with the presence of the weapon, or by fired but missed shots), we are still successful.

In offensive scenarios, which are peculiar to military special operations or some law-enforcement and counter-terror operations, the key objective is to disable the adversary completely as quickly as possible. If the adversary is not disabled immediately (even if he dies seconds after being shot), he may defeat the entire task of the offensive team, by calling an alarm (in military operations behind enemy lines), or by shooting a hostage or setting off a bomb (in counter-terror operations). So far, the only effective and fast way to disable an adversary is to kill him with one or two well-placed shots, which are thus mandatory for offensive operations. In defensive operations, the shooter may or may not have a chance to fire accurately, and, especially in some military situations where there are no friendly personnel or innocent bystanders around (to avoid collateral casualties), even unaimed fire in the general direction of the enemy may save the day.

Usually, the handgun is considered as a purely defensive weapon. However, in the early stages of handgun history, some of them were regarded as offensive weapons, mostly by cavalry troops and naval boarding parties. Several early semi-automatic pistols included such specific features, requested by the cavalry, as single-hand cocking or enhanced safety when handling the loaded firearm on horseback. One of the last documented military campaigns to see significant use of pistols as offensive weapons was the 1916 raid of US cavalry forces in Mexico, against

Pancho Villa's guerillas. Semi-automatic pistols, along with revolvers, also saw limited use in offensive, as well as defensive, actions in the trenches of the First World War, where their compact size and manoeuvrability was at a premium. However, by the end of the Great War the introduction of the much more effective sub-machine guns relegated pistols to a purely defensive status for many decades.

Still, there are some properties specific to pistols which are essential for some modern offensive situations. These are compact size, high manoeuvrability even in the most cramped areas, and ease of silencing (fitting a sound-moderating device). Such pistols are very usable by various Special Forces, which are often required to operate at short ranges and with as much stealth as possible. The classic example of such an offensive handgun is the American Mk23 mod.0 pistol, developed for US Special Operations Command by the famous Heckler & Koch Company of Germany.

We can summarize some key attributes of offensive and defensive weapons in one table (see below) and compare their relative importance for each class.

Let's look at these attributes more closely.

High reliability is obviously a mandatory requirement for either class of weapon.

Minimum weight is important for a defensive weapon, as a heavy weight penalty will make carrying the weapon uncomfortable, especially if the gun is carried every day for long hours in a purely defensive capacity. A defensive weapon must not interfere with the primary functions of its owner. On the other hand, offensive pistols are usually carried only during specific missions and they are primary equipment for operators, be they Army recon (reconnaissance) or police SWAT (Special Weapons and Tactics) operators. Increased weight also provides the benefits of lesser recoil and faster second shot recovery, both essential for accurate combat shooting.

High first shot readiness is essential for a defensive handgun, as almost all defensive scenarios start very suddenly for the defending party; even a small fraction of a second required to make the gun operational may cross the thin line between life and death. Offensive scenarios are almost always planned in advance and each operator usually has enough time to get his weapon to ready status before the hostilities begin.

High accuracy is essential to an offensive weapon – a missed first shot at the enemy sentry or at the terrorist holding a hostage at gunpoint may cause the entire operation to fail. In some defensive operations the 'spray and pray' technique is quite justifiable and, in some instances, the only course of action possible. Of course, each and every shooter must use aimed fire instead of 'spray' fire as much as possible, but this is not essential for success in at least some defensive situations.

Minimum handling requirements are more important for defensive weapons, as excessive routine maintenance requirements will impair the primary functions of the owner. Complicated maintenance procedures also require more training and the chances are that the users will avoid the complications to save time and effort, which will eventually result in degraded reliability of the weapon. Offensive weapons are usually primary tools for their operators, so sufficient training will not be an issue, nor will the necessary time for routine periodic maintenance of the weapons.

The same logic applies to the **maximum ease of operation** criterion. The simpler the operation of the gun, the less likely are various operator errors both in

Key attributes of offensive and defensive weapons

Attribute	Offensive weapon	Defensive weapon
Reliability	Important	Important
Minimum weight	Less important	Important
High first shot readiness	Less important	Important
High accuracy	Important	Less important
Minimum handling requirements	Less important	Important
Maximum ease of operation	Less important	Important

emergency defensive situations and during necessary routine maintenance, especially considering how seldom most users actually handle their defensive pistols outside of the holster.

Finally, there is some confusion regarding so-called 'assault' pistols. From the logical point of view, any pistol which is used for offensive actions can be called a 'weapon of assault'. But, mostly due to the 'Assault Weapons Ban' enforced in the USA between 1994 and 2004, the term 'assault pistol' was coined to describe the large, semi-automatic only, clones and copies of sub-machine guns, or 'sub-machine gun wannabees'. These weapons, originally listed in the ban by their names, in most cases represented crude and simple designs with 'bad' (from politicians' point of view) military looks. The most famous of these weapons were guns like the Intratec TEC-9 (later TEC-DC9) and its variations, Goncz GA-9 and its Claridge copies, various semi-automatic-only clones of the Ingram MAC-10 and UZI sub-machine guns, and others. These weapons are of little use for any 'service' users, be they in law enforcement or the military, as they can use standard sub-machine guns of the same size but with much more firepower. For civilian users, these were mostly 'fun' weapons, as most of these 'assault pistols' were far too bulky and heavy for concealed carry and were often too unreliable for someone to stake their life on them.

Military and Police Pistols – What's the Difference?

Regardless of the fact that many pistols are used for both police and military applications (weapons like the Beretta 92F or Glock 17 usually come to mind first), some of the essential requirements for pistols are quite different between military and law enforcement.

When considering the defensive role of pistols, the key requirements of high reliability, low weight, ease of handling and maintenance are the same for both military and police guns. However, the tactical situations in which these weapons are most often used are quite different. A typical police officer rarely spends more than one shift out of his 'base', each day returning to the warmth of his station, with its armourers and cleaning facilities. The typical soldier, such as an artillery crewman or tank driver, could spend many weeks in the middle of the desert, or in cold winter mountains, or in tropical forests. Military pistols are much more likely to see weeks of rain or days of sand storms without any cleaning or other maintenance.

On the other hand, the hostilities which a typical LEO (Law-Enforcement Officer) may expect are much wider in their nature than typical military ones. Many LEOs worldwide are killed each year by their own pistols, taken from their holsters by surprise. For the typical police patrolman it is almost impossible to distinguish a potential threat before it becomes actual, as criminals almost never wear uniforms, nor do they carry their weapons in open sight. Because of this, police weapons ideally need to be inoperable, at least immediately, if withdrawn from the holster by 'the other guy'. The simplest way is to use one or more less-than-obvious manual safeties, although this approach directly conflicts with the requirement for ease of operation. For many years various manufacturers and designers have tried to achieve the 'approved user only' effect by using magnetic rings, key locks and, recently, electronic bio-authentification devices (which read and discern fingerprints, for example). The problem is that none of these devices has so far achieved the level of reliability necessary for use in a modern combat handgun.

Another aspect, peculiar to police usage of pistols (and other weapons), is that during most police shootings there are not only police officers and criminals, but also innocent bystanders, who must not be endangered. This raises the issue of bullet over-penetration – bullets must not penetrate the body of the target and must not produce dangerous ricochets, in the interests of those who happen to be standing behind or to the side of the criminal. These requirements strongly encourage the use of various expanding (hollowpoint or softpoint) bullets in police weapons. Military operators are prohibited from using such expanding projectiles by the 1899 International Peace Conference at The Hague. Also, typical adversaries of the military these days often wear body armour and helmets, and in these situations the enhanced penetration of some types of bullets, completely unacceptable by police standards, is very desirable.

Special Purpose Handguns – Silenced

From the earliest stages of human history, several aspects of warfare have called for stealth operations – stealthy movements and the stealthy disposal of enemies such as sentries, lone soldiers and officers. For many centuries the 'cold arms' (knives) were weapons of choice for such operations, even after the advent of firearms. But by the end of the nineteenth century several designers developed 'sound moderators', generally known as silencers – devices which significantly decrease the sound volume of a gunshot. Among the first successful designs was one by Sir Hiram Maxim, the designer of the first practical machine gun. At first, silencers were advertised for hunting, allowing hunters to fire multiple shots at the game without disturbing it and scaring it off. But very soon the virtues of sound moderators were discovered by criminal elements, so in many countries the civilian ownership of silenced firearms was prohibited or severely limited.

Military elements began to use silenced weapons more or less extensively during World War II. Soviet 'special purpose groups' of the NKVD, which operated behind German lines, used Nagant gas-sealed revolvers fitted with 'Bramit' silencers; German ABWEHR (military intelligence) and Gestapo (secret police) operatives used various 7.65mm (.32) and 9mm pistols fitted with detachable silencers; British commandos used Welrod pistols, one of the first purpose-designed integrally silenced weapons; American OSS (Office of Strategic Services, the predecessor of the CIA) operatives used integrally silenced .22 calibre High Standard pistols. In most cases these weapons were used behind enemy lines, to take out sentries and guard dogs, or for the assassination of important people, without drawing unnecessary attention to the shooters.

In the post-war period silenced weapons remained in use by various military special operations units around the world. Several law-enforcement organizations also used silenced weapons, which can be useful in many operations against terrorist organizations or drug dealers and smugglers. The goal of these weapons is the same – to take out guards and suppress any other unexpected resistance without alarming other 'bad guys' in the vicinity.

In most cases, silenced weapons are the same as service weapons (pistols), but fitted with extended barrels which have some means by which to mount the silencers securely (such as threads or lugs). Sometimes such weapons are fitted with slide locks which, when engaged, block the slide, turning the pistols into single-shot, manually reloaded weapons. This is useful when extreme stealth must be maintained, as the sound of the steel slide slamming backward and forward is much louder than the sound of the gunfire through the silencer. Only two countries, China and the former Soviet Union, have taken the trouble to develop and mass-produce purpose-designed semi-automatic, integrally silenced pistols. China developed the 7.62 × 17mm Type 64 and improved Type 67 pistols.

The USSR developed the 9 × 18 PB pistol and, later on, several unique silent pistols which do not require any type of silencer to avoid the sound of the gunshot. These pistols use specially designed integrally silenced ammunition, which locks the expanding powder cases (a major source of the gun sound) within the strengthened cartridge case, behind a special piston. It must be noted that a similar type of ammunition was developed in the USA during the Vietnam War, to be used in 'tunnel revolvers', based on the S&W Model 29 .44 calibre revolvers. Unlike Soviet designs, American 'tunnel ammo' used buckshot, loaded into a container, instead of a single bullet. A very limited number of the tunnel revolvers was made for US Special forces, but this concept has not been pursued in the USA since then. A similar idea is also used in the German HK P11 underwater pistol, but that covers all such firearms known to the general public, which are silent yet do not require a special silencer. The design and basic principles of silencers are described in Appendix Two.

Special Purpose Handguns – Underwater

The key problem with underwater shooting is that the density of water is about 500 times greater than the density of air. This results in greatly increased drag on the bullet, so a typical pistol bullet can travel only a few metres in water before falling to the bottom harmlessly. To decrease the drag, several countries have developed long, needle-like projectiles which are fired from purpose-designed underwater weapons, a few of which are described below in the text. These weapons are issued to various special naval warfare units like US Navy SEALs, or Soviet/Russian Navy PDSS (*Protivo-Diversionnye Sily i Sredtsva Flota* – the Soviet Navy's anti-diversion forces) and others. They are also used by frogmen (combat divers) against other frogmen and various less dangerous creatures which may be encountered under the sea, such as smaller sharks or other similar species.

However, several pistol manufacturers widely advertise their standard 'above water' pistols as capable of firing when submerged. From what is said above, it may be of very little value for most users, but the point of this advertising is that these pistols are actually capable of firing with the barrel filled with water without any subsequent damage to the gun. And once you remember that rain will not care if the police officer watches for danger with his gun muzzle up, and the river crossing is a routine thing for most infantry

ABOVE: US-made High Standard HD Military .22 calibre pistol was used by OSS (the predecessor of the CIA) operatives during WWII and also saw some action behind enemy lines in South Asia.

RIGHT: Czechoslovak Vz.27 pistol, chambered for the 9 × 17mm Browning Short (.380 ACP) cartridge, fitted with an extended barrel and a detachable silencer. German agents used similar pistols during the Second World War.

and special operations units, you will see that it may be very reassuring for most users to know that their sidearm will not explode or otherwise fail in the moment of danger because of water in the barrel.

Special Purpose Handguns – Automatic

The first attempts to make pistols fire automatically (for as long as the trigger is pulled and the magazine is not empty) were made during the First World War, with the intention of providing troops with compact 'trench sweeper' weapons. However, these attempts were hardly worth the effort, because the low mass of the recoiling parts resulted in an excessive rate of fire, which, combined with the small magazine capacity plus increased wear and rapid barrel overheating, resulted in little success. The advent of purpose-designed sub-machine guns, relatively small carbine-like weapons which fired pistol-calibre ammunition from large capacity magazines, made the prospects of the fully automatic pistols even less rosy.

Nevertheless, during the 1930s several European companies produced small numbers of selective-fire pistols, often equipped with detachable shoulder stocks/holsters. The most famous of these weapons were the *Schnellfeuer* (rapid-firing) versions of the Mauser C96 pistols, commercially designated the Model 712. These pistols featured detachable ten- or twenty-round magazines and shoulder stocks/holsters and were fitted with fire-mode selectors at the left side of the frame. These pistols were significantly smaller

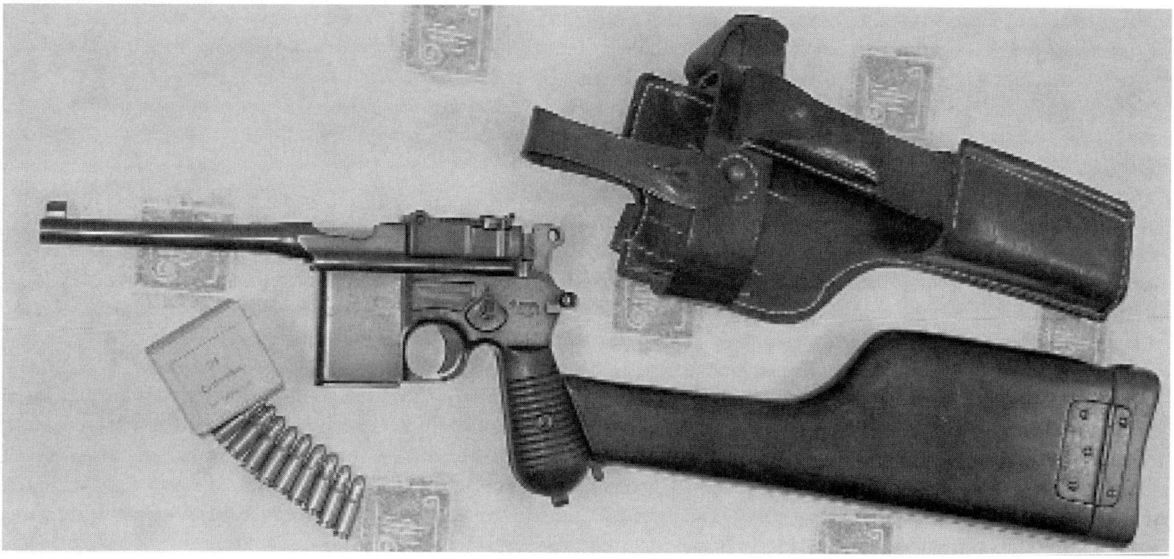

Mauser C96 Model 712 Schnellfeuer; the fire-mode selector is visible on the left side of the frame.

than contemporary sub-machine guns, but were not capable of providing similar firepower because of their lighter barrels and higher rate of fire, both resulting in rapid overheating.

Astra produced similar weapons in Spain, as the Model 904 and Model F, the latter being fitted with a rate-of-fire reduction device, to reduce overheating and to make automatic fire more controllable. Astra Modelo F pistols have seen some use with the Spanish *Guarda Civil* (Civil Guard) law enforcement organization. Both Mauser and Astra pistols were also exported to China during the 1930s, where they were quite popular for urban warfare.

Glock 18 automatic pistol.

After the Second World War a few more selective-fire pistols appeared on the scene. The first was the Soviet Stechkin APS, a relatively large pistol with a twenty-round magazine and a detachable shoulder stock. It fired a moderate-power 9 × 18mm PM cartridge. Originally intended as a self-defence weapon for front-line officers and vehicle crews, this pistol turned out to be very expensive to make and cumbersome to carry and was soon shelved. Beretta of Italy produced two 9mm designs, first the Model 951A and a couple of decades later the Model 93R. These pistols were intended mostly for use by *Carabineri* and other special purpose law enforcement and VIP protection units, whose officers would require extremely compact yet effective sidearms. The key idea was to improve the effectiveness (and lethality) of pistols by firing quick bursts instead of single shots, so as to achieve multiple hits on a single target in the shortest period of time. These pistols, like the later Glock Model 18, could not provide suppressive burst fire because of their rapid overheating and small magazine capacity. So, with the development of compact sub-machine guns such as The Beretta 12S, Heckler-Koch MP5k, Mini-Uzi or Steyr MPi69, the fully automatic pistols gradually became obsolete. Today, only one major manufacturer worldwide offers a fully automatic pistol in its line of products, the Austrian Glock Company.

Chapter Two

Overview of the Development of Combat Pistols Before 1945

If we look at the history of semi-automatic pistol development throughout the world, we will see that most activity in this field prior to 1945 occurred in Europe. This is no surprise, as the development of pistols, as well as any other military small arms, has always been fuelled by the prospect of war (both in European countries and in their colonies), and by the experience gained through these wars. Probably the first war to see any use of semi-automatic pistols was the Boer War in South Africa. Boers, buying whatever arms they could, made some use of Mauser C96 pistols. While opposing British forces were mostly armed with Webley & Scott revolvers, we know at least one famous account of a British man using a semi-automatic. This was no less than Winston Churchill, who at the time was armed with a Mauser C96 pistol, bought with his own money.

Not surprisingly, one of the major centres for small arms development was Germany, which sought to reshape Europe during both world wars and thus needed the most modern and efficient arms. The other centre of development and manufacture, Belgium, was more oriented towards exporting its products to many of the smaller and newly formed European countries (such as the Baltic States or Yugoslavia), which also needed modern arms, if not for war at least as a matter of a pride.

Overseas, there was much less prospect of a coming war, so, once the US military had found itself a suitable (and excellent) semi-automatic pistol in the Colt M1911A1, it saw no reason to finance any further developments. Police forces at the time were much less confident in the 'complicated and unreliable' semi-automatic pistols and also saw no reason to switch from proven and reliable revolvers. Of course, there was a civilian market, but at the time it was satisfied with the same pistols that were used by the Army and with the smaller pocket pistols, which had little service value. Other than the USA, only a few countries on both American continents were able to afford the development and manufacture of their own designs, and in most cases these were more or less clones of the American M1911, such as the Argentinian Ballester-Molina or the Mexican Obregon.

European Combat Pistols Before 1945

Development of the first successful semi-automatic pistols became possible by the end of the nineteenth century thanks to several major technological achievements, including the invention of smokeless powder (which produced significantly less fouling per shot than earlier black powder), and improvements in steel processing and treatment, which allowed for the manufacture of relatively small parts of intricate shape, yet strong enough to withstand high pressures and rapidly repeating recoil.

Among the very first self-loading pistols ever produced on an industrial scale was the Borchardt C93 pistol, developed by 1893 in Germany by the American engineer, Hugo Borchardt. Several thousands of these pistols were produced before and just after the turn of the century by the Ludwig Loewe and, later on, DWM companies. This big and heavy pistol had an ungainly appearance and, despite being tested by numerous armies around the world, did not achieve any military sales. Before the end of the nineteenth century, several other designs appeared in Germany. The most notable of those was the Mauser C96 pistol, developed in 1894–6 by the three Federle brothers, who were employees at the Mauser factory. This pistol used a more powerful version of the 7.63mm Borchardt cartridge and saw considerable sales over the next four decades. It never became the mainstream in military circles, but was issued in significant numbers to German troops during both world wars (mostly chambered for the 9 × 19mm German service ammunition), and was bought by several other military customers worldwide.

After the First World War, significant numbers of the short-barrelled C96s were bought by Bolshevik Russia to arm its newly formed Red Army and NKVD. Generally known as 'Bolo' Mausers (Bolo – from *Bolshevik*), these pistols resulted in the adoption of a slightly modified version of the 7.63mm Mauser cartridge by the Soviet Red Army for its own semi-automatic pistols and sub-machine guns. This type of ammunition is generally known as 7.62 × 25mm TT.

There were other pistols developed and manufactured before 1900. For example, the German company Theodor Bergmann AG produced several semi-automatic pistols designed by Louis Schmeisser. These pistols went through many variations, starting in 1895–6 with blowback-operated, clip-fed pistols with integral magazines, through to locked-breech, recoil-operated pistols with detachable box magazines of the 1903 and later patterns. Bergmann pistols deserve their place in history due to the fact that they were adopted by the Spanish and Dutch armed forces, chambered for the relatively powerful 9 × 23mm Bergmann cartridge (better known by its Spanish name, 9mm Largo). Most of the Bergmann pistols produced after 1906 were actually manufactured under licence in Belgium, by Pieper.

The turn of the century brought two of the most notable and famous pistol designs of the period – the Parabellum pistol by Georg Luger and the FN Mle.1900 pistol by John Browning. Georg Luger was an engineer and a sales representative for the DWM factory. He was responsible for sales of the Borchard C93 pistol. As the C93 failed to attract many customers due to its large size and poor balance, Luger redesigned its basic mechanism, by 1899 creating a much more elegant and compact pistol. It must be noted that even by modern standards the grip shape and overall balance of Luger's pistol is often considered among the best. This pistol was tested and subsequently adopted by the Swiss Army in 1899, and in the following year DWM started producing the Luger pistol for Switzerland. In the same year Bulgaria also ordered a significant number of Luger pistols. The well-known nickname, 'Parabellum', usually associated with this pistol and its ammunition, comes from DWM's trademark, which, in turn, is a part of the ancient Latin saying *Si vis pacem, para bellum* ('If you want peace, prepare for war.'). In modified form, Luger pistols served with the German Army through both world wars; but the most important impact of the Parabellum was the 9mm ammunition, developed by DWM in 1902 at the request of the German Navy. This cartridge, universally known today as 'the 9mm', has the metric designation of 9 × 19mm and also bears the names Luger and/or Parabellum. Being just about the ideal compromise between power, size and recoil, after 100

7.63mm Mauser C96 pistol, shown in its 1912 modification, with detachable shoulder stock/holster.

Two of the Bergmann-made pistols, developed by Louis Schmeisser – top model 1896 with integral end-block clip-fed magazine; bottom model 1897 with detachable box magazine.

plus years of service worldwide, it still remains the most popular centrefire pistol cartridge in the world.

The American gun genius John Browning had developed his first semi-automatic pistols by 1895–6, offering these designs to several American companies. The Colt Company purchased rights for the more powerful locked-breech design and then Browning sold his less powerful, blowback-operated design to the Belgian Fabrique Nationale (FN) company. The first FN pistols of Browning design were made in 1899, chambered for the moderately powerful 7.65 × 17mm SR Browning cartridge (also known as .32 ACP [Automatic Colt Pistol]). The next year, FN engineers slightly shortened the Browning pistol and

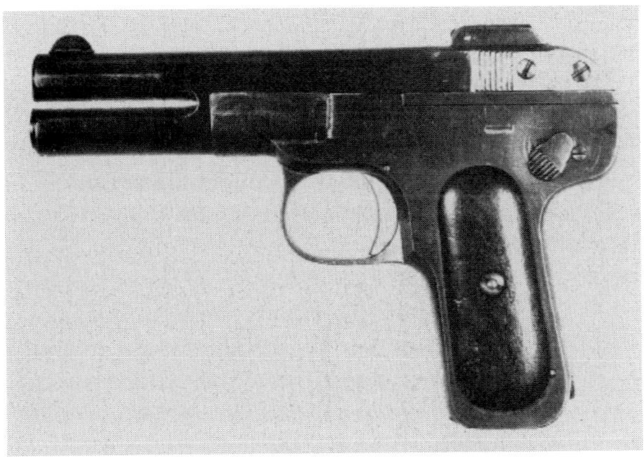

Prototype 7.65mm blowback pistol, made by Browning at the Fabrique Nationale factory in Belgium, circa 1898.

7.65mm FN Mle.1900 pistol, also known as the Browning No.1, became the prototype for most pocket pistols in the world. Made by the million, it has been widely used by police forces in Europe and was also very popular as a civilian self-defence pistol.

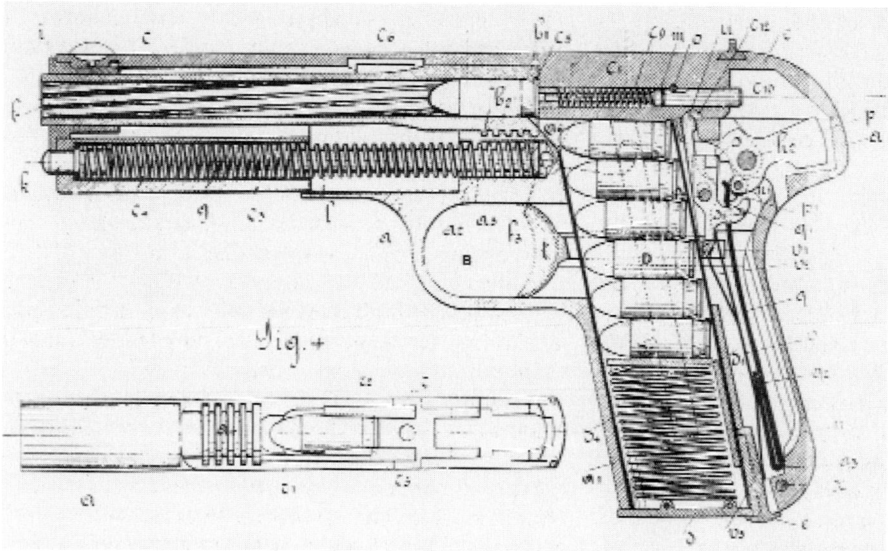

LEFT: Diagram of the FN Browning Mle.1903 pistol.

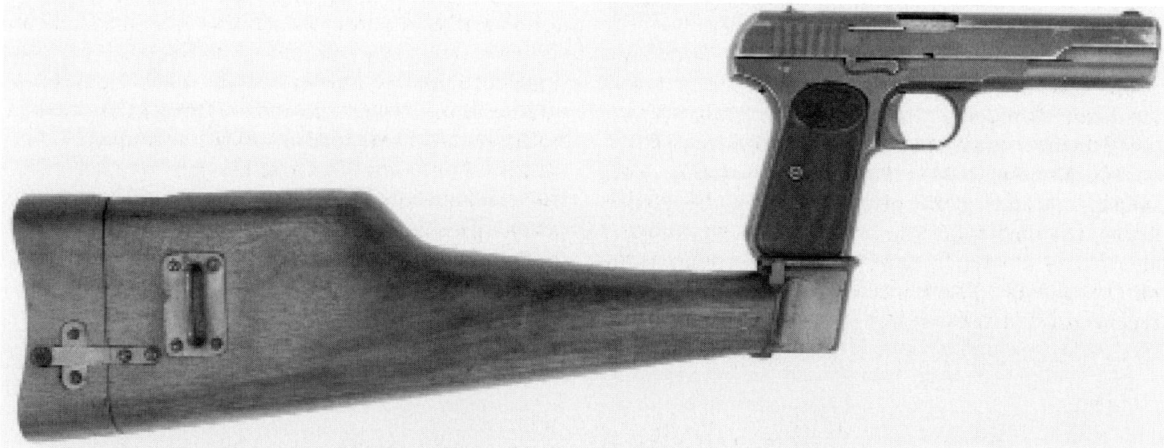

ABOVE: Military FN Mle.1903 9mm pistol with detachable shoulder stock/holster and extended ten-round magazine.

LEFT: The Browning Mle.1910 pistol, one of the most successful pocket pistols ever made; it was favoured by many European police and security organizations between the world wars and for a couple of decades afterwards.

this became the Mle.1900, or the Browning No.1 pistol. The Number 1 was a great success from the start, made by the million before the First World War. In fact, the very first shot of the Great War was made from an FN Browning Mle.1900 pistol, when Serbian nationalist Gavrilo Princip shot and killed the Archduke Ferdinand, thus creating the formal *casus belli.* Following the initial success of the Mle.1900, Browning developed several other designs that were later manufactured by FN. These were the Mle.1903 (a relatively large blowback-operated military-type pistol, adopted by the Belgian and Swedish Armies), and the Mle.1910, one of the most successful pocket pistols of the era, which was manufactured and sold in large numbers for more than sixty years.

During the pre-war period FN introduced another cartridge developed by Browning. The 9 × 17mm Browning Short, also known as the 9mm *Corto* (Italian) or 9mm *Kurz* (German), was first offered in Europe around 1912, along with the slightly modified FN Browning Mle.1910 pocket pistols. However, this cartridge was originally developed by Browning around 1908 in the USA as the .380 ACP, for the modified Colt Browning M1903 'hammerless' pistol, which, in turn, was more or less a scaled-down copy of the FN Browning Mle.1903 (or vice versa). This cartridge improved the effectiveness of the pocket pistols, reaching its heyday between the world wars, when it was adopted by several European countries for military service. These included the armies of Italy, Czechoslovakia and Yugoslavia, as well as Germany (as a substitute standard) and a large number of European police agencies. Most armies switched from the 9mm Short to the more powerful 9 × 19mm ammunition soon after the Second World War ended; however, many European police forces used this ammunition well into the 1970s.

There were several other European pre-war designs worth mentioning, such as the Austrian Mannlicher self-loading pistols, as well as the Roth-Steyr M1907 and Steyr-Hahn M1911, designed by Karel Krnka, also in Austria. The Roth-Steyr M1907 can be distinguished as the first semi-automatic pistol adopted by a major European military force – the Cavalry of the Austro-Hungarian Empire. The Steyr-Hahn M1911 was adopted by the Austrian Army, as well as exported to other countries, most notably Chile. Both of these pistols, developed at the famous Steyr arms factory, shared some similar properties, such as a recoil-operated locked-breech design with a rotating barrel and an integral magazine in the grip which was reloaded using stripper clips. The Steyr M1911 pistol saw significant use in both world wars. During the Second World War many of the Steyr M1911 pistols were rebarrelled by the Germans from its original 9 × 23mm Steyr cartridge to the 9 × 19 Luger round and issued to various German forces. There was also a rather rare version of the M1911, known as M1911/16 or M1916. This pistol had an elongated grip which held sixteen rounds in its integral magazine; the grip was cut for a shoulder stock/holster, and, above all, it had the option to fire in fully automatic mode. Unsurprisingly, a clip-fed pistol

9mm Parabellum pistol by Georg Luger, in its Pistole 08 version (1908), used by the German Army through both world wars.

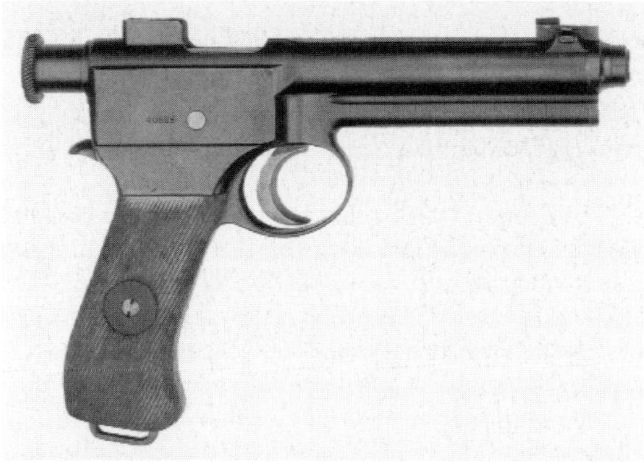

Roth-Steyr M1907 cavalry pistol.

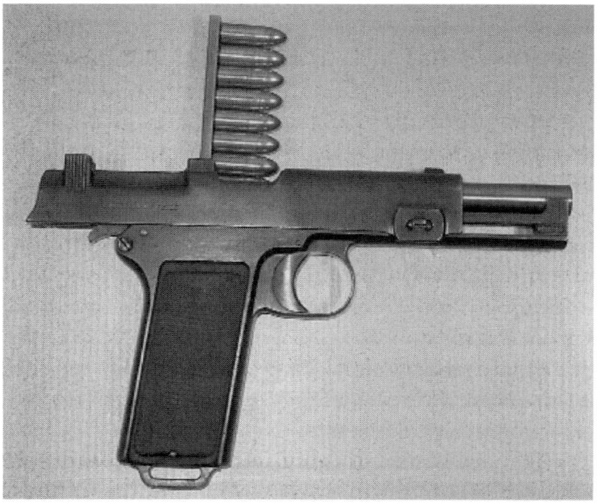

9mm Steyr M1911 pistol. Shown is the export version with the Chilean crest; note how the pistol is loaded from stripper clips with the slide held open.

was not the best platform for an automatic weapon and this did not prosper.

The Steyr factory was not the sole source of arms development in the Austro-Hungarian Empire. Rudolf Frommer in Budapest produced a number of complicated designs, featuring long-recoil locked-breech operation, which was rather unnecessary for the moderate-power ammunition used. Most famous of these designs was the Stop, which had appeared by 1912. It was a neat little pistol, well made and cleverly designed; however, it was far too complicated, especially considering the marginally powerful 7.65 × 17mm Browning cartridge it fired. Nevertheless, significant numbers of 7.65mm Stop pistols were made for *Honved* (the Hungarian military organization within the Austro-Hungarian empire) and it served until the late 1920s, when it was replaced in service by a much simplified blowback pistol of the same calibre.

In Britain, the famous Webley & Scott arms maker developed several self-loading pistols of roughly similar appearance but of different size and calibre. Starting circa 1904, Webley & Scott produced several working designs in .25 (6.35mm), .32 (7.65mm) and .38 (9mm) calibres, which were sold to civilians as well as issued to various police forces across the British Empire. In 1912 the Royal Navy tested the biggest of the Webley & Scott designs, chambered for .455 calibre ammunition. This was a huge, blocky pistol with a locked-breech, recoil-operated action. In 1913, the Navy adopted this weapon as the 'Webley & Scott self-loading pistol, Mark I Navy, calibre .455'. Pistols of similar design were also issued to Royal Flying Corps and Royal Artillery officers. Limited numbers of these pistols served in the Navy until 1945, but the main body of the British armed forces was still equipped with revolvers as standard sidearms until the 1950s. There were a few other attempts to produce military-type self-loading pistols in the UK, but none was successful.

The First World War, which started with the shots from the FN Browning Mle.1900 pistol in Sarajevo, effectively underlined the necessity and usefulness of self-loading pistols for military use. The interwar period was the time when self-loading pistols became mature and several important designs appeared on the scene during that time.

Between the wars, development of self-loading pistols in Europe was centred in two countries – Germany and Belgium. The most famous of the Belgian developments at that time was the FN Browning Grand Puissance Mle.1935 (or FN GP-35 in short), better known under its English name, FN

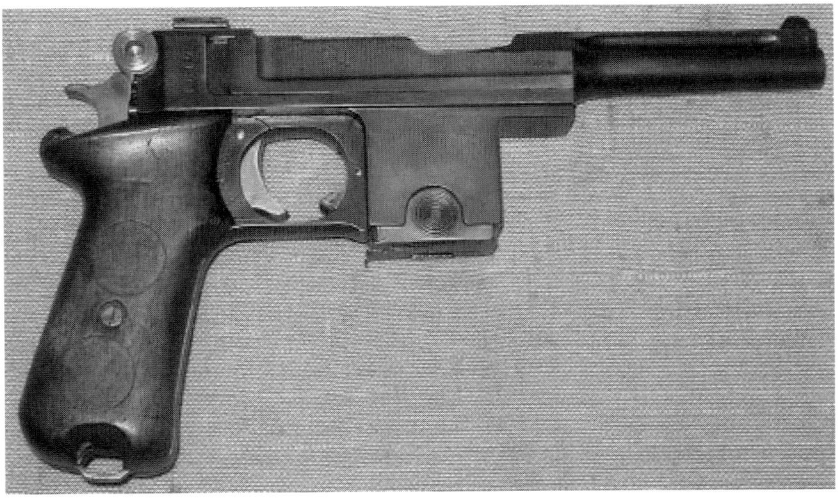

Danish 9 × 23mm Bergmann-Bayard M1910/21 pistol, designed in Germany, manufactured in Belgium and later upgraded in Denmark.

Browning High Power. Initially developed by John Browning for the Belgian company at the request of the French Army, the design was carried on by Browning's assistant and FN chief designer, Dieudonné Saive, after Browning's death in 1926. Browning built this pistol around the modified short-recoil system of his own design and a high-capacity magazine, developed by Saive. Finalized some nine years after the death of Browning, the High Power pistol was an instant success. Adopted by several countries, including Belgium itself before the start of the Second World War, it saw considerable use by both sides of the conflict, being made for German forces in occupied Belgium and for Allied forces in Canada. Still being made in several variations some seventy years after its introduction, the High Power remains the standard against which all other 9mm pistols are measured.

The achievements of German engineers in the inter-war period were no less significant and made the most impact on post-war developments. The famous

The Frommer Stop pistol has been made in 7.65 × 17mm (.32 ACP, military version) and 9 × 17mm (.380 ACP, civilian version). It was complicated and expensive, considering its marginally powerful ammunition.

Webley & Scott Mark I Navy pistol of .455 calibre (adopted 1913) was huge and heavy; it was in many ways inferior to the American Colt M1911 pistol, supplied to Britain as a military aid during both world wars.

Webley & Scott 9mm pistol of the 1909 pattern, which was used by various police units across the British empire, most notably in South Africa.

FN GP-35 (Browning High Power) pistol, made in Belgium during the German occupation.

company founded by Carl Walther is responsible for developing the first *successful* semi-automatic pistol with a double-action (DA) trigger. While double-action revolvers were known for centuries and the first double-action semi-automatics appeared in Europe prior to World War I, it was only with the introduction of Walther's *Polizei Pistole* (PP) in 1929 that the double-action pistol was brought to the attention of police and military users. Two years later Walther introduced an even more compact double-action pistol, the *Polizei Pistole Kriminal* (PPK), which also became a major success. Both of these pistols were widely used by various European police and military agencies, with the result that these designs were carried on into the post-war period, starting to disappear from the holsters of European police officers only in the mid-1970s. The Walther Company is also responsible for making the first military-type pistol with a double-action trigger. Walther's *Heeres Pistole* (HP) first appeared in 1938, being purchased by the Swedish Army. In 1940, Hitler's *Wehrmacht* (German armed forces) adopted the HP as the Pistole 38 (P38). This pistol also saw significant use in the post-war period, remaining in service (in slightly modified form) up until the 1990s. The success of the P38 introduced many military experts to the usefulness of double-action pistols and, not surprisingly, many armies at least tried to adopt double-action pistols as a result of war-time experience.

The double-action Walther PP pistol was a real success from the start; not surprisingly, it spawned a number of imitations and clones in the post-war period.

In 1938 another well-respected arms maker from Germany, the J.P. Sauer & Sohn Company, produced the Sauer 38H pistol. This relatively small blowback pistol featured a double-action trigger with manual safety, supplemented by one very handy addition – a manual cocking/decocking lever. This lever was especially useful because the Sauer 38H had a concealed hammer. While this pistol was not the first to have manual decocking (the Polish VIS P.35 pistol that was introduced in 1935 has a somewhat similar arrangement in the form of the lever, located on the slide), it was quite successful; However, the impact of the new design became visible only in the mid-1970s, when the same company teamed with the Swiss SIG company to produce the line of SIG-Sauer pistols, most of which were fitted with frame-mounted manual decocking levers.

Both the Walther and Sauer companies also experimented with lightweight pistols, trying to use aluminium alloys for pistol frames instead of steel. Several compact pistols in 7.65mm and 9mm Kurz were made with Duralumin frames, but again, only the post-war achievements in metallurgy permitted the wider use of lightweight aluminium alloys in the construction of semi-automatic pistols (and other small arms).

The Italian Army adopted its first semi-automatic pistol in 1910. This was the Glisenti, developed by the designer Bethel Abiel Revelli, who was better known for his designs of sub-machine and machine guns. However, the Glisenti pistol, as well as its 'improved' Brixia version, were of somewhat weak construction and used a down-loaded version of the German 9 × 19mm Luger cartridge. It should be considered a very bad idea to try to fire the 9 × 19mm Luger ammunition from either of these pistols.

One of the oldest gun-making companies in the world, the Pietro Beretta Co., entered the world of semi-automatic pistols in 1915, producing a blowback pistol chambered for 9mm Glisenti ammunition. After several improvements, made by Beretta's chief designer, Tulio Marengoni, Berreta finally produced a very good and strong, if slightly underpowered, design, the Model 1934. Adopted in 9mm Corto (9 × 17mm) by the Italian Army and in 7.65 × 17mm by the Italian Navy (as the Model 1935), this pistol served as a platform for several post-war designs by the same company.

The Soviet Union had begun development of a military-type semi-automatic pistol by the late 1920s, in an attempt to replace obsolete M1895 Nagant revolvers. However, it should be noted that the first semi-automatic pistols had been produced in the USSR several

7.65mm Sauer 38H pistol. The decocking lever is clearly visible on the frame, behind the trigger. *(Courtesy of Cherry's Fine Guns)*

The Polish VIS-35 pistol in many respects was based on the American Colt M1911 design; however, it had several original features, including a slide-mounted hammer decocking lever (quite an unusual feature for a single-action pistol).

9mm Glisenti M1910 pistol.

A Beretta M1923 which belonged to the Buenos Aires police (Argentina).

The Beretta M1934, a blowback pistol of simple and strong design, was the standard sidearm of the Italian Army. *(Courtesy of Cherry's Fine Guns)*

years before, when the designer Korovin, in Tula, developed a 6.35mm TK pistol – a pocket-type weapon, issued to high-ranking Communist Party officials and Red Army officers. Initially, the Red Army requested two military pistols – one of more or less standard type and another of 'pistol carbine' type, with a longer barrel, high-capacity magazine and detachable shoulder stock. Obviously, the latter concept was prompted by the Mauser C96 pistols which were bought in significant numbers from Germany during

6.35mm TK (Tula, Korovin) pistol of 1926. This pocket handgun was the first semi-automatic pistol to be made in the USSR. Firing a slightly more powerful version of the 6.35mm Browning (.25 ACP) cartridge, it was issued to top-ranking officials of the Red Army and the Communist Party.

the 1920s. The same C96 pistol inspired the new ammunition, requested for new Red Army sidearms. In fact, the new 7.62mm Russian pistol ammunition was the same as the 7.63 × 25mm Mauser except for a larger extractor groove and a larger primer.

After extensive trials, held by the Red Army Artillery Board in the late 1920s, the 'pistol carbine' concept was abandoned and the design of Fedor Tokarev was adopted for general service in 1931. In 1933 a slightly improved version of the Tokarev pistol appeared, universally known as the TT-33. It was a good and powerful pistol, but not without certain flaws, and by 1939 the Red Army started another contest for the next military pistol. By 1941 the pistol of designer Voevodin was considered a favourite, with its high-capacity magazine that held eighteen rounds of 7.65 × 25mm ammunition. The outbreak of the Great Patriotic War against Hitler's occupation buried this design – the Red Army fought this war with Tokarev's TT-33 pistols and Nagant revolvers.

During 1942, Tokarev attempted to improve his pistol by introducing a double-stack magazine with a capacity of fourteen 7.62mm rounds in a thicker and slightly redesigned grip. Only a handful of these modified pistols was made, very few of which survive to this day. The fate of the 'genuine' TT-33 was much better. Being made by the million in the USSR until the mid-

Handguns that were tested in 1928–9 by the Red Army, from top to bottom: designs by Tokarev, Prilutskij and Korovin.

Pistols that were tested by the Red Army in 1938–9, from top to bottom: designs by Rakov, Voevodin and Tokarev. All were chambered for the standard 7.62 × 25mm ammunition, although, the pistol by Voevodin had the most chance of being adopted.The slender barrel profiles of these pistols were caused by the requirements of Red Army Tank command, which demanded that pistols must be able to fit into the small firing ports made in the turrets of pre-war Soviet tanks for close-in defence.

1950s, it was declared obsolete in 1951 but remained in service until the 1970s. In the post-war period, manufacturing licences and documentation for the TT-33 were supplied to several pro-Soviet countries, including China, Hungary, Poland and Yugoslavia. China and Serbia (the former Yugoslavia) still manufacture the TT-33 clones today, as the Type 51 and M57, respectively. Both of these countries also make variations of the basic TT design, both in its 'native' 7.62 × 25mm calibre and in the much more popular 9 × 19mm.

France, while showing significant interest in the Browning's pistol developed by FN in Belgium, decided to adopt a 'native' design in 1935. This was developed by the Swiss designer Charles Petter, who at the time worked for the French SACM arms factory. The pistol, initially known as the Mle.1935, had an elegant appearance, but it fired the somewhat underpowered 7.65 × 20mm *Longue* (long) ammunition. The construction of this pistol was influenced by the Browning designs, most notably the Colt M1911. Production difficulties resulted in the appearance of the simplified and less 'good-looking' MAS Mle.1935S variation, with the original pistols renamed Mle.1935A. The patents of Charles Petter, especially for the 'en-bloc' hammer and sear unit, were later bought by the Swiss company SIG and served as the basis for the post-war SIG P210 pistol. It must be noted that the Russian designer Tokarev had used a similar idea of the modular hammer unit (in his TT pistol) five years before Petter.

7.62mm Tokarev TT-33 pistol, manufactured in 1941.

Spain, another major manufacturer and exporter of handguns before and during World War I, supplied the world with endless numbers and marks of so-called 'Ruby' pistols, which, in most respects, were clones of Belgian FN/ Browning designs. During the interwar period, Spain also supplied many countries with quality clones of foreign designs, such as the Mauser C96 (produced by the Astra and Azul companies in various calibres and sometimes with full-auto capability). Others included clones and variations of the American Colt M1911 pistols, made in a wide variety of calibres (7.63 and 7.65mm, 9mm and .45) and sizes by the Star and Llama companies. There also were a few indigenous designs, such as the Astra 400 – a mili-

7.62mm modified Tokarev TT pistol with high-capacity fourteen-round magazine, manufactured circa 1942.

French SACM M1935A pistol, designed by Charles Petter.

(Courtesy of Cherry's Fine Guns)

The 7.65mm (7.65 × 17 Browning) Gabilondo 'Plus Ultra' pistol, designed during World War I, was a typical Spanish-made 'military' pistol of the era, derived from smaller, Browning-inspired weapons by lengthening the barrel and magazine.

The 7.65mm (7.65 × 17 Browning) Unceta Victoria Model 1911 pistol was also quite popular as an officer's sidearm during World War I. It is a typical example of the so-called 'Ruby' breed, although one of the better quality ones.

tary-type pistol with a simple blowback mechanism. This pistol was most famous for its ability to fire many types of 9mm ammunition, from 9 × 17 Short to 9 × 19 Luger and the standard Spanish 9 × 23 Largo. A slightly smaller version of the Model 400 was the Model 600/43, developed by 1943 at the request of Germany, which needed more and more 9mm pistols for the war. The Model 600 was chambered for 9 × 19 Luger ammunition only. Some of these pistols served with West German police units until the mid-1970s.

Czechoslovakia, which became an independent state after the fall of the Austro-Hungarian Empire, developed into one of the major arms suppliers of the interwar period. While most famous small arms to bear the CZ logo were machine guns and rifles, several pistols were produced in Czechoslovakia in the 1920s and 30s, before it was occupied by Germany in 1939. These pistols showed a great deal of fine workmanship

9mm Largo Astra Model 400 (also known in Spanish service as the Modelo 1921), blowback-operated pistol, which was capable of digesting various types of 9mm ammunition without any changes – at least, in theory.

9mm Largo Star Model 1920 pistol. This was the first in the long line of Colt M1911 clones made under the Star trademark.

The 7.63mm Astra Model 900 was a clone of the German Mauser C96, less expensive than the original. It was quite popular among customers in China and South America.

in construction and finish, but had few practical improvements over existing designs. Germany used many of the Czechoslovak pistols, made during the occupation, for its armed forces and police. Most of these pistols were relatively simple blowback designs, chambered either for 7.65 × 17 or 9 × 17 ammunition.

With the start of World War II, most of the development of combat handguns was frozen, with efforts being concentrated on the ever-increasing production of existing designs necessary for the fighting troops. Remaining developments mostly focused on production issues, in attempts to make pistols more quickly, with fewer resources and less money. When Hitler's fortunes began to change in late 1943, many German arms makers started to work on mostly simplified and crude weapons, intended to be used by ordinary people recruited into the *Volkssturm*. The typical *Volkspistole* (people's pistol) was a crude weapon, made from

7.65 × 17 Vz.27 pistol, made in Czechoslovakia under German occupation. German forces issued these pistols under the designation of P.27(t).

The 9 × 17 Vz.37 pistol was unusual in having a tipping-up barrel, a feature used for disassembly; another, less popular feature of this design was the double-action-only trigger with a very heavy pull.

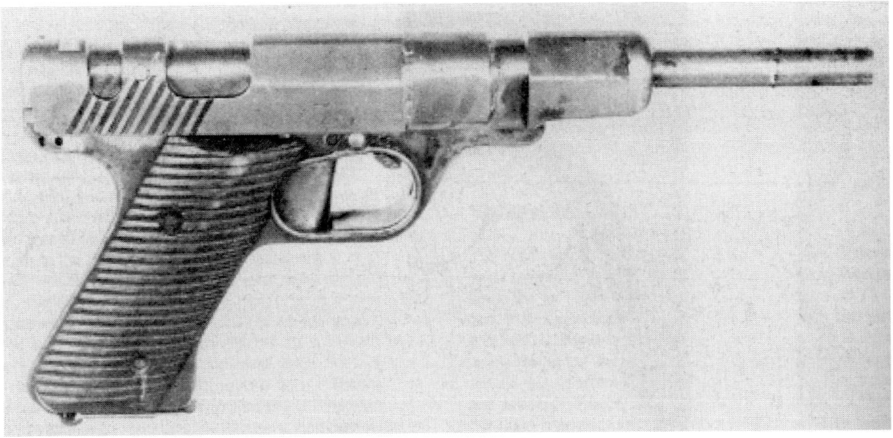

One of the German late-war Volkspistole *designs, made at the Mauser Werke. This prototype weapon was almost entirely made from steel stampings.*

steel stampings and ordinary steel, with hardly any finish and a miserable expected service life. However, in attempts to make these weapons even simpler, German designers developed several new operating systems, mostly of delayed (retarded) blowback type.

Of those systems, two made significant impacts on the post-war developments. The first was the design of the engineer Barnitzke, who developed a gas-retarded blowback system at the Gustloff Werke. This system, initially used in *Volksgewehr* semi-automatic rifles and several *Volkspistole*, was later used in pistols by Steyr-Daimler-Puch. The second design, developed at the Mauser Werke by the designer Vorgrimler, used a pair of rollers to delay the opening of the unlocked breechblock. This system saw extensive use in the post-war designs of the German company of Heckler & Koch, in all types of small arms, from pistols through assault rifles to machine guns.

American Combat Pistols Before 1945

John Moses Browning is regarded as the icon of the American handgun world. He had developed his first working pistols by 1895–6, immediately covering four major operating principles with his patents.

Colt M1902 .38 ACP (9mm) calibre pistol designed by Browning.

(Courtesy of Cherry's Fine Guns)

Colt M1903 Hammerless .32 ACP calibre (7.65mm) pistol. (Courtesy of Cherry's Fine Guns)

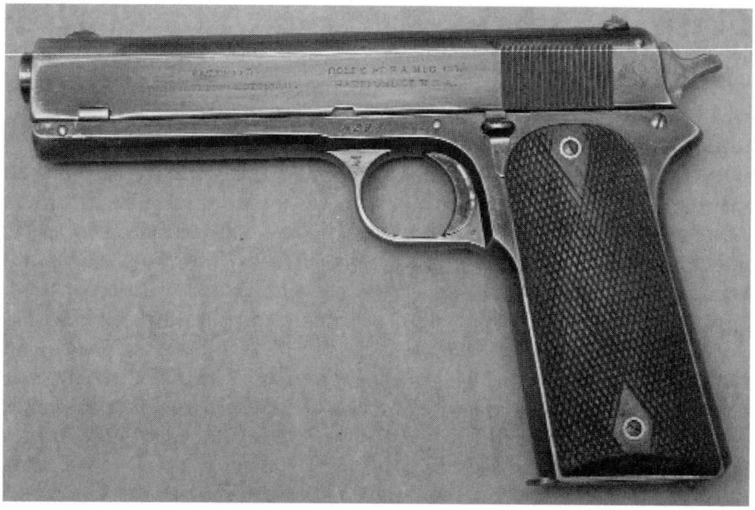

Colt M1905, the first Colt-Browning design to be made in .45 calibre (11.43mm). (Courtesy of Cherry's Fine Guns)

'Pistol, automatic, .45 calibre M1911', made by Colt. (Courtesy of Cherry's Fine Guns)

Modified M1911A1 .45 calibre pistol, as adopted by the US Military in 1926. *(Courtesy of Cherry's Fine Guns)*

He later extensively exploited and improved two of these systems (the short-recoil system with tilt-barrel locking and the simple blowback), and left two others idle (these were the short-recoil-operated system with rotating barrel locking and the gas-operated locked-breech system). By 1900, Browning had made arrangements with two companies, one American and one European, shaping the world of the semi-automatic handgun for ever. The European company was FN of Herstal, Belgium; the American one was the Colt's Patent Fire Arms Co., of Hartford, Connecticut.

Colt produced the first Browning pistol in 1900. This was a relatively large, recoil-operated weapon, firing 9mm ammunition of significant power, at least by the standards of the time. During the succeeding years, Browning improved these pistols in a number of ways. In 1903, Colt also introduced the blowback-operated Model 1903 'hammerless' pocket pistol (also known as the Colt Type M). This pistol was more or less a copy of the Belgian-made FN Browning M1903 military pistol (or vice versa), but with some modifications.

After initial examination of available pistols supplied by Colt, as well as by European makers such as DWM (Luger pistols) and Mauser (C96 pistol), the American Army decided to go ahead with a service pistol to replace the existing revolvers. After extensive studies on the terminal effect of various bullets and calibres, made in 1904 by the officers Thompson and La Garde, the US Army settled on a bullet diameter of 0.45in (11.43mm) as the future calibre for military sidearms. The year 1905 saw the start of lengthy and extensive trials to find the best available pistol for American troops. Two major contenders in these trials were the Colt Company with pistols designed by John Browning and the Savage Company with the pistol designed by Elbert Searle. There were a few other designs, such as the Grant Hammond recoil-operated and Philips gas-operated pistols, but these did not survive the trials. Several foreign designs were also briefly tested, the most famous of which was the Luger pistol in .45 calibre.

The outcome of the trials is well known – in 1911 the US Army finally chose the improved Colt-Browning design as the winner. This started the era of 'the 1911' – the US Pistol, 45 calibre, semi-automatic M1911 became the most successful semi-automatic pistol ever made. Slightly modified in 1926 to became M1911A1, this pistol is also known as the 'Government model', when sold on the civilian market. Serving as the official sidearm to the US Military until 1985, this pistol is still widely used by US Military and law-enforcement forces. Copies and clones of the M1911, made either under licence or without it, were and still are manufactured in many parts of the world, including Europe and South America. Today, almost 100 years since its introduction, the M1911 is still made by dozens of manufacturers, from the largest companies down to smaller outfits and individual gunsmiths.

During the first three decades of the twentieth century Colt continued to make its earlier .38 calibre

Savage model 1907 pistol in .380 calibre (9 × 17), a scaled-down version of the prototype military pistol.

pistols of Browning design, as well as smaller blowback pistols of the 1903 pattern. The latter, also known as the Colt M1903 'Hammerless' (because of its concealed hammer), or 'Colt Model M', was purchased by the US Military and issued to general officers as a self-defence weapon.

The .45 calibre Savage-Searle pistol entered in the trials was never put into production. However, it spawned a number of smaller variations, chambered for .32 (7.65 × 17) and .380 (9 × 17) ammunition. First offered in 1907, Savage pocket pistols were among the very first pistols to have double-stack detachable magazines with a ten-round capacity, being advertised as 'ten shots quick'. Manufactured for nearly two decades, these pistols were favoured by civilian users. A few European armies also bought numbers of Savage pistols before and during the First World War. Military customers of the Savage pistols included the Portuguese and French armies.

Other companies and individual designers also attempted to enter the pistol market, but without much success. Smith & Wesson produced its .35 calibre Model 1913 pocket pistol, based on the European Clement design, but it was not very impressive. Designer John Pedersen, working for the Remington Arms Company, produced a large, delayed-blowback .45 calibre semi-automatic pistol at the request of the US Navy in around 1917, but the war ended before this handgun could be manufactured. Because of the Armistice, the Navy decided to select the already adopted M1911 pistol. With no need for another military-type pistol, Remington went the same route as Savage had done before, redesigning the Pedersen system into a smaller pocket-type pistol. Known as the Remington Model 51, this finely crafted pocket pistol was made for some time, but it was too complicated for the .32 (7.65 × 17mm) and .380 calibre (9 × 17mm) ammunition it used and was finally discontinued.

With the start of World War II, the US Government ordered M1911A1 pistols from several manufacturers

Prototype .45 calibre pistol, designed by John Pedersen for Remington, circa 1917.

Remington M51 pistol, made in .32ACP (7.65 × 17SR) and .380ACP (9 × 17) calibres in-between 1918 and 1927, was a direct offspring of the experimental .45 calibre military pistol. It was finely made, but too expensive.

Prototype .45 calibre pistol, made during the Second World War from stamped and formed steel.

The single-shot .45 calibre Liberator pistol was the least expensive 'military' pistol ever made.

other than Colt. There were few attempts to develop any new pistols, except for the exotic and unsuccessful Liberator pistol, a smooth-bored single-shot weapon of .45 calibre. These weapons, made almost entirely from steel stampings at the Guide Lamp division of the General Motors Co., were to be used at point-blank ranges by resistance fighters in Europe. For this reason, Liberator pistols, along with ten cartridges and comic-strip-type manuals, were packed in cartons to be dropped via parachute from aeroplanes into the occupied territories.

There also were attempts to make M1911-type pistols using as much steel stamping as possible. The resulting prototype weapons were bulky and heavy and, as the existing technologies proved sufficient for war demands, this project never went into production.

Chapter Three

Trends in the Development of Service-Type Pistols Since 1945

It must be noted that, despite much work over the last sixty years, most modern pistols have no significant advantages over the best handguns developed before World War II. Most 'modern' pistols still fire the same ammunition and are the same size. Of course, modern pistols usually are lighter, hold more rounds, have better sights and better-placed controls, and sometimes they are also more durable than the 'old-time' pistols. But, there's no task that a typical modern gun will accomplish that an 'old' pistol will not. Speaking simply, the well-trained person armed with a 1911-vintage Colt Government model pistol, or with a 1935-vintage Browning High Power (assuming that either gun is in fine mechanical state), would be at no critical disadvantage against a person armed with the most modern sidearm such as a Glock or Beretta.

Certainly, modern pistols have better sights and about twice the magazine capacity of many older pistols, but every advantage has its own downside. For example, lighter pistols usually have stronger felt recoil. Also, a larger magazine capacity may result in false hopes of suppressing the enemy by a large amount of 'barrage fire'. This may work in certain defensive situations, but the smaller magazine capacity emphasizes the requirements of aimed fire (and necessary training), as every round counts.

However, it would be entirely wrong to disparage all modern developments, as in many situations even a slight advantage may cross the thin line between life and death. For example, at night, a shooter armed with a 'modern' pistol equipped with luminous night sights will have an edge over the shooter with the old-style pistol with simple iron sights.

So, let's look at the major improvements that have appeared during the last sixty years.

First, we must note the appearance of new materials and finishes. For the first half of the twentieth century, most pistols were made from machined steel, with wood or plastic used for grip panels, and usually finished in blue or with nickel plating. There were several main goals in the use of new materials: the first was to make guns lighter and thus more comfortable to carry for long hours and days; the second, to make the guns less expensive; and the third, to make guns more durable and less sensitive to the elements.

The first goal was pursued with the introduction of lightweight aluminium alloys. First tried by German manufacturers such as Sauer and Walther in several pocket pistols during late 1930s, aluminium became important during the 1950s, when two major American gunmakers, Colt and Smith & Wesson, brought out several pistols and revolvers with aluminium frames. Colt offered its 'Commander', based on their famous M1911A1/Government model pistols. Smith & Wesson offered its Double Action Automatic, since 1957 also known as Model 39. Both pistols were the results of the unsuccessful US Army trials held in the early 1950s with the intention of replacing the venerable M1911A1 with a lighter 9mm pistol. During that period, the aluminium frames were made in exactly the same manner as steel ones – by cutting and machining the solid bars of forged metal. Later on, several companies, such as the US-based Sturm, Ruger & Co., developed the precision casting techniques which allowed both steel and aluminium parts to be made with minimum machining. There were originally a lot of concerns about the strength and durability of the cast parts, but nowadays these parts are generally no less durable and strong than the older, machined parts, while being significantly less expensive.

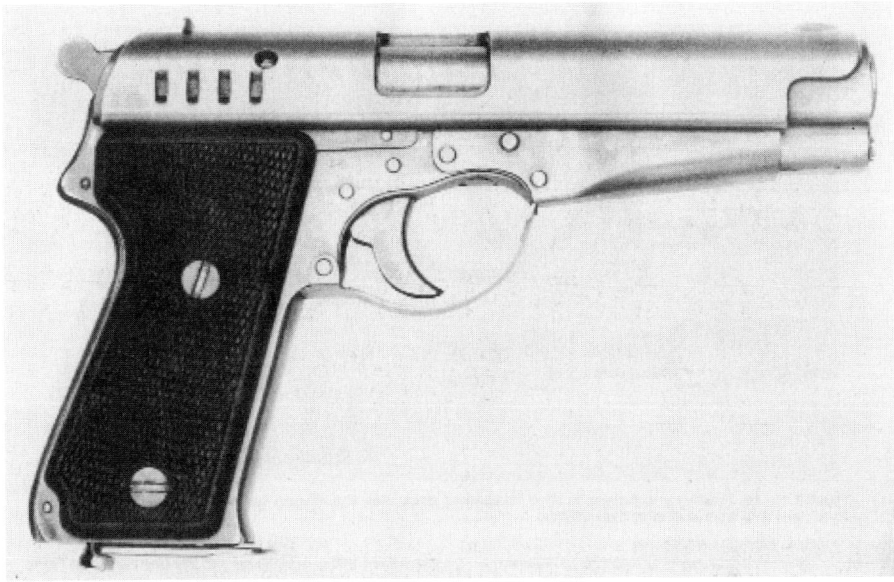

Experimental 9mm pistol, made in Germany by Walther in the late stages of the Second World war. Both frame and slide were made from stamped-steel parts.

Developments in the chemical industry resulted in the next improvement in materials – in 1970 the German company Heckler & Koch introduced the first pistol with a frame made from impact-resistant polymer. This pistol, the VP70, was not a success and was made in very limited numbers. It took another ten years before polymer-framed pistols took the market by storm. The commercial success of the Austrian Glock pistol, developed for the Austrian Army trials, was enormous – by the late 1980s Glock pistols were all the rage in the military, law-enforcement and civilian segments of market. The one-piece polymer frame was much lighter than steel or aluminium and significantly cheaper to make and take care of, as the polymer is not corrosive and is self-lubricating. There was a lot of media hype about 'all-plastic' pistols, invisible by X-ray and magnetic detectors, but about 60 per cent of the mass (and most of the parts) of the Glock, as well as any other polymer-framed pistol, is made from high-grade ordnance steel and therefore easily detectable by any appropriate equipment.

Another way to make weapons less expensive and more suitable for mass production was to use steel stampings. German companies experimented widely with steel stamping and forming during the Second World War. However, the first mass-produced pistols to use major parts made from steel stampings appeared only in the 1960s, from Heckler & Koch. HK P9 pistols used stamped slides and frames, as well as a few other stamped parts. But the most famous pistols to use slides stamped and formed from sheet steel were SIG-Sauer pistols, starting with the extremely successful P220 pistol, which appeared in the mid-1970s. These pistols combined aluminium alloy frames with stamped-steel slides and once were called 'expensive pistols made by affordable technologies'. However, today many pistols still use machined or precision-cast frames and slides, as this offers more strength and durability, especially when firing modern high-energy cartridges such as the .40 S&W or .357SIG.

Another modern material that improved the handling and simplified the daily maintenance of pistols was stainless steel. The first mass-produced handgun to be entirely made from stainless steel was the Smith & Wesson Model 60, 'Chief's Special Stainless' revolver. By 1968, Smith & Wesson were producing a Mark 22 Model 0 pistol for US Navy SEALs, which had a stainless-steel barrel and slide. However, it took more than ten years to develop and put into production an all-stainless-steel pistol. Since 1980, the Smith & Wesson Company has offered its semi-automatic pistols in stainless steel, as well as in carbon steel and in steel slide/aluminium frame versions. Plenty of other makers also now offer their guns made from stainless steel, entirely or in part.

On many modern pistols, 'old-school' finishes like bluing or phosphate coating have been succeeded by modern polymer coatings such as Teflon or Brunition; combined with special surface treatments, these finishes offer excellent resistance to wear and rust. Some of

Colt Government model pistol fitted with SureFire™ tactical flashlight using an add-on adapter.

these coatings have the additional benefit of a self-lubrication effect, further minimizing maintenance requirements.

There were also certain developments in manufacturing techniques which enabled less expensive, yet more precise and uniform, production. The most important of these were the metal precision casting process (usable for both steel and aluminium) and CNC (computer numerically controlled) machining. Precision casting processes are used for the production of major parts, such as frames and slides, while smaller parts, such as triggers and levers, are often manufactured using a metal-injection moulding (MIM) process. Barrel manufacturing has also improved, especially with the introduction of the cold-hammer forging process, which allows barrels to be made in one process, complete with rifling and firing chamber, without separate bore drilling and subsequent rifling and chamber reaming.

A lot of effort has been spent on the development of the trigger and safety systems. As we have seen from the previous chapter, the German military and police accepted the double-action trigger mechanism in self-loading pistols during the 1930s. However, after the Second World War most of the world was still using single-action (SA) pistols, although after thorough examination of the captured German pistols, many armies showed interest in the double-action pistol, as a safer way to carry a gun loaded and ready for action. The fact that the early double-action pistols usually had a long and heavy trigger pull did not concern the military, as Army pistols are usually regarded as purely defensive weapons.

The first to adopt a double-action semi-automatic was the Soviet Union, which in 1951 introduced the compact Makarov PM pistol and the relatively large Stechkin APS automatic pistol, both chambered for the indigenous 9 × 18mm ammunition. These pistols showed the significant influence of German developments, specifically the Walther PP. While many European police forces used double-action Walther pistols, most armies continued to use single-action handguns well into the 1970s. The only other major Army to issue a double-action pistol during the early post-war decades was the *Bundeswehr* – the West German Army – which adopted a slightly modified version of the wartime Walther P38 as the P1. American military trials of the early 1950s also brought to the market the double-action 9mm Smith & Wesson Model 39, but it was not until late 1960 that this pistol was adopted by the relatively large police department of the state of Illinois. The range of double-action pistols had expanded by the mid-1970s, when several companies, fuelled by requirements to replace ageing military and police sidearms of pre-war design, turned out service-type 9mm pistols. These designs, which included 'military-type' full-size pistols like the American S&W M59, Italian Beretta 92, Swiss SIG P220 and Czechoslovak CZ 75, plus the more compact 'police-type' pistols such as the Walther P5, Heckler & Koch P7 or SIG-Sauer P225, allowed their users to carry pistols with a live cartridge in the chamber and with the hammer lowered, decreasing the chances of accidental or negligent discharge during the carrying and handling of weapons.

The next 'improvement' in trigger systems came in around 1982, when the Austrian Glock Company introduced its 'safe-action' trigger. This system, with no manual safeties and with a pre-cocked firing-pin spring, was actually an adaptation of the 1907-vintage Roth-Steyr trigger. This design, combined with several automated safeties, made the gun safe against accidental discharges, while being extremely simple in manipulation. Of course, there were several drawbacks to this system. First, it does not allow making a quick second strike to the primer in the case of misfire – one must first pull the slide slightly back to pre-compress the firing-pin spring. Second, the pre-set trigger pull weight, along with the relatively short trigger travel and lack of any manual safeties, resulted in a number of negligent discharges, mostly among the US police officers. The latter fact resulted in the advent of the so-called 'New York trigger pack' for Glock pistols, which included new, stronger springs to make the trigger pull heavier.

This experience also resulted in increased interest in double-action-only (DAO) trigger systems and there are a few recent developments which permit only true double-action firing (with second-strike capability), with manageable yet 'fool-safe' trigger pull weight and travel. Of cause, most of the DAO pistols are considered as purely defensive weapons and many companies still make similar pistols available either with traditional DA or with DAO triggers, to suit the particular requirements of the buyer. There are also a few pistols which are available in DAO or SA versions. Of those, the most popular is the FN Five-seveN, which in its basic version is available as a DAO 'military defensive' pistol, while the Five-seveN Tactical version is available as an SA 'special forces offensive' pistol, which can also have adjustable sights for maximum accuracy.

There have also been several improvements in the sights of combat pistols. Pre-war military and police pistols usually had either low-profile fixed sights, or (adjustable for range) tangent sights, rather optimistically marked up to 500m (as on FN High Power pistols) or even the totally unrealistic 1,000m (as on Mauser C96). However, real-life experience showed that most of the pistol shooting in combat is done at ranges of 25m or less, so the complicated and expensive range-adjustable sights are of little value. The little fixed sights, on the other hand, were found to be less suitable for rapid aiming. The availability of various types of ammunition with different ballistics resulted in removable fixed sights, often dovetailed into the slide. This also simplified the zeroing procedure, as the windage adjustments can be made by drifting the front or rear sight in its dovetail and vertical adjustments by replacement of the sight blade.

As not every shooting incident occurs in daylight, there is a clear need for high-visibility sights. The simplest way is to paint the front and rear sight with bright (white, yellow or orange) stripes or dots, to make sights more visible against dark backgrounds. For night shooting, simple paint is often replaced by the self-luminous inserts, usually filled with tritium. For those special purpose units, which may use various types of ammunition in their pistols (such as sub-sonic ball rounds for use with silencers, or +P JHP for use in 'non-silent' combat), several manufacturers offer pistols with low-profile adjustable rear sights. These sights are pre-set for a desired type of ammunition and are used in combat in the same way as any other fixed sight.

Other sighting aids include tactical flashlights and laser aiming modules (LAMs). Before World War II the Germans tried flashlights attached to pistols, but at the time batteries were too heavy and this practice was extremely rare. Advances in the fields of chemical power sources and small light bulbs have resulted in the appearance of small yet powerful lights, which can also withstand the violent forces of recoil. Currently, most lights are mounted under the barrel, using either a special add-on mount or integral mounting rails on the frame of the gun. The laser aiming modules are more recent additions, starting to appear on combat

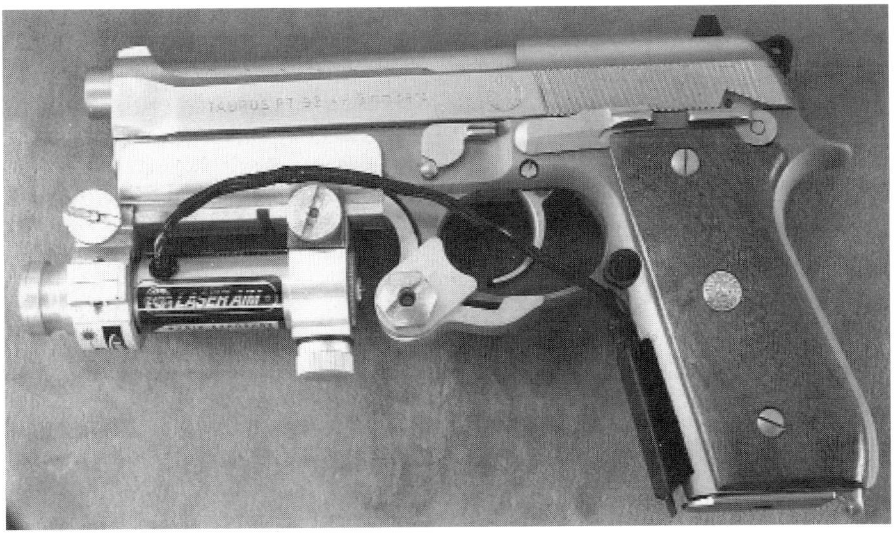

Early model (circa mid-1990s) LAM, mounted on the Taurus PT92 pistol using a special adapter, clamped onto the trigger guard. *(Courtesy of Cherry's Fine Guns)*

pistols by the mid-1990s. The LAMs are used to project a bright aiming spot onto the target, making snap-shooting much easier, with an additional psychological effect on the target which could be very useful for certain police operations, such as arrests. However, in search operations, both light and laser beam can be used against the operator, as each gives away the operator's position and current point of aim. Both instruments must therefore be used with care, and it should be considered a bad idea to make 'Hollywood-style' entries into a danger zone with laser beams waving around all the time. Several Special Operations Forces also make use of invisible infrared (IR) lasers and illuminators, when the aiming spot can be seen only through night-vision goggles or other IR-diapason equipment.

The next element to be given attention by manufacturers over the last few decades was the overall ergonomics and handling characteristics of the pistol. It may be surprising how much actual shooting results can depend on such 'minor' matters as grip size and shape, layout of controls and so on. The first well-accepted move to improve the ergonomics of combat pistols was to make at least some of the controls ambidextrous. In the simplest cases, manufacturers just put the safety levers on either side of the pistol. Other

The modern 'LaserGrips' LAM from the US-based Crimson Trace Company is built into the grip panel of the Smith & Wesson revolver. The button at the front strap of the grip automatically switches on the LAM, once the gun is properly held in the hand. This set-up is very useful for defensive handguns,and is available for a wide variety of semi-automatic pistols and revolvers. (Courtesy of Crimson Trace)

'southpaw-friendly' options may include a magazine release button that can be reinstalled by the user on either side of the frame, or an ambidextrous slide-stop lever. Several of the most modern pistols have magazine release buttons or levers on both sides of the frame simultaneously.

The next step in the direction of better ergonomics was to produce pistols with grips of adjustable size. The correct grip size is essential for both trigger handling (and accuracy) and for recoil management, plus people tend to have hands of very different sizes. For many years, grip adjustment options were mostly limited to aftermarket grip panels and, on pistols like the Colt M1911, grip backstraps of different shapes, for example straight, as in the 1911 pattern, or arched in the 1911A1 pattern). The first pistol to offer interchangeable grip backstraps of different shapes and sizes was the Walther P99, which appeared in the mid-1990s. Each P99 pistol was shipped from the factory with three backstraps of a different shape and size, so users could adjust the grip for their own preference. Since then, several other manufacturers have also offered their pistols with interchangeable backstraps or entire grip panels of different sizes. However, there are still plenty of modern pistols that could not be easily tailored for at least some shooters. For example, the otherwise very fine Beretta 92 pistols have a quite thick grip, which is less comfortable for people with smaller hands. Another example of a pistol which may require some aftermarket, non-factory grip tuning are larger calibre Glocks (such as the .45 calibre Model 21 or 10mm Model 20). These pistols have relatively thick and blocky grips, so some users that have to carry these pistols send them to a gunsmith who will reduce the grip circumference by cutting 'excessive' plastic from the rear part of the grip.

It is well-known that the factor which most affects the combat effectiveness of every firearm is the ammunition it fires. There have been about a dozen pistol cartridges used as service ammunition by military and police forces worldwide. These range in calibre from 7.62mm to 11.43mm (0.3 to 0.45in), as well as varying in power. However, only three cartridges have dominated the world. Historically and arithmetically, the first is the famous 9 × 19mm cartridge, developed by the DWM factory for the Luger pistol in 1902. Universally known as the 9mm Luger or 9mm Parabellum, or simply as the 'nine millimetre', this cartridge was adopted as a NATO standard during the first post-war decade. By 1985, the United States military had also adopted the 9mm as its new sidearm ammunition, and in 2001–4 this cartridge, although in modified and intensified loadings, was approved for service by the Chinese police and Russian military and law enforcement organizations.

Second in this list comes the venerable .45 ACP cartridge, developed by US Army Ordnance experts in 1905 and adopted for US service in 1911. It has seen significant use on both American continents and is still considered amongst the best 'man-stoppers'.

The third is the smallest of the three, the Soviet 9 × 18mm PM cartridge, sometimes also called the '9mm Soviet' or '9mm Russian'. Adopted by the Soviet Army in 1951, by the early 1980s it had become the standard pistol ammunition throughout the Warsaw Pact countries. It has also been widely used in China and a few other countries in South-East Asia and Africa.

It must be noted that the difference between 1902-vintage DWM-made 9mm and modern NATO-standard 9mm ammunition is even more insignificant than the differences between the 1902-dated Luger pistol and any current service 9mm pistol. There are certain differences in the chemical mixtures of primer compositions and powders, plus a few differences in the materials used, but the terminal (damage) effect on the target is about the same, as long as we consider military-issue ammunition. Limited to non-expanding projectiles by the Hague Convention, armies still issue lead-core full metal jacket bullets (FMJ) which are almost the same as 100 years ago. Only during the last decades have we seen certain developments in this field, fuelled by the ever-spreading use of body armour. Several countries have developed special armour-piercing (AP) bullets for pistol cartridges, with the intention of issuing such ammunition to military personnel, as well as for some special law enforcement users. In 2003, the Russian Army adopted domestically developed 9 × 19mm ammunition with AP bullets as the future standard for all military sidearms, but most armies still continue to issue standard 'ball' ammunition for their troops.

The first attempt to radically improve terminal effects on protected military personnel was made by the Belgian FN Company during the late 1980s and early 1990s. During the development of the 'project 9.0 personal defence weapon', FN produced the well-publicized P90 sub-machine gun and specialized 5.7mm ammunition, which combined low recoil, good accuracy and high body armour penetration. A few years later FN produced a 5.7mm military-type pistol, the FN Five-seveN, but this is yet to be adopted by any major military force. There have been a few more developments in the small-calibre niche, such as now-obsolete British .224 BOZ and the German 4.6 × 30mm HK (which is currently in use by German and British armed services along with the HK MP7 PDW/compact sub-machine gun), or the recent 5.8 × 21mm Chinese.

One of the most unobtrusive versions of the LAM is offered by the American LaserMax Company; this LAM unit replaces the standard recoil spring guide rod and does not change the outline of the firearm to any extent. The aiming point is also very close to the bore axis, which helps the accuracy. Shown here is the SIG-Sauer P229 pistol with the LaserMax LAM installed; similar units are available for many other popular combat pistols. (Courtesy of LaserMax)

Police forces, on the other hand, have no such legal limitations as the military, but they do have their own concerns, including liability issues. Since the 1960s, there has been constant development of various types of projectiles, intended to increase the wounding and incapacitating capabilities of issued weapons. There are two major routes for this increase – the first is to make the bullet expand on impact with the target and the second is to design the bullet to fragment. Either way results in faster energy transfer to the target and causes more damage when compared with the standard FMJ bullet. Added benefits of such projectiles are limited penetration and a lesser probability of ricochets.

It is interesting to note that, while some experts back in the 1970s and 1980s promised the advent of increasingly powerful service ammunition, this has never happened on a large scale. The reason is that the power of the bullet is directly related to the recoil and weight of the firing platform. If you make more powerful ammunition for a gun of limited weight, there will be increased recoil and the gun will eventually become uncomfortable and even uncontrollable to fire. You also cannot increase the weight of the weapon beyond certain reasonable limits, nor can you use more complicated recoil-mitigating designs (such as the long-recoil, locked-breech system) because of cost and reliability issues.

However, there have been a few successful attempts to develop more powerful service type ammunition over the last decades. These are the 10mm Auto (10 × 25), the .40 S&W (10 × 22) and the .357SIG (9 × 22) cartridges, all developed in the USA in the last decades of the twentieth century. The 10mm was an attempt to design an 'ideal' combat pistol round which would outclass both the 9mm and .45, while combining the larger magazine capacity of the former with the bigger terminal effect of the latter. The .40 S&W was developed from the 10mm in an attempt to provide smaller and lighter guns, while retaining the decent stopping power of the 10mm bullet. The .40 S&W has recently become very popular as police issue ammunition in the USA and in some South American countries, most notably Brazil. The .357SIG, which was based on a necked-down .40 S&W case, was an attempt to improve penetration when firing standard 9mm expanding bullets and also became quite popular with American law enforcement agencies. The only other comparable development is the Russian 9 × 21 ammunition, originally produced in AP loading for the FSB (Russian Federal Security Service). Approximately similar in ballistic performance to the .357SIG, the 9 × 21 Russian is now available in a variety of loadings, which include AP, tracer and expanding bullets. The rest of the world seems to be content with the ubiquitous 9 × 19 ammunition.

The Future of Combat Pistols

It is clear that combat handguns, like most other firearms, have reached their peak of development and even the most expensive research will not bring any significant increase in overall combat effectiveness. The only way to make a quantum leap in effectiveness would be to adopt an entirely new principle of bullet acceleration or an entirely new damage mechanism. The first can be achieved, for example, by introducing so-called 'Gauss guns', which can accelerate projectiles to significant velocities using electromagnetic fields. The second can be achieved by using entirely different kinds of energy for damage effect on the target, such as thermal energy, generated at the target by laser beam or by electric shock and delivered to the target using thin wire or plasma channels generated through the air. In either case, weapons based on such principles, while possible in theory, in reality would be far too big and expensive to be considered for any type of small arms, at least today. There could be other systems and principles, but none of these is likely to appear in small arms in the coming decade or two.

Considering the above, we should expect only gradual and, in most cases, minor improvements. The only field that can see some significant improvements is the safety of weapons, via the introduction of technology that will not permit the use of the weapons by unauthorized persons. Today, systems that grant access by reading and distinguishing fingerprints are built into laptop computers and mobile phones, but their reliability not yet reached the level necessary for combat firearms – tools that are used in moments of grave danger. But, once these systems become mature, they will significantly improve the safety of many law-enforcement officers, as well as ordinary citizens who are permitted to own and carry firearms.

Part II

Modern Combat Pistols

Argentina

The Argentinian Army was among the first on both American continents to adopt semi-automatic pistols, starting with the purchase of Austrian Mannlicher M1905 pistols. From 1927, the Mannlicher M1905 was gradually replaced in service with a more modern and effective pistol, the Colt M1911A1, known in Argentina as *Pistola Sistema Colt modelo 1927*. Early pistols were purchased from the USA. Later on, the local government-owned arms factory, *Fabrica Militar de Armas Portatiles* (FMAP), started to manufacture the same pistols under licence. American-made pistols had the usual Colt markings on the slide and 'Government model' rollmark on the frame; Argentinian-made pistols had a large DGFM (FMA.P) rollmark on the left side of the slide. Except for the markings and finish, these pistols were exactly the same as standard US M1911A1 pistols. During the late 1930s, the HAFDASA (*Hispano Argentina Fabrica de Automoviles Sociedad Anonima*) company produced a cheaper .45 calibre pistol, initially known as Ballester-Rigaud, and – since 1940 – as the Ballester-Molina. Broadly based on the same Colt M1911 design, this pistol was more similar to the Spanish-made Star Modelo P. Ballester-Molina pistols have seen significant use by Argentinian military and police forces and also were exported to Britain during World War II. HAFDASA stopped the production of Ballester-Molina pistols in around 1953.

In 1968, the General Directorship of Military Factories (*Direccion General de Fabricaciones Militares* – DGFM) purchased the manufacturing rights for the famous FN High Power pistols from Belgium. Production of the High Power pistols was started in 1969 at the *Fabrica Militar de Armas Portatiles Domingo Matheu* (FMAP 'DM') and is still continuing. The Argentinian Army was the major customer for these pistols, which were also exported to neighbouring countries such as Bolivia, El Salvador and Uruguay. Argentinian versions of the High Power are usually known as 'FM GP' (to distinguish them from Belgian-made FN GP pistols) and are available in several versions and configurations.

.45 calibre Sistema Colt modelo 1927 pistol, marked with the Argentinean Army crest.

.45 calibre Ballester-Molina pistol with Argentinian police markings.

The privately owned Bersa SA company started to manufacture smaller, pocket-type pistols during the 1960s. Since the early 1990s Bersa has also made several full-size and compact pistols, suitable for service purposes. Recently, the Argentinian Federal Police and the Buenos Aires Province Police have adopted the Bersa 'Thunder Nine' pistols as standard-issue sidearms. These pistols, along with the FM GP, are also widely exported from Argentina, being fairly popular because of their inexpensive prices and good workmanship.

Bersa Thunder 9

Trigger type: Double/single action
Calibre/ammunition used: 9 × 19mm or .40 S&W
Weight: 870g (30.5oz) empty
Length: 192mm (7.49in)
Barrel length: 110mm (4.29in)
Magazine capacity: 17 (9mm) or 13 (.40) cartridges

The Bersa Thunder-9 and Thunder-40 pistols are similar except for the calibre and magazine capacity. The design of these pistols was apparently inspired by the German Walther P88 pistol. The frames of the Bersa Thunder pistols are made from high-strength

Bersa Thunder 9 pistol.

aluminium alloy, the barrels and slides are steel. The pistols use a locked-breech, short-recoil-operated system of the modified Browning type. The barrel is locked to the slide by a single large lug, which engages the ejection port in the slide. Unlocking is caused by the cam-shaped underbarrel extension. The trigger is of the double-action type, with an exposed hammer and an ambidextrous frame-mounted safety, which decocks the hammer when engaged and then locks the slide and sear. The action also features a firing-pin safety, which blocks the firing pin until the trigger is pressed. The slide release lever is also ambidextrous and the magazine release button can be easily mounted on either side of the grip. Magazines are of the double-stack type. Sights are fixed, with the front sight integral to the slide and rear dovetailed to the frame. Grip panels are made from polymer.

Bersa also makes compact versions of the Thunder pistols, which are available in 9mm, .40 S&W and .45 ACP.

Manual safety Ambidextrous levers on the frame, above the grip panels. Push levers up for safe; push down for fire. The 'ready-to-fire' position is marked by red dots on the slide, which are covered by the levers when on safe.

Field stripping procedure 1) remove the magazine by pressing the magazine release button; 2) check that the chamber is empty; 3) turn down the disassembly lever, which is located at the left side of the frame, above the trigger guard; 4) pull the slide forwards until it slides out of the frame; 5) remove the barrel and recoil spring from the slide.

Reassemble in reverse order.

FM GP Modelo Militar.

FM GP/Browning Modelo Militar/ Detective

	Modelo Militar	Modelo Detective
Trigger type	Single action	
Calibre/ ammunition used	9 × 19mm Para	
Weight	929g (32.5oz) empty	916g (32.1oz) empty
Length	197mm (7.68in)	172mm (6.7in)
Barrel length	118mm (4.6in)	93mm (3.63in)
Magazine capacity	14 cartridges	14 cartridges

The FM GP/Modelo Militar is basically the same as the Belgian-made FN Browning High Power pistol, so for a detailed description please refer to the respective

FM GP Modelo Detective.

entry. Current production commercial pistols, marked as FN GP M95, are similar to FN High Power Mark III pistols and also have ambidextrous extended safety levers. The FM GP Modelo Detective pistols are exactly the same as standard FM GP pistols, except for a shorter barrel and slide. Conversion kits are available for standard FM GP pistols to convert them into Detective configuration. These kits consist of a new barrel, slide and recoil spring with guide.

AUSTRIA

After liberation from German occupation, Austria retained significant industrial potential. During World War II, most Austrian factories manufactured ammunition and arms of German design and in the early post-war period the Austrian *Bundesheer* (Armed forces) was armed with many German weapons, such as the Walther P38 pistols or the MP 40 sub-machine guns. Austrian police also used some pistols of German origin, such as the Walther PP and PPK, some of them being made before the war and some after. Austrian police also purchased Browning High Power pistols from Belgium.

In 1980, the Austrian Army announced trials with the intention of obtaining a new military handgun, to replace the obsolescent Walther P38 pistols. The basic requirements for a new pistol were as follows:

- calibre: 9 × 19 Parabellum
- magazine capacity: not less than 8 rounds
- ambidextrous handling
- field stripping and reassembly without special tools
- maximum number of parts: 58 (same as in the P38)
- complete parts interchangeability
- reliability: no more than 20 stoppages per 10,000 rounds fired
- drop resistance: pistol must remain intact and must not fire when dropped onto a steel plate from a 2 metre height.

Participants in the trials included such designs as the Steyr GB, Heckler & Koch P7M13, SIG-Sauer P220 and P225, Beretta 92S and the Glock 17. In 1982 the Austrian Army officially adopted the brand new Glock 17 as the Pistole (P80) and starting from 1986 Glock pistols gradually replaced the older P38 in military service. In the following years, thanks to their unorthodox design and aggressive marketing, Glock pistols became one of the most popular combat handguns, being adopted by numerous military and law-enforcement forces worldwide.

GLOCK

The creation of the Glock 17 pistol is a very interesting and successful attempt to develop a firearm using the 'clean sheet' concept. Instead of adopting an existing design to meet the requirements, Gaston Glock, owner of a relatively small company, Glock GmbH, called for the best designers and experts and asked them to create an ideal pistol. It was clear that the Austrian Army wanted a defensive pistol – simple, reliable, lightweight and inexpensive. The Glock team accordingly created what would be called the most utilitarian pistol of recent decades. The low weight was achieved by using a polymer frame – a solution probably borrowed from the Austrian Stg.77 (Steyr AUG) assault rifle, adopted in 1978. Simplicity of operation combined wiith a minimum number of parts was achieved by using a patented 'safe action' trigger system, which has no manual safeties.

After initial success with the basic Glock 17 pistol and its smaller counterpart, the Glock 19 (which appeared in 1988), the company developed and brought to market a large number of models, available in every major calibre from 9 × 17 Browning up to 10mm Auto and .45 ACP. There are three generations of production Glock pistols. The first generation lasted between 1986 and 1990 and included only the Glock 17 and 19 pistols. Second-generation pistols, which also included pistols chambered for .40 S&W, 10mm Auto and .45 ACP, had slightly different frames, with textured grips. The third generation had entered production by the mid-1990s, when frames received integral accessory rails and finger-grooved grips. So far, Glock pistols have been adopted by the armies of

9mm Glock 17 pistol, first generation. *(Courtesy of Cherry's Fine Guns)*

Glock Pistols Chambered for 9 × 19mm Luger/ Parabellum Ammunition

Model	**17&18 (selective fire)**	**19**	**26**	**34**
Length	186mm (7.3in)	174mm (6.8in)	160mm (6.2in)	207 mm (8.15in)
Barrel length	114mm (4.5in)	102mm (4in)	88mm (3.4in)	135mm (5.32in)
Weight empty	625g (21.9oz)	595g (20.8oz)	560g (19.6oz)	650g (22.96oz)
Magazine capacity	17, 19, 33 rounds	15, 17, 19, 33 rounds	10, 12, 15-33 rounds	17, 19, 33 rounds

Glock Pistols Chambered for .357 SIG and .40 S&W

Model (calibre)	**22 (.40) 31 (.357)**	**23 (.40) 32 (.357)**	**27 (.40) 33 (.357)**	**35 (.40)**
Length	186mm (7.3in)	174mm (6.8in)	160mm (6.2in)	207mm (8.1in)
Barrel length	114mm (4.5in)	102mm (4in)	88mm (3.4in)	135mm (5.3in)
Weight empty	650g (22.8oz)	600g (21oz)	560g (19.6oz)	695g (24.3oz)
Magazine capacity	15, 17 rounds	13, 15, 17 rounds	9, 11, 13+ rounds	15, 17 rounds

Glock Pistols Chambered for 10mm Auto and .45 ACP

Model (calibre)	**20 (10mm) 21 (.45)**	**29 (10mm) 30 (.45)**	**36 (.45)**
Length	193mm (7.5in)	172mm (6.7in)	172mm (6.7in)
Barrel length	117mm (4.6in)	96mm (3.7in)	96mm (3.7in)
Weight empty	785g (27.5oz)	700g (24.5oz)	570g (19.6oz)
Magazine capacity	15 (10mm) rounds 13 (.45) rounds	10 rounds	6 rounds (single stack magazine)

Austria and Sweden, as well as by large numbers of law-enforcement organizations worldwide.

It must be noted that while many experts describe Glock pistols as having a 'revolutionary design', in fact the design was the result of carefully considered evolution, incorporating previously known solutions into one extremely practical, rugged package. For example, polymer frames had been known for ten years before the introduction of the Glock 17; automated safeties built into the trigger were known from the 1930s; and pre-cocked firing pins from the first decade of the twentieth century. While extreme ruggedness, affordable price and ease of use made Glock pistols top sellers for about two decades, by the start of the twenty-first century their position had somewhat declined. In the view of the author, there are two key problems that

.45 ACP Glock 21 pistol, second generation. (Courtesy of Cherry's Fine Guns)

the Glock Company has so far failed to address. The first is ergonomics – the Glock pistol was designed for Army use and was intended to be a 'one size fits all' proposition. In fact, the square-shaped grip of the typical full-size Glock pistol is of less than ideal shape, at least for many shooters, the author included. The second issue is the famed 'safe-action' trigger. There have been too many accidental and negligent discharges of Glock pistols, because of the lack of any manual safeties, combined with a relatively low trigger pull weight. Upgrades such as the 'New York trigger pack', which included heavier springs, did help a little, but in this configuration there were few advantages over the other pistols with DAO triggers, which were as safe to fire as upgraded Glocks, but allowed a rapid second strike in the case of a misfire. Of course, the basic reason for these accidents was insufficient training of the users, but it is customary for most people to blame tools for operator errors.

Glock pistols are short-recoil-operated, locked-breech weapons, except for two models (Glock 25 and 28), which are chambered for low-power 9 × 17 Browning ammunition and use a blowback action. The barrel is locked into the slide by a single large lug, which engages the ejection window in the top of the slide. Unlocking upon recoil is achieved by a single cam-shaped lug, which interacts with the steel insert, moulded into the polymer frame. The same insert also holds the slide rails, which cannot be replaced in the frame in the case of wear or damage. The patented 'safe-action' trigger system has a firing pin, with the mainspring being preloaded during the reloading cycle. The full cocking of the striker is achieved only by the pull of the trigger. There are no manual safeties on Glock pistols. Automated safeties include the trigger safety, which blocks the trigger if it is not pulled properly, the firing-pin block and the out-of-battery disconnector, which does not permit fire if the barrel is not fully locked to the slide.

Double-stack magazines are made from polymer, the standard magazine capacity for the Glock 17 being seventeen 9mm rounds, with extended magazines available in nineteen- and thirty-three round capacity. Glock 18 pistols, which are about the only full-auto

Three 9mm third generation Glock pistols of all available sizes, top to bottom – models 17, 19 and 26.

ABOVE: *Glock 18 automatic pistol. Note the fire mode selector on the slide.*

RIGHT: *Glock 17, partially disassembled.*

capable pistols currently made in the world, have slightly different dimensions of slide-to-frame rails, so the Glock 18 slide with its fire-selector mechanism cannot be installed on any other 9mm frame. However, several companies now offer add-on modules, which can convert any Glock pistol to selective fire. Such modules are usually installed at the rear of the slide, replacing the rear slide plate, while original Glock 18 pistols have a fire mode selector mounted on the left side of the slide. The standard sights are fixed, with the rear sight dovetailed to the frame. Latest generation pistols also feature integral accessory rails on the frame, below the barrel.

Steyr Pi-18 pistol, prototype model. Note that it had a manual safety instead of simple decocking.

Manual safety None.

Field stripping procedure 1) remove the magazine by pressing the magazine release button; 2) check that the chamber is empty; 3) while pointing the pistol in a safe direction, pull the trigger to release the firing-pin spring; 4) retract the slide for about 3mm (0.1in) and hold it, then push the disassembly buttons at both sides of the frame (above the trigger); 5) while holding the disassembly buttons down, pull the slide assembly forwards and out of the frame; 6) remove the barrel and return the spring from the slide.

Reassemble in reverse order.

STEYR GB

Trigger type: Double/single action
Calibre/ammunition used: 9 × 19mm Luger/Parabellum
Weight: 845g (29.6oz) empty; 1,285g (45oz) loaded with 18 rounds
Length: 216mm (8.42in)
Barrel length: 136mm (5.3in)
Magazine capacity: 18 rounds

The famous Austrian company Steyr-Daimler-Puch AG (now Steyr-Mannlicher) began the development of a new, high-capacity 9mm pistol in 1969, at the request of the Austrian military. The first prototype pistols, designated Steyr Pi-18 (Pistolc 18) had appeared by 1974 and it is believed that some of the prototypes were made as selective-fire weapons. However, for various reasons the government suspended the project and Steyr sold the licence and documentation for the Pi-18

Steyr GB pistol.

Same pistol; slide is locked back by a slide stop.

to its US-based importer and distributor, Rogak Inc. During the late 1970s and early 80s Rogak produced about 2,000 P-18 pistols, which were very roughly made stainless-steel copies of the Austrian pistol. In 1980, with the holding of new military pistol trials in Austria, Steyr revived the Pi-18 project, supplying modified GB pistols to the trials. As the trials were won by the Glock pistol, Steyr instead began commercial production of its pistol, while also submitting it for US trials for a new 9mm military sidearm. The GB failed in these trials, which were won by the Italian Beretta 92 pistol, and in 1986 the Steyr company ceased the manufacture of the GB. It is believed that a total of between 15–20,000 GB pistols were made. Most of these pistols were sold on the commercial market, where the GB was relatively popular due to its large magazine capacity and good accuracy.

Steyr GB, partially disassembled. Note the gas cylinder/slide bushing removed from the front of the slide.

The Steyr GB is a delayed blowback pistol with a fixed barrel, based on the WWII-era Barnitzke system. The opening of the slide is delayed by the pressure of the gases, which are bled from the middle of the barrel via two gas ports. The gas cylinder is located around the front part of the barrel and is attached to the removable front slide bushing. While the bullet is in the barrel and pressure is high, it delays the opening of the slide by acting on the front part of the gas cylinder/slide bushing. Once the bullet leaves the barrel, the pressure drops and the slide then recoils free, completing the reloading cycle. The frame of the GB is made from stamped and formed steel, welded together where necessary. The trigger guard is a separate assembly, permanently fixed to the frame. The double-action trigger unit has an external hammer and a slide-mounted decocking lever, combined with an automatic firing-pin block, which does not permit the movement of the firing pin if the trigger is not completely pulled.

Manual safety None; a lever at the left side of the slide safely drops the hammer when pressed, then returns to the original position when released.

Field stripping procedure 1) remove the magazine by pressing the magazine release button; 2) check that the chamber is empty; 3) turn down the takedown lever, located above the trigger at the left side of the frame; 4) twist and remove the barrel bushing; 5) remove the captive return spring assembly; 6) pull the slide back and off the frame.

Reassemble in reverse order.

Steyr M and M-1A

Trigger type: Double-action only, pre-cocked
Calibre/ammunition used: .40 S&W, 9 × 19mm Para, .357 SIG
Weight: 780g (27.3oz) empty
Length: 180mm (7in)
Barrel length: 101mm (3.9in)
Magazine capacity: 12 (.40 or .357) and 15 (9mm) rounds

The Steyr M pistol was introduced to the shooting public in 1999, along with its smaller version, the Steyr S. In January 2004 the Steyr-Mannlicher company introduced an updated version of the M-series pistol, designated Steyr M-A1, which replaced in production the earlier M model; Steyr S pistols were discontinued in 2003. The Steyr M pistols are among the most ergonomically designed handguns of recent decades, with excellent pointing characteristics and recoil management. These pistols have several unusual features, such as trapezoidal sights and an unorthodox manual safety, but they are quite comfortable and require not much familiarization time. So far, the only known government users of Steyr M pistols are the Malaysian police forces, which bought significant stocks of these pistols to replace the Taurus .38 calibre revolvers.

The Steyr M is a short-recoil-operated, locked-breech pistol. It uses a Browning-type tilting barrel system. Barrel movement is controlled by a cam-shaped lug under the chamber; slide locking is achieved by the one large lug on the barrel, which engages the ejection port in the slide. The frame of the pistol is made from impact-resistant polymer. The barrel unlocking surface in the frame and the slide rails are combined in a single removable unit, made of steel, which is easily replaceable if worn or damaged. The trigger unit is of the DAO type, with the striker being partially pre-cocked upon cycling the slide.

Original version of the Steyr M pistol, in 9mm. (Courtesy of Cherry's Fine Guns)

Current production M-1A pistols are available in two versions, one with a manual safety and one without. The manual safety comprises two 'engage safety' buttons, located at both sides of the frame,

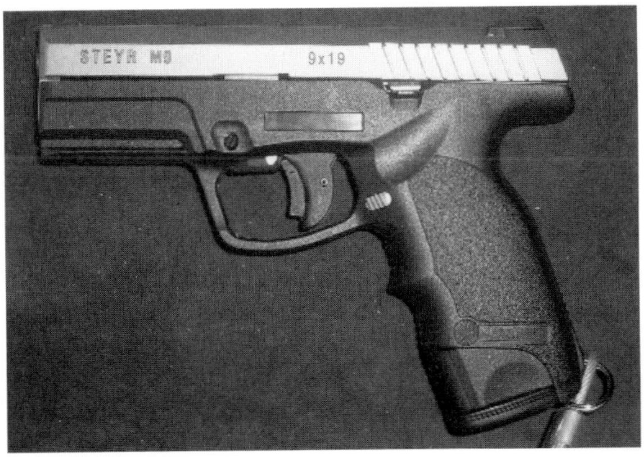

Current production Steyr M-A1 pistol, 9mm. The safety is on and the safety release button (marked with a white dot) is visible inside the trigger guard, just ahead of the trigger.

.40 S&W Steyr M-A1 pistol; current production model with no manual safety.

The trapezoid sights of the Steyr M pistol are most unusual, but quite comfortable once you get used to them. The round object just below the sights is the firing-pin cocking indicator.

above the trigger, and one 'disengage safety' button, located inside the trigger guard, just ahead of the trigger. There is also an external key lock, located at the left side of the frame. Once engaged, the key lock blocks the firing mechanism, locks the slide and does not permit disassembly of a pistol. All Steyr M pistols are also fitted with an automated trigger safety. The standard open sights are of unusual form, with a triangular-shaped front sight and the rear sight with a trapezoidal opening (instead of the more traditional V- or U-shaped opening). Both front and rear sights have white marks for better contrast and both are dovetailed to the slide, so, if necessary, the user can replace the factory sights with aftermarket ones. The bottom of the frame under the barrel is shaped as an integral accessory rail, with proprietary dimensions on pre-2004 Steyr M pistols and with a Picatinny-style rail on current production pistols.

Manual safety (applicable only to versions with safety installed) To set on safe: press inwards simultaneously two buttons, located above the trigger; the push-button will appear inside the trigger guard, with white dots visible on either side.

To disengage safety and set the pistol to ready: push up the button inside the trigger guard with the index finger.

Integral key lock Located next to the disassembly lever at the right side of the frame, above the trigger. Once locked with an external key, this completely blocks the operation of the pistol, except for magazine insertion/removal.

Field stripping procedure 1) remove the magazine by pressing the magazine release button; 2) check that the chamber is empty; 3) while pointing the pistol in a safe direction, pull the trigger to release the firing-pin spring; 4) press down the lock button, which is located next to the disassembly lever at the right side of the frame, then pull the lever down; 5) pull the slide assembly forwards and out of the frame.

Reassemble in reverse order.

Belgium

For several centuries Belgium has been one of the leading manufacturers and exporters of small arms in Europe, a position it still retains, mostly thanks to the famous *Fabrique Nationale Herstal* (FN), located near the city of Liège. After the liberation of Belgium in 1945 FN continued its operations, supplying the world with high-quality small arms, from pistols to heavy machine guns.

During the post-war decades, Belgian police were armed mainly with FN/Browning Mle.1922 pistols, usually in 9 × 17 Browning Short calibre. The Belgian military continued to purchase and issue the famous FN GP-35 (FN Browning High Power) pistol, which still remains the standard sidearm for the Belgian armed forces. Belgian police during the 1980s switched from the older 7.65mm and 9 × 17mm pistols to the more potent 9 × 19 pistols. Since there's no single standard sidearm approved for police service in Belgium, each local police department buys whatever pistols they find appropriate (and affordable).

During the post-war rearmament and NATO standardization of the 1950s, FN produced large numbers of 9mm GP-35 pistols for export worldwide. FN also continued to manufacture smaller pistols, intended for police and civilian use, including versions of the FN Browning Mle.1910 and Mle.1922 pistols, as well as even smaller 6.35mm FN Baby 'vest-pocket' pistols. During the early 1950s, the FN's chief designer, Dieudonné Saive, decided that high-capacity, 9mm double-action pistols had a great future. He then quickly developed a nice-looking military-type pistol, which combined basic GP-35-type action and a high-capacity magazine with a double-action trigger. Unfortunately, FN management rejected this prototype, generally known as FN M1952DA, on the grounds that it would impact on the sales of the High Power pistol, the flagship of the FN pistol line. Had the executives at FN the ability to see a little further ahead, the 'Wonder Nine' revolution could have happened about two decades earlier than it did in reality. In fact, at the time there was only one other high-capacity, double-action, 9mm combat pistol – the Soviet-made Stechkin APS – which was, obviously, unavailable to Western buyers. While the GP-35 still remains in production and is in use by several military forces worldwide, the double-action pistols generally took

ABOVE: 1952 prototype 9mm FN double-action pistol, designed by D. Saive.

RIGHT: 9 × 17mm FN Model 125 pistol, a post-war modification of the FN Browning Mle.1922. This particular gun has been used by Dutch police.

over the military and police markets during the 1970s and 1980s and, unfortunately, FN entered this market niche a little too late.

FN first tried to market its new double-action pistol to the US military, submitting the newly developed FN BDA9 (Browning Double Action 9mm) pistols to the XM9 trials during the early 1980s. FN also submitted a modified FN GP-FA pistol, a single-action version of the venerable GP-35, fitted with the patented 'Fast Action' trigger. Neither FN design succeeded in the trials and FN decided to shelve the GP-FA but bring the BDA9 to the market. Sales of the BDA9 line of pistols (which included standard DA or DAO pistols in full size, medium size with standard grip and shorter barrels and compacts with cut-down barrels and grips) were less than satisfactory, so in 1987 FN discontinued this line. During the same period of time, the Browning Arms Company, a US subsidiary and importing agent to FN, felt the need for a compact double-action pistol. Browning Arms entered into contract with the famous Italian arms maker, Pietro Beretta, and the Italian company began production of the Browning BDA380 pistol, also known in Europe as FN Model 140DA. This pistol, which was sold to civilian markets in the US and also saw limited use by several European police and security forces, is an adaptation of the Berretta Model 84 pistol which is described in detail in the Italy section.

Today, the BDA380 is still being produced in Italy for Browning. However, this relatively small pistol has no place in the modern military and law-enforcement market, and by 1990 FN had reintroduced a slightly modified line of 9mm double-action pistols. Known as Browning BDA9 in the civilian market, this pistol is offered for government and police buyers as FN GP-DA or HP-DA. This pistol has seen limited use by various European police agencies and also has been adopted by the Finnish Army as a standard military sidearm. It must be noted that the 'Browning BDA9' designation had previously been used for an entirely different pistol, as during the late 1970s Browning Arms Co., then an independent company, imported into the USA some German-made SIG-Sauer P220 pistols (in 9mm, .38 Super and .45 ACP), and marketed them as the Browning BDA. Once the SIG-Sauer conglomerate (now SIGARMS company) had established its own subsidiary in the USA, it took over the import of SIG-Sauer pistol and now SIG/Browning pistols are a rarity.

With all of these decades of conservative thinking (at least in the field of pistols), in 1995 FN surprised the world with its revolutionary Five-seveN military handgun. This was the defensive military pistol in its essence – lightweight, extremely simple to operate and maintain, with low recoil and a large magazine capacity. Specially designed ammunition of 5.7mm calibre, available in various projectile types with an armour-piercing bullet loading being standard, allowed for decent effectiveness against enemy military personnel wearing body armour and steel helmets. Today, about twenty years since the introduction of 5.7 × 28 ammunition (which emerged during late 1980s along with the FN P90 sub-machine gun/personal defence weapon), its terminal effectiveness is still questioned by many experts. However, there is a very simple concept which makes the 5.7mm family of weapons very promising for military use. This concept is encapsulated in old American saying: 'It is better to hit the target with a .22, than to miss with a .45.' Indeed, the low recoil and large magazine capacity greatly increase the chances of inflicting at least some damage to opponents during a hot firefight. And, in most cases, even a lightly wounded enemy soldier will put much less effort into fighting. However, the prospect of the introduction of an entirely new calibre to the already complicated military chain of supply, as well as the conservative thinking of some experts and decision-makers, still delays the wide adoption of 5.7mm weapons as a military issue.

Surprisingly, both the P90 sub-machine gun and the Five-seveN pistol have attracted a good deal of attention from special purpose elements of law-enforcement agencies in several countries, including the USA. These people appreciate the light weight, large magazine capacity, serious penetration and great accuracy potential (once some necessary changes have been made to the weapons). Recognizing this new niche, by the late 1990s FN had produced an 'offensive' version of the Five-seveN pistol, designated 'Five-seveN Tactical'. This pistol replaced the 'foolproof' DAO trigger of the original Five-seveN with a single-action trigger, combined with a manual safety. This set-up is much more conducive to accuracy, but requires more training and attention when operating the pistol. Today, several special purpose law enforcement units in the world, including a few US agencies, are using the Five-seveN Tactical pistols along with FN P90 sub-machine guns.

5.7mm Five-seveN pistol of early manufacture.

As the Five-seveN had a slow start, FN decided to produce a more traditional pistol, intended mostly for the police market. First shown in 1999, the new FN Forty-Nine pistol was also a typical 'defensive' arm, with lightweight construction thanks to a polymer frame and with a simple DAO trigger. Designed around the relatively new .40 S&W cartridge, which is very popular in the USA, the Forty-Nine is also available in a 9mm version. However, this pistol received little attention from the shooting public, neither civilian nor police, and in 2003 FN introduced its most modern pistol, the FNP9. This is a very versatile pistol, because of its modular trigger unit and grip and it can be effectively used for all purposes, be that civilian self-defence, police work, or military issue. So far, the only forces to adopt the FNP9 are the Belgian police and Luxembourg military, but it is possible that this pistol will see more widespread use in the near future.

FN *Grande Puissance* Mle. 1935/Browning High Power/FN HP-SA

Data for Current Production (Mark III) Pistols

Trigger type: Single action
Calibre/ammunition used: 9 × 19mm Para, also .40 S&W
Weight: 885g (31oz) empty; 1,080g (37.8oz) loaded with 13 rounds
Length: 200mm (7.8in)
Barrel length: 118mm (4.6in)
Magazine capacity: 10 (.40) or 13 (9mm) rounds

John Browning developed the first high-capacity 9mm semi-automatic pistol of military type in 1922. This development was based on French requirements for a new military sidearm, submitted to FN in 1921. In this work, John Browning was supported and supplemented by his son, Val Browning and, especially, by his assistant and apprentice, Dieudonné Saive. Monsieur Saive was responsible for the development of a working high-capacity magazine for the new pistol and, after the death of John Browning, he assumed the responsibility of finalizing the design. In fact, most of the work for the 'Browning High Power' pistol had been done under the Saive's supervision, with John Browning's key role in this development being the modified barrel locking/unlocking system, based on his early patents. Browning had submitted the patent application for a new pistol in 1921, but the US patent was issued only in 1927, several months after his death. It must be noted that the design went through many changes between 1922 and 1934, and in 1935 FN started the production of the *Grande Puissance* (High Power) pistol.

Typical FN GP-35 pistol of post-war manufacture.

FN Browning High Power Mark II (HP Mk2).

If there is a single most successful 9mm combat pistol, the GP-35 is the one. Still being manufactured, it remains the yardstick against which all other 9mm pistols are measured. GP-35 pistols are known for excellent reliability and durability; they are also quite accurate and have one of the most comfortable grips of all combat-type pistols. There are certain complaints about the trigger pull, which is to some extent impaired by the magazine safety, but actually this is really only an issue for target shooters. GP-35 pistols are still in use by Argentinian, Belgian, British, Canadian, Indian, Israeli and several other armed forces. Clones and copies of this pistol are or were made in Argentina, Bulgaria, Canada, China, Hungary and Israel. Fabrique Nationale also continues to produce modern versions of the GP-35, which are sold under the 'Browning' trademark for civilian markets and under the FN trademark for law enforcement and military markets.

Of course, the basic design has evolved with age. The tangent sight and shoulder stock options were dropped soon after World War II, based on wartime experience. More changes had appeared by the mid-1960s, when FN engineers replaced the original internal extractor with a more reliable and accessible external extractor and also replaced the ring on the hammer with a spur. In 1982, FN introduced the 'High Power Mark II' (Mk2) pistol, which had improved plastic grip panels of better shape, as well as extended and ambidextrous safety levers. Other improvements included a small ring on the front of the slide, which

FN Browning High Power Mark III (HP Mk3). (Courtesy of FN)

FN Browning High Power Safety Fast Shooting (GP-SFS). (Courtesy of FN)

FN GP-35, partially disassembled.

was intended to protect the barrel crown, as well as an improved feeding ramp, necessary for reliable shooting with hollow-point ammunition.

At about the same time, FN designers developed a proprietary 'Fast Action' system, with the intention of allowing the safe carry of a single-action pistol with the hammer lowered on a chambered round. Pistols with 'Fast Action' triggers were submitted to the US XM9 trials of the mid-1980s, but did not succeed. For unknown reasons FN shelved the 'Fast Action' system for almost twenty years, reintroducing it only in around 2002 as the 'Safety Fast Shooting' (SFS) system in a modified High Power pistol. In the meantime, FN had introduced a third generation of High Power pistols in 1987, as the 'High Power Mark III' (Mk3). The most important change was in the method of manufacture – instead of expensive machining, FN turned to a precision casting process and also strengthened some parts for enhanced durability. In the mid-1990s FN also introduced the .40 calibre High Power pistol, which is more popular among civilian shooters, thanks to its accuracy and relatively mild (for its calibre) recoil. The slide of the .40 S&W model was beefed up to handle safely the increased recoil impulse of the heavier round.

The High Power pistol is a short-recoil-operated, locked-breech pistol of all-steel construction. The classic locking scheme, developed by Browning, utilizes two (three for .40 calibre models) lugs, machined on the top of the barrel just ahead of the chamber area. These lugs lock into the recesses on the inside of the slide. To lock and unlock, the High Power uses a cam-shaped lug under the breech of the barrel, which interacts with the cross-pin, permanently installed in the frame. The standard trigger is of the single-action type, with an exposed hammer. A manual safety lever is located on the frame (ambidextrous on Mk2 and Mk3 models) and there are three other automatic safeties – one is a magazine disconnect (which blocks the trigger if the magazine is removed), another is a disconnector which prevents firing if the slide is not fully closed and locked to the barrel and, finally, there is a safety half-cock notch on the hammer. The Mk3 versions also have an additional automatic firing-pin safety.

In the 'Fast Action' and 'Safety Fast Shooting', the gun is so designed that the hammer can be lowered onto the loaded chamber without the trigger being pulled, just by pushing the hammer forward. This movement forces the 'safety' lever to pop up, locking the sear and slide so that the gun can be carried safely in 'condition two', with the hammer down but the main spring cocked. To get the gun into the 'ready' position, the shooter pulls the lever down, unlocking the slide and sear; the hammer automatically pops into the cocked position and the pistol then can be fired in single-action mode.

Most post-war High Power pistols were made with fixed sights, or with target-type adjustable sights for civilian use. Pre-war-type set-ups with tangent rear sights, adjustable for a range of between 50 and 500m, were available for some time in the 'High Power Captain' version, as well as in a few military contract guns. The standard magazine holds thirteen rounds, although there are some aftermarket magazines with capacities up to thirty rounds. The pistols are available in a wide variety of external finishes, from practical Parkerizing or bluing up to polished chrome and elaborate engravings, with wooden or plastic grip panels.

Manual safety (for all versions except the GP-SFS) To set on safe: push up the lever at the rear of the frame. Note: the safety works only with the hammer in full-cock position.

To disengage safety: pull the lever down.

Field stripping procedure 1) remove the magazine by pressing the magazine release button; 2) check that the chamber is empty; 3) pull the slide all the way back and lock it with the safety lever; 4) remove the slide stop first by pushing its axis pin at the right side of the frame, then by pulling it completely out to the left; 5) while holding the slide back, disengage the safety and then ease the slide and barrel group forwards and out of the frame; 6) remove the return spring by pushing its head slightly forwards and then pulling it out of the detent in the barrel lug; 7) lift up the barrel from the slide.

Reassemble in reverse order.

FN 140DA/Browning BDA-380

Trigger type: Double/single action
Calibre/ammunition used: 9 × 17 Browning
Weight: 640g (22.4oz) empty
Length: 172mm (6.7in)
Barrel length: 81mm (3.2in)
Magazine capacity: 13 rounds

The FN 140DA/Browning BDA-380 pistol is based on the Beretta model 84 pistol and is made in Italy for FN and Browning. BDA-380 is a blowback pistol with an aluminium alloy frame and a steel slide. The barrel does not recoil, but unlike most other blowback pistols it is not fixed permanently to the frame and the return spring is located below the barrel. The double-action trigger unit has an external hammer, with ambidextrous safety/decocking levers located on the slide. The external extractor also acts as a loaded chamber indicator and protrudes from the slide if a cartridge is chambered. The double-stack magazine holds thirteen rounds and is interchangeable with that for Italian Beretta 84 pistols. Sights are of the fixed type, with the rear sight dovetailed to the slide.

Manual safety To set on safe: push up the lever on the slide. If the hammer is in the cocked position, it will be lowered (decocked) automatically. To disengage the safety: pull the lever down; a red dot will appear on the slide, the hammer will remain in the lowered position.

Browning BDA-380 pistol

Field stripping procedure 1) remove the magazine by pressing the magazine release button; 2) check that the chamber is empty; 3) pull the slide all the way back and lock it with the slide-stop lever; 4) depress the takedown lever stop, which is the small button located on the left frame above the trigger guard; 5) while holding the button down, rotate the takedown lever (located at the right side of the frame) downwards for about 90 degrees ; 6) set the safety to the 'off' position; 7) pull the slide slightly back to disengage the slide stop, then carefully push it all the way forwards and out of the frame; note that the barrel will be removed along with the slide and return spring; 8) remove the return spring by pushing its head slightly forwards and then pulling it out of the detent in the barrel; 9) lift up the barrel from the slide.

Reassemble in reverse order.

FN HP-DA, HP-DAO, HP-DAC/FN GP-DA/ Browning BDA9

	HP-DA	HP-DAC
Trigger type	Double/single action or double-action only	
Calibre/ ammunition used	9 × 19mm Luger/Parabellum, .40 S&W	
Weight	905g (35.3oz) empty	710g (24.9oz) empty
Length	200mm (7.8in)	173mm (6.8in)
Barrel length	118mm (4.6in)	96mm (3.7in)
Magazine capacity	15 rounds	10 or 15 rounds

FN HP-DA pistols are short-recoil-operated, locked-breech weapons of all-steel construction. The locking system is basically similar to that of the FN GP-35, described above. The trigger is of the double-action or double-action-only type, depending on the model (HP-DA or HP-DAO, respectively). In both variants, there is an automatic firing-pin block safety; in the HP-DA, there are ambidextrous decocking levers, located on both sides of the frame, above the grip panels. It must be noted that decocking systems on first-generation models (made 1983–7) were different from those of the second generation (made since 1990). On first-generation HP-DA pistols, these levers were of slightly arched shape and had to be pushed up to decock the hammer, whereas current production HP-DA pistols have flat levers that must be pushed down for

Early production 9mm FN GP-DA (BDA9) pistol, made between 1983 and 1987.

9mm FN GP-DA (BDA9) of current manufacture.

9mm FN GP-DA Compact (BDA9C) of current manufacture. Insert shows part of FN GP-DAO Compact (without the decocking lever and with the spurless hammer). *(Courtesy of FN)*

decocking. HP-DAO pistols have no such levers and also have spurless hammers. HP-DA pistols made between 1983–7 were available in three sizes – standard (full-size), Medium (HP-DAM, with shortened barrel and standard grip) and compact (HP-DAC, with shortened barrel and grip). Current production pistols are available in two sizes only, standard and compact (HP-DAC). While compact pistols have a shorter grip and are equipped with shorter magazines of smaller capacity, these pistols will also accept large-capacity, full-size magazines from standard model pistols. All HP-DA pistols are fitted with fixed sights as standard, with both front and rear sights being dovetailed into the slide.

Manual safety None; ambidextrous decocking lever mounted on the frame above the grips.

Field stripping procedure 1) remove the magazine by pressing the magazine release button; 2) check that the chamber is empty; 3) remove the slide stop first by pushing its axis pin at the right side of the frame and then by pulling it completely out to the left; 4) release the slide and barrel group forwards and out of the frame; 5) remove the return spring by pushing its head slightly forwards and then pulling it out of the detent in the barrel lug; 6) lift up the barrel from the slide.

Reassemble in reverse order.

FN Five-seveN

The Five-seveN pistol (its designation is always written in words, not in digits such as '5-7') had been developed by 1995 as a military-type defensive pistol, with emphasis on ease of use, ease of shooting with minimum recoil and high probability of effective hits against enemy personnel wearing body armour. However, to date the Five-seveN pistol has mainly found a niche in the hands of special purpose police and security forces, such as various SWAT units and the US Secret Service. The only military force that had adopted the Five-seveN pistols by early 2005 was the Belgian Air Force, which replaced its venerable 9mm FN Browning GP-35 pistols with the standard (DAO) version of the Five-seveN.

The key to the capabilities of the FN Five-seveN is its ammunition, a purpose-designed 5.7 × 28mm

FN Five-seveN

	Five-seveN	Five-seveN Tactical
Trigger type	Double-action only	Single action
Calibre/ammunition used	5.7 × 28mm	
Weight	610g (21.4oz) empty 750g (26.3oz) loaded with 20 rounds	620g (21.7oz) empty 760g (26.6oz) loaded with 20 rounds
Length	208mm (8.1in)	
Barrel length	122mm (4.8in)	
Magazine capacity	20 rounds	

cartridge, which looks like a scaled-down 5.56 × 45mm NATO assault rifle round. Fitted with high-velocity pointed bullets, this ammunition provides low recoil, flat trajectory and good penetration against body armour. The rapid tumbling of the long bullets in soft body tissue provides the effectiveness against 'soft' targets. Initially developed as a pure 'government-issue only' firearm, since 2004 the FN Five-seveN has also been sold on the US civilian market. Since the armour-piercing ammunition is not permitted for civilian use in most jurisdictions, only 'training' soft-core ammunition is sold for civilian shooters.

The FN Five-seveN pistol is a short-recoil, retarded (delayed) blowback-operated pistol. The barrel of the FN Five-seveN is allowed to recoil in the frame for a very short distance. The captive return spring is mounted around the barrel and there is a U-shaped lever, which is located in the frame, below the barrel. When the slide is in battery, the raised hands of the lever are connected to the notches in the slide. When the pistol is fired, the slide begins to recoil, turning the lever back. This movement of the lever carries the barrel back, adding its mass to the mass of the recoiling slide, thus slowing down (retarding) the initial opening of the slide. As soon as the barrel reaches its rearmost position in the short recoil, the U-shaped lever turns down enough to disengage its arms from the slide, thus allowing it to recoil freely and complete the reloading cycle.

The FN Five-seveN pistol has a polymer frame with

ABOVE: *FN Five-seveN Tactical pistol, 2004 production. This one is fitted with low profile fixed sights and with a Picatinny-type accessory rail on the frame. Note the ambidextrous safety lever just above the trigger.* (Courtesy of FN)

RIGHT: *FN Five-seveN, partially disassembled.* (Courtesy of FN)

steel inserts and steel slide rails; the slide is formed from stamped steel, with a steel breechblock insert and polymer coating. The trigger unit of the standard FN Five-seveN pistol is striker fired, DAO, so the firing pin (striker) is always at rest, except for the moment the trigger is pulled. There is an automatic firing-pin block for added safety. The FN Five-seveN Tactical pistols, on the other hand, have single-action triggers, also striker fired. To provide the necessary safety when the pistol is carried, there are ambidextrous safety levers, located above the trigger on the frame. FN Five-seveN pistols are fitted with low-profile, combat-type fixed sights, dovetailed to the slide; models sold to the civilian market can be fitted with low-profile, sport-type adjustable rear sights. The front part of the frame is shaped as an accessory rail to accept tactical lights and/or laser aiming modules. Magazines are made from polymer and hold twenty rounds of ammunition.

Manual safety None on the standard FN Five-seveN pistols; ambidextrous safety levers mounted on the frame above the trigger on FN Five-seveN Tactical pistols. Push it up to set on 'Safe'; pull down to disengage safety; a red dot will appear on the frame above the lever.

Field stripping procedure 1) remove the magazine by pressing the magazine release button; 2) check that the chamber is empty; 3) pull back the takedown lever, which is located at the left side of the frame, above the front of the trigger guard; 4) pull the slide back slightly, then release it forwards and out of the frame; 5) take the barrel with the integral return spring assembly out of the slide.

Reassemble in reverse order.

FN FORTY-NINE

Trigger type: Double-action only
Calibre/ammunition used: 9 × 19mm Para, .40 S&W
Weight: 730g (25.6oz) empty
Length: 197mm (7.7in)
Barrel length: 108mm (4.2in)
Magazine capacity: 16 (9mm) or 14 (.40) rounds

The FN Forty-Nine pistol has been developed especially for the US market. It is unavailable in European FN catalogues and is being manufactured in the US-based FN plant. For this reason, it incorporates the most desirable features for such a niche, like light weight (thanks to its polymer frame), safety and simplicity of operation (thanks to its DAO trigger) and serious terminal effectiveness and stopping power, due to the .40 S&W ammunition, selected as the primary chambering for this pistol. The first prototypes of the FN Forty-Nine appeared in 1997 and mass production began in 1999, with the 9mm version of the pistol appearing a little later. However, despite all of its positive properties, plus the traditional robustness and quality typical of an FN product, the Forty-Nine has not been a great success in the market. It is hard to judge why, as the only known drawback of this pistol is a heavier-than-average trigger pull (about 5kg/11lb).

The FN Forty-Nine is a short-recoil-operated,

Forty-Nine pistol in .40 S&W calibre. (Courtesy of FN)

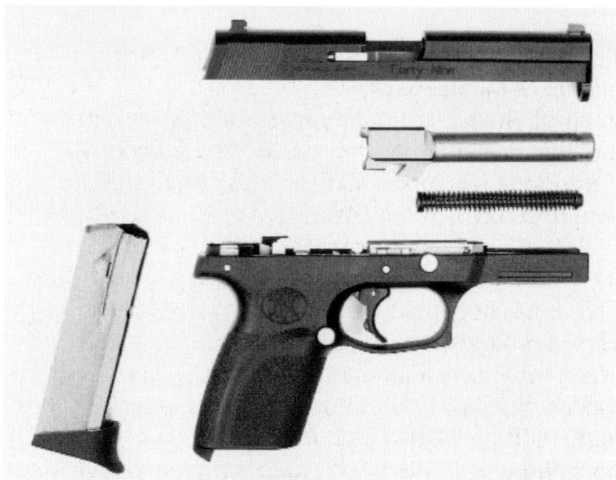

Forty-Nine pistol, partially disassembled. (Courtesy of FN)

locked-breech pistol. The barrel is locked to the slide with a single large lug, which engages the ejection port. Unlocking is controlled by the cam-shaped lug under the barrel, which interacts with a steel insert, set in the polymer frame. The same insert also carries the slide rails and could be replaced by any competent gunsmith if worn or damaged. The trigger is of the double-action type, striker fired. There is an automatic firing-pin block safety, but no manual safety is provided. Both front and rear sights are fixed, with contrasting inserts and dovetailed into the slide. The bottom of the frame, below the barrel, is shaped as an accessory rail.

Manual safety None.

Field stripping procedure 1) remove the magazine by pressing the magazine release button; 2) check that the chamber is empty; 3) pull the slide all the way back and lock it there with the slide stop; 4) rotate the takedown lever downwards (located at the left side of the frame, above the trigger guard); 5) pull the trigger, retract the slide slightly to disengage the slide stop, then release the slide forwards and off the frame, while still holding the trigger pulled; 6) remove the return spring and the barrel from the slide.

Reassemble in reverse order.

FN FNP-9/Browning PRO-9

Trigger type: Double/single action
Calibre/ammunition used: 9 × 19mm Para, .40 S&W
Weight: 710g (24.9oz) empty; 890g (31.2oz) loaded with 16 rounds
Length: 200mm (7.8in)
Barrel length: 100mm (3.9in)
Magazine capacity: 16 (9mm) or 14 (.40) rounds

The FN FNP-9 pistol, which is also sold on the civilian market as the Browning PRO-9, first appeared in 2003 in an obvious attempt to provide a more flexible and modern design than the relatively unsuccessful FN Forty-Nine described above. While retaining the low weight and large-capacity magazine of the Forty-Nine, FNP-9 offers a much more adaptable design, thanks to the modular trigger unit and modular grip. So far, the FNP-9 has not seen much use, but available reports are very favourable – users especially like the good ergonomics and smooth trigger pull both in DA and SA modes. The first FNP pistols were offered in 9mm, now both FNP and Browning PRO pistols are offered in both 9mm and .40 S&W calibres, as FNP-9/PRO-9 and FNP-40/PRO-40 respectively. FNP/PRO pistols are manufactured in Belgium and also in the USA

9mm FNP-9 pistol. (Courtesy of FN)

at FN Manufacturing Inc., the US subsidiary of FN Herstal.

The FN FNP pistol is a short-recoil, locked-breech pistol. The barrel is locked to the slide with a single large lug, which engages the ejection port. Unlocking is controlled by the cam-shaped lug under the barrel, which interacts with a steel insert set in the polymer frame. The same insert also carries the slide rails and it could be replaced by any competent gunsmith if worn or damaged. The trigger unit is modular, hammer-fired and could easily be replaced when the pistol is disassembled. Trigger units are available in SA/DA with manual safety, SA/DA with decocker only, and DAO variations. Both front and rear sights are fixed, with contrasting inserts, and dovetailed into the slide. The bottom of the frame, below the barrel, is shaped as an accessory rail.

Manual safety Ambidextrous levers above the grip or decocking levers or none (depending on the trigger unit installed). Push the lever up for 'safe'; pull it down for fire; a red dot will appear on the frame.

Field stripping procedure 1) remove the magazine by pressing the magazine release button; 2) check that the chamber is empty; 3) pull the slide all the way back and lock it there with the slide stop; 4) rotate the takedown lever downwards (located at the left side of the frame, above the trigger guard); 5) pull the trigger, retract the slide slightly to disengage the slide stop, then release the slide forwards and off the frame, while still holding the trigger pulled; 6) remove the return spring and the barrel from the slide.

Reassemble in reverse order.

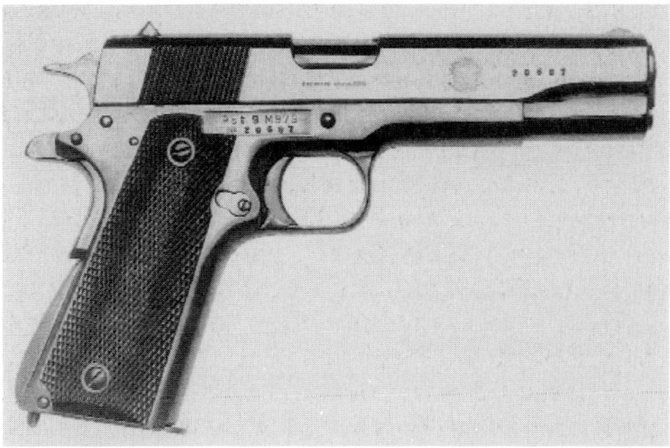

9mm IMBEL M973 pistol with Brazilian Army crest.

45 calibre IMBEL M911A1 pistol, as imported by Springfield Armory. (USA) *(Courtesy of Springfield Armory)*

Brazil

Brazil has one of the strongest arms industries in South America and produces a lot of military equipment, including small arms, but mostly of foreign design. During the post-war period, the standard issue sidearm of the Brazilian military was the American Colt M1911A1 pistol, made under licence by the state-owned IMBEL (Industria de Material Belico do Brasil) factories. Initially produced and used in its original .45 ACP loading, in the early 1970s it was rechambered for 9 × 19mm ammunition and issued to military forces as the *Pistola 9 M973*. The .45 calibre M1911 copies are still manufactured under the designation of *Pistola .45 M911A1*, for domestic use and export. IMBEL also produces M1911A1 copies in other calibres, including .380 ACP and .40 S&W, with single or double-stack magazines and different barrel lengths. IMBEL pistols are imported and distributed in the USA by the Springfield Armory and are known for good quality fit and finish.

In 1977, the Brazilian military decided to complement the 'obsolete' M1911A1 and M973 pistols with a more modern weapon. That year, Brazil contracted with the famous Italian company Pietro Beretta for 40,000 9mm M92 pistols, the newest creation of the oldest Italian gunmaker. According to the contract, all of the pistols had to be manufactured in Brazil, with Beretta building a factory for this contract. Once the contract was completed in the early 1980s, the Brazilian company Forjas Taurus SA purchased Beretta's factory in Brazil, along with all manufacturing rights for the Model 92 pistol. With this purchase, Taurus complemented its line of double-action revolvers with a modern double-action semi-automatic pistol which, in modified form, is still being produced as the Taurus PT92. It must be noted that the Brazilian engineers decided to develop the basic Model 92 design in a different way, when compared with the original Italian pistols. The most notable difference between the modern Taurus PT92 and the Beretta Model 92 is the placement of the manual safety – Brazilian pistols retained the original location of the safety on the frame, while the Italians soon put it on the slide. From the author's point of view, the Brazilian set-up is more comfortable to operate and is also more versatile, because of the different safety/decocker combination.

During the last decades of the twentieth century, Taurus produced a wide variety of semi-automatic

.45 calibre IMBEL M911A1 Micro pistol, as imported by Springfield Armory (USA). *(Courtesy of Springfield Armory)*

pistols, including full-sized and compact models, either based on the original Model 92/PT92 design or on proprietary designs. The latest offering from Taurus is the PT 24/7 pistol, a modern-looking polymer framed DAO weapon, suitable for both police and civilian use, which appeared in 2004. Available in all major calibres (9mm, .40 S&W and .45 ACP), it offers a large magazine capacity along with excellent handling and good safety. According to available information, these pistols are already in limited use by several local police organizations in Brazil.

Up to the present, the Brazilian Army still uses older 9mm Beretta 92 and Taurus PT92 pistols, as well as 9mm IMBEL M973s. Police officers use mostly .40 calibre Taurus pistols such as the PT100 and PT940, or .45 calibre pistols from Taurus (PT945) and IMBEL (various versions of the US M1911A1).

Taurus PT 92/PT 99/PT 100/PT 101

Trigger type: Double/single action
Calibre/ammunition used: 9 × 19mm Para, .40 S&W
Weight: 960g (33.6oz) empty
Length: 216mm (8.4in)
Barrel length: 127mm (4.95in)
Magazine capacity: 15 or 17 (9mm) or 11 (.40) rounds

Early (original) versions of the PT92 pistol were exactly the same as the original Beretta 92 pistols, that is, with the frame-mounted manual safety and with the magazine release button located at the bottom of the left grip panel. In the 1985 the original PT92 was replaced in production by the PT92AF version, which had ambidextrous safety levers (on both sides of the frame) and the magazine release button at the base of the trigger guard. Since 1992, the PT92AF-D replaced the previous version, the key improvement being the addition of the decocking mechanism to the manual safety. The new safety/decocker allowed the pistols to be carried either in 'condition one' (with the hammer cocked and the safety on) or in 'condition two' (with the hammer down on a loaded chamber and the safety either on or off). At the same time, Taurus also produced similar versions of the PT99 pistol, which differ only in the type of rear sights – PT92 models are fitted with fixed sights and PT99 with adjustable sights. In the mid-1990s Taurus introduced versions of the PT92/PT99 pistols in the .40 S&W calibre, designated as PT100 (fixed sights) and PT101 (adjustable sights). These .40 calibre pistols are widely used by Brazilian police. Taurus offers the PT92 line of pistols in various finishes, with aluminium alloy frames and slides and barrels made from carbon or stainless steel.

All Taurus PT92-series pistols are short-recoil-operated, locked-breech weapons. They use the Walther P38-type locking system, with the tilting locking piece

Taurus PT 111 'Millennium Pro' pistol in 9mm. This compact, lightweight polymer framed locked-breech pistol is well suited for personal defence and as a police back-up sidearm. (Courtesy of Taurus)

Taurus PT 24/7 pistol in 9mm. It is a DAO pistol, but it also has a manual safety and a built-in key lock. With a polymer frame, it weights only 750g (26.3oz) and its magazine holds seventeen rounds of 9mm ammunition. (Courtesy of Taurus)

located under the barrel. In the forward position this piece engages the notches in the slide with two locking lugs, securely locking the barrel to the slide. Upon recoil, the locking piece swings down and releases the slide, allowing it to recoil freely and commence the reloading cycle. The trigger is of the double-action type, with an exposed hammer and a frame-mounted safety (ambidextrous on the current production pistols). The sights are dovetailed to the slide, with either fixed or adjustable rear depending on the model. The double-stack magazines of the 9mm pistols hold either fifteen (earlier production) or seventeen (current) rounds, .40 S&W magazines hold eleven rounds.

Manual safety To set on safe: push up the lever on the frame. The hammer will be locked regardless of its position (cocked or lowered). To disengage the safety: pull the lever down to the horizontal position; a red dot will appear on the slide. To decock (safely lower) the hammer, push the safety lever down from the 'safety off' position; the hammer will be lowered and the safety lever will return to the 'fire' (safety off) position automatically.

Field stripping procedure 1) remove the magazine by pressing the magazine release button; 2) check that the chamber is empty; 3) depress the takedown lever stop, which is the small button located right of the frame

9mm Taurus PT92AF-D pistol with manual safety/decocker. (Courtesy of Taurus)

9mm Taurus PT92AF pistol. (Courtesy of Cherry's Fine Guns)

Early version of the Taurus PT99 pistol with adjustable rear sight. Note that the magazine release button is located at the bottom left side of the grip.

.40 calibre Taurus PT101AF-S pistol, with slide and barrel made from stainless steel. Note that it has adjustable sights. (Courtesy of Taurus)

above the trigger guard; 4) while holding the button down, rotate the takedown lever (located on the left side of the frame) downwards about 90 degrees; 5) push the slide all the way forwards and out of the frame; 6) remove the return spring by pushing its head slightly forwards and then pulling it carefully out of the detent in the barrel; 7) press the locking-piece plunger (small stud projecting from the bottom part of the barrel towards the breech) and lift up the barrel from the slide.

Reassemble in reverse order.

9mm Taurus PT 911 pistol. (Courtesy of Taurus)

TAURUS PT 909/PT 911/PT 940/PT 945

The first of the Taurus 900 series pistols, the .40 S&W calibre PT 940 was introduced in 1992 in an attempt to provide both civilian and police shooters with a pistol substantially more compact than the full-size PT 92/PT 100 series pistols, yet big and powerful enough to be used as a primary duty or self-defence weapon. Two years later, Taurus introduced a .45 calibre version, which was slightly longer and heavier due to the larger round. A 9mm version, the PT 911, appeared in 1997 and In around 2005 Taurus introduced the newest pistol in its 900 line, the PT 909. The PT 909 is mechanically similar to the PT 911 but has a different grip shape (with a curved backstrap, like the PT 945) and an integral Picatinny-type accessory rail on the frame below the barrel.

Taurus 900 series pistol are short-recoil-operated, locked-breech weapons. Frames on all PT 900 series models are machined from forged aluminum blocks, while slides and barrels are made either from carbon or stainless steel. Locking is of the modified Browning type, with a tilting barrel that engages the slide with a single locking lug at the top of the chamber area, which enters the ejection window. Unlocking is controlled by the cam-shaped lug below the barrel. The trigger is of the double-action type, with an exposed hammer and

.40 S&W Taurus PT 940 with stainless-steel slide. (Courtesy of Taurus)

TAURUS PT 909/PT 911/PT 940/PT 945

	PT 909, PT 911	PT940	PT 945
Trigger type		double/single action	
Calibre/ammunition used	9 × 19mm Para,	.40 S&W	.45 ACP
Weight, empty	817g (28.6oz)	820g (28.7oz)	855g (29.9oz)
Length	184mm (7.2in)	178mm (6.9in)	190mm (7.4in)
Barrel length	102mm (4in)	92mm (3.6in)	108mm (4.2in)
Magazine capacity	15 or 17 rounds	10 rounds	8 rounds

.45 ACP Taurus PT 945 pistol. *(Courtesy of Taurus)*

ambidextrous safety/decocking levers. Magazines are double stack on 9mm and .40 S&W models and single stack on .45 ACP models. The magazine release button is located at the base of the trigger guard. The sights on service models are of the fixed type, with both front and rear sights dovetailed into the slide. Pistols are available from the factory with standard or night (with luminous inserts) sights.

Manual safety To set on safe: push up the lever on the frame. The hammer will be locked regardless of its position (cocked or lowered). To disengage the safety: pull the lever down to the horizontal position; a red dot will appear on the slide. To decock (safely lower) the hammer, pull the safety lever down from the 'safety off' position; the hammer will be lowered and the safety lever will return to the 'fire' (safety off) position automatically.

Field stripping procedure 1) remove the magazine by pressing the magazine release button; 2) check that the chamber is empty; 3) depress the takedown lever stop, which is the small button located on the right of the frame above the trigger guard; 4) while holding the button down, rotate the takedown lever (located at the left side of the frame) downwards about 90 degrees; 5) push the slide all the way forwards and out of the frame; 6) remove the return spring by pushing its head slightly forwards and then pulling it out carefully of the detent in the barrel; 7) lift up the barrel from the slide.

Reassemble in reverse order.

BULGARIA

Soon after the end of World War II, Bulgaria joined the Warsaw Pact (a now-defunct counterpoise to NATO in a post-war bipolar world) and produced small arms of Soviet design. For many years, Bulgarian armed forces and police were armed with Makarov PM pistols, made there under Soviet licence. After the fall of the Iron Curtain and the disintegration of the Warsaw Pact, Bulgaria turned to 'Free World' values and, among other changes, permitted its citizens to own handguns for sport and personal defence. However, most of the output of Bulgarian handgunmakers is exported. There are just two major small arms makers in Bulgaria – 'Arcus'and 'Arsenal'. Arcus makes several types of handguns, pistols and revolvers. Arsenal makes a wide spectrum of small arms, including Kalashnikov-type assault rifles, machine guns and grenade launchers. All handguns made by Arsenal are based on the Soviet Makarov PM pistol, with minor variations in external appearance, finish and sights.

Weapons from both factories are well accepted in the world, thanks to good quality and low prices.

ARCUS 94 AND 98DA

	Arcus 94	Arcus 98DA
Trigger type	Single action	Double/ single action
Calibre/ ammunition used	9 × 19mm Para	9 × 19mm Para
Weight	970g (34oz) empty	950g (33.3oz) empty
Length	203mm (7.9in)	203mm (7.9in)
Barrel length	118.5mm (4.6in)	118.5mm (4.6in)
Magazine capacity	13 or 15 rounds	15 rounds

The Model 94 pistol has been manufactured by Arcus since 1994. It is an almost exact copy of the Belgian FN Browning High Power pistol, described above, with minor differences in the shape and dimensions of some parts. All functioning and disassembly procedures are the same as with the Belgian pistol. The Arcus 94 will accept all Browning High Power compatible magazines.

The Arcus 98DA is a further development of a Model 94. The key difference is the trigger, which is of double/single-action type, with a frame-mounted safety (ambidextrous safety levers are optional). Operation of the safety is similar to the FN GP-35 and

Arcus 94, except that it can be applied with the hammer either in cocked or lowered position. No means for safe decocking are provided. The Arcus 98DA is also available in a compact version, with a shortened barrel and slide. Arcus 98DA pistols are now the standard issue sidearms of the Bulgarian military and police.

Manual safety To set on safe: push up the lever at the rear of the frame. Note: for Model 94 the safety works only with the hammer in full-cock position; for Model 98DA the safety locks the hammer whether cocked or not. To disengage the safety, pull the lever down.

Field stripping procedure 1) remove the magazine by pressing the magazine release button; 2) check that the chamber is empty; 3) pull the slide all the way back and lock it with the safety lever; 4) remove the slide stop first by pushing its axis pin at the right side of the frame and then by pulling it completely out to the left; 5) while holding the slide back, disengage the safety and then ease the slide and barrel group forwards and out of the frame; 6) remove the return spring by pushing its head slightly forwards and then pulling it out of the detent in the barrel lug; 7) lift up the barrel from the slide.

Reassemble in reverse order.

Canada

The first combat handgun to be produced in Canada was the famous FN Browning High Power pistol. In late 1943 the Canadian government obtained production rights and sample FN HP pistols from FN managers, who managed to escape Hitler's occupation of Belgium. During the following two years, the Canadian company John Inglis produced more than 150,000 High Power pistols. Out of these, about 50,000 were acquired by the Canadian military and about 10,000 went to the Chinese forces which were fighting the Japanese occupation. The rest were shipped to British and other Commonwealth forces, as well as Denmark (during the early post-war years). To the present, the FN High Power pistols, both of Inglis wartime manufacture (refurbished) and of Belgian post-war manufacture, have remained the standard issue sidearms of the Canadian military. During the early 1990s Canadian armed forces also acquired some SIG-Sauer P225 pistols and issued these as non-standard equipment.

During the late 1940s and early 1950s a small Canadian company called North American Arms Corporation of Canada (NAACO), located in the city of Toronto (this should not be confused with the US-based company of the same name), developed and offered for all interested buyers a new, extremely powerful pistol, known as 'Brigadier'. This gun looked much like an enlarged FN High Power pistol except for the slide-mounted safety and large calibre bore. The new pistol was built around a new, proprietary cartridge of .45 calibre and extreme power, known as the .45 NAACO. This ammunition, which generated muzzle energy in the 1,700–1,800 Joules range (about three to four times more than typical 9 × 19 Luger or .45 ACP loadings), was too powerful to be considered for military purposes. The NAACO 'Brigadier' pistol was extremely heavy (about 1.9kg [4.2lb] empty) and held only eight rounds in the magazine. Not surprisingly, this monster has never been put into production.

9mm Inglis High Power Mk1 pistol of wartime production.*

Experimental Inglis High Power pistol; lightweight model of early post-war manufacture.

NAACO Brigadier pistol, an upscaled version of the famous FN High Power pistol.

In the late 1980s, a previously unknown company called Para-Ordnance introduced a 'high-capacity conversion kit' for US M1911A1-type pistols. This kit consisted of an updated frame with a thicker grip, which accommodated a double-stack magazine (also supplied with the kit), effectively doubling the available round count for 1911 shooters. The kit also included a new trigger assembly with suitable dimensional changes to fit into the widened grip frame. With the initial success of these kits, Para-Ordnance soon introduced a line of complete M1911A1-type pistols with high-capacity magazines, available in 9mm, .40 S&W and .45 ACP calibres. Compact and semi-compact models followed and Para-Ordnance pistols found buyers among both civilian and police shooters. In 1999, Para-Ordnance introduced the new line of LDA (Lightweight Double Action) pistols, which combined the established 1911-type general design with the DAO type of trigger, for improved safety. At the present time, Para-Ordnance offers its pistols with either standard single-action or patented LDA triggers, with single or double-stack magazines, and in a variety of sizes and finishes. It is believed that Para-Ordnance LDA pistols have already found some customers among the police agencies of the USA and Canada.

Para-Ordnance P7-45LDA .45 calibre pistol, with single-stack seven-round magazine and patented LDA trigger.

Para-Ordnance P14-45 and P14-45LDA

Data for Full-Size P14-45 Pistols
Trigger type: Single action or double-action only (LDA)
Calibre/ammunition used: .45 ACP, also .40 S&W and 9 × 19
Weight: 1,120g (39.2oz) empty
Length: 216mm (8.4in)
Barrel length: 127mm (4.95in)
Magazine capacity: 14 rounds (16 rounds for .40, 18 rounds for 9mm models)

Early Para-Ordnance pistols were almost exact copies of the Colt M1911A1 pistol, except for the wide-body frames; in fact, these early Para-Ordnance frames accepted standard M1911A1 slides and barrels without any major modifications. However, with time, the Para Ordnance design team, led by Ted Szabo, introduced several new features, other than different sizes of frame and barrel. First came the LDA modification, which was designed to allow the gun to be safely carried with a round in the chamber and the hammer down and to avoid a heavy double-action trigger pull when bringing the gun into action in fast-pacing defensive scenarios. First appearing in 1999, the Para-Ordnance line of LDA pistols now includes all sorts of guns – full-sized and compact, with single- and double-stack magazines. Another step forward from the basic M1911 design was the patented 'Power Extractor', or PXT. Since 2004, all Para Ordnance pistols are manufactured with PXT-type extractors, which are said to improve extraction reliability.

All Para Ordnance pistols are short-recoil-operated, locked-breech weapons. The classic Browning locking system consists of two lugs at the top of the barrel, which engage the cuts on the inside of the slide. The unlocking action on the recoil is caused by the swinging link, which connects the barrel breech with the frame. Most Para Ordnance pistols are made with steel frames; however, there were certain versions with

.45 calibre P14-45 pistol with single-action trigger.

.45 calibre P14-45 LDA pistol with double-action-only trigger.

aluminium alloy frames. Standard Para Ordnance pistols have a single-action trigger with an exposed hammer and a manual safety on the frame. There is also a typical M1911-style automatic grip safety and a disconnector, which does not permit the gun to be fired unless the slide is fully closed and the barrel is locked.

However, the LDA models have an entirely different trigger set-up. The key for the 'Lightweight Double Action' is the hammer cam, which is interposed between the main spring and the hammer. During the cocking cycle the hammer is pulled back; this movement causes the cam to rotate and to compress the main spring. Once the spring is compressed, the hammer cam is held cocked by the main sear. The hammer, which is not directly connected to the main spring, is then returned forwards by its own, relatively soft hammer spring. When the trigger is pulled, it raises the hammer until it connects to the cocked hammer cam; at this moment, the sear is depressed and the hammer cam, powered by the main spring, forces the hammer forwards and towards the firing pin; at the same moment, the automatic firing-pin safety is released to allow the firing pin to hit the primer. This system, which is obviously based on FN's 'Fast Action' trigger of the early 1980s, allows for a very light and smooth trigger pull, since the hammer spring is very light and the stronger main spring is cocked only during the reload. However, LDA could not be considered as a 'true double-action-only' design, as it does not permit a fast second strike in the case of misfire; in this case, the shooter must manually recock the main spring and hammer cam by pulling the hammer back manually (however, the Para-Ordnance manual does not recommend that; instead, it advises cycling the slide and ejecting the failed round, which will also recock the main spring automatically).

Another issue is that while the gun looks uncocked at all times except when the trigger is being pressed, the main spring may still be cocked and under tension. There's no way to safely release the main spring and hammer cam, except by a trigger pull; so, after unloading and before storage, the shooter must point the pistol in a safe direction and dry-fire it once (pull the trigger completely and then release it). Otherwise, the hammer cam will remain cocked and the main spring will be under tension, which could impair pistol reliability in the long term. Another unusual feature of the LDA is that the automated grip safety locks the slide when not disengaged properly by the shooting hand.

Para-Ordnance pistols are available in a variety of magazine capacities; the maximum magazine capacity and calibre is usually indicated by the model designation, that is, P14-45 is a .45 calibre pistol with fourteen-round capacity and P16-40 is a .40 S&W calibre pistol with sixteen-round capacity. Para-Ordnance pistols can be fitted with either fixed or adjustable sights, with the rear sight dovetailed into the slide.

Manual safety A lever, located at the side of the frame above the grip panel (ambidextrous levers on the 'Limited' series pistols). Push the lever up to set on 'safe', pull the lever down to set on 'fire'. Note: on the single-action pistols, the safety will engage only when the hammer is cocked; on the LDA series of pistols, the hammer is always uncocked and the safety is applied with the hammer down. It is not advised by the factory manual to cock the hammer directly on the LDA pistols.

Field stripping procedure (relevant only for full-size single action and LDA pistols of all calibres; for compact and semi-compact models, please refer to the Owner's manual) 1) remove the magazine by pressing the magazine release button; 2) check that the chamber is empty; 3) with the slide in the forwards position, pull the trigger once to drop the hammer; 4) put the pistol on the table with the muzzle up, with the grip safety resting on the table; 5) press inwards the recoil spring plug (located just below the muzzle), and rotate the barrel bushing clockwise for about a quarter of a turn (use the special bushing wrench if available); 6) gently allow the recoil spring plug to extend out of the slide; 7) push the slide back until the disassembly notch on the slide is aligned with the slide stop lug. Note that the grip safety must be depressed to do so on LDA pistols; 8) push the rounded end of the slide stop pin inward, then completely remove the slide stop, pulling to the left and out of the frame; 9) push the slide forwards and out of the frame; 10) lift the return spring guide out of the slide; 11) remove the barrel out of the slide.

Reassemble in reverse order.

China, People's Republic of

The Chinese Communist Party established the People's Republic of China (PRC) in 1949. Not surprisingly, Mao Zedong and Josef Stalin, the leaders of China and Soviet Union, quickly established very tight and warm relations between their countries. For a few years, the Soviet Union transferred a lot of military technology and knowledge to China and during this time the PLA (People's Liberation Army of China) was armed with weapons of Soviet design. However, with the death of Stalin in 1955 and the resulting political changes in the USSR, Mao quickly turned from his former ally due to differences in Communist ideology. Further development of military equipment has therefore been done by the Chinese in their own way (but not without looking over others' shoulders, be that the USSR or the USA).

The first military sidearm to be produced in the PRC was the 7.62mm Type 51 pistol, an exact and licensed copy of the Soviet Tokarev TT-33 pistol. As its name implied, it was adopted by the PLA in 1951. The same pistol, when exported from the PRC via the state-owned NORINCO (China North Industries Corporation), is often designated the Type 54. Later in the same decade the PLA adopted the 9mm Type 59 pistol, also an exact copy of the Soviet Makarov PM pistol. In the following years, various Chinese state-owned arms factories produced many clones and copies of the most well-known foreign weapons, which were widely exported to all parts of the world. Chinese-made copies of the Tokarev TT, either in the original 7.62mm Type 54 configuration, or in various 9mm versions, often with minor changes in design, can be encountered anywhere in Asia and Africa, especially in the hands of various left-wing movements and rebels. The same pistols were, and sometimes still are, sold on commercial markets.

China also produced and exported through

The 7.62mm Type 54 pistol is an exact copy of the Soviet Tokarev TT-33 pistol, manufactured for export.

The 9mm M213 pistol is an 'improved' version of the Type 54, fitted with a manual safety on the frame, above the left grip panel.

NORINCO various copies of the US Colt M1911 pistols (as the NORINCO M1911 Government), Czechoslovak CZ 75 and CZ-85 pistols (as NORINCO NZ-75 and NZ-85 respectively), SIG-Sauer P226 (NORINCO NC-226/NP-22) and P228 (NORINCO NP-34), FN Browning High Power Mk3 (NORINCO NP-36) and many others. There are also a few domestic designs, which are produced for export only, such as 9mm Type 77B and NP-20 pistols. Most of these pistols, while advertised in NORINCO literature as 'improved versions', are in fact cheaper imitations, usually built from good materials but with sub-standard fit and finish.

It is interesting to note that since the introduction of Type 51 and Type 59 pistols, the PLA has preferred weapons of indigenous design, leaving the copies of foreign weapons to foreign users. During the 1960s and 1970s the PLA and the Chinese police adopted several pocket-type pistols, as well as a few specialized weapons – two integrally silenced pistols (Type 64 and improved Type 67) and one machine pistol, the Type 80. By the early 1990s the PLA and police were armed with a mix of obsolete and mostly underpowered pistols, so in 1994 it was decided to develop a new, indigenous pistol of modern design, suitable for police and military use. Closely watching international trends in small arms development, Chinese experts decided to adopt two calibres – the world's standard 9 × 19mm for police use and the indigenous 5.8 × 21mm for military use. The 5.8mm ammunition is, obviously, based on the same concept as the Belgian 5.7 × 28mm ammunition – low recoil, flat trajectory and high penetration are the key properties. The 5.8mm calibre has been selected to ensure commonality of production machinery with the 5.8 × 42mm ammunition, adopted in the late 1980s by the PLA for the new assault rifles and other infantry weapons (sniper rifles and machine guns included). Both 9mm and 5.8mm ammunition is used in versions of the same pistol; however, there are certain external and internal differences between 'police' 9mm and 'military' 5.8mm versions of the newest Chinese QSZ-92 pistol. At the present time, the QSZ-92 pistol has entered production and has already

9mm (9 × 18) M59 pistol, a close copy of the Soviet Makarov PM.

The 9mm M201C pistol is a further improvement over the Type 54, with added manual safety and grip panels of a more comfortable shape.

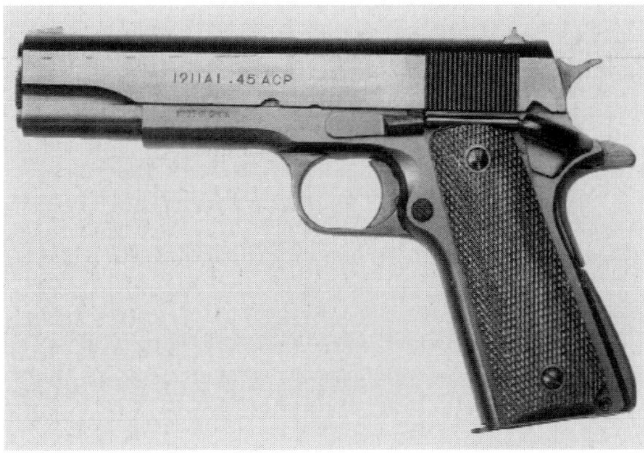

.45 calibre NORINCO M1911 pistol; it is somewhat popular among civilian shooters in the USA as a platform for inexpensive customized 1911-style pistols.

ABOVE: QSZ-92 pistol in both versions – 9mm 'police' top and 5.8mm 'military' bottom. (Courtesy of Small Arms magazine [China])

RIGHT: 7.62mm Type 64 silenced pistol. The slide locking/ unlocking button is visible in the cuts made at the top of the slide.

been observed in the holsters of some police units in mainland China; the 9mm version of this pistol is also offered for export by NORINCO as the M9/NP-42.

Type 64 and Type 67 Silenced

	Type 64	Type 67
Trigger type	Single action	
Calibre/ ammunition used	7.62 × 17 Type 64	
Weight	1,800g (63oz) empty	1,050g (36.8oz) empty
Length	222mm (8.7in)	226mm (8.8in)
Barrel length	95mm (3.7in)	95mm (3.7in)
Magazine capacity	9 rounds	9 rounds

The Type 64 silenced pistol is a specially designed weapon for clandestine operations. A few of these pistols have seen action in South-East Asia, in the hands of North Vietnamese soldiers. Type 64 is designed around indigenous 7.62mm ammunition, developed especially for that gun, but later used in few other Chinese pistols. The Type 64 cartridge is apparently based on the older 7.65 × 17SR Browning/.32 ACP case, but, unlike the original, it is of rimless design and thus is not interchangeable with the Western round. It is also loaded to relatively low pressures, to decrease the sound of gunfire. This results in a very limited effective range and low penetration of the bullet, requiring more skill to disable targets instantly. The Type 64 is extremely heavy and bulky for its low-powered ammunition and by the late 1960s it had been replaced by the Type 67, which is similar to Type 64 except for a much slimmer silencer and lower weight. It is believed that Type 67 pistols are still in use with special operations units of the PLA and Chinese police.

The Type 67 pistol is a simple blowback design, with a fixed barrel which is enclosed by an integral cylindrical 'sound moderator', or silencer. The rear part of the silencer body, around the barrel, is filled with a thin wire mesh; the front part of the silencer, ahead of the muzzle, contains several baffles, which are used to slow further and cool down the expanding muzzle gas. The barrel is ported close to the chamber to let some powder gases escape into the interior of the silencer. The silencer is screwed onto the frame and held in place by the swinging-down trigger guard. The short and bulky slide contains the rotating locking piece, with two lugs. The cross-bolt-type button located in the slide controls the rotation of this piece.

7.62mm Type 67 silenced pistol.

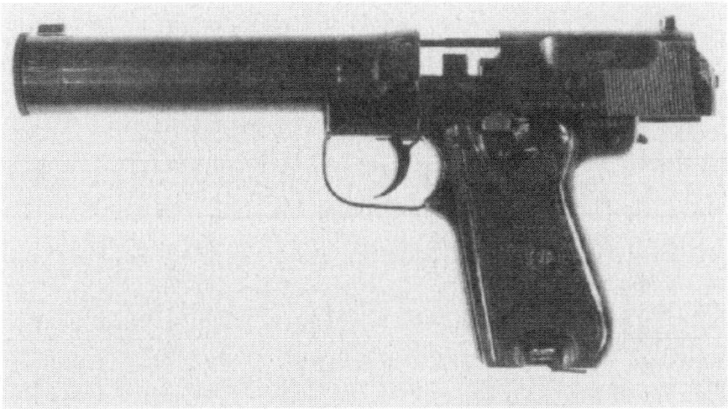

7.62mm Type 67, with slide locked back.

The purpose of this arrangement is to lock the slide manually to the barrel, thus turning the pistol into a manual repeater and avoiding the sound of the slide cycling back and forward. The return spring is located within the slide, above the barrel axis and around the spring guide rod. The trigger is of a single-action type, with an external hammer and a manual safety, located at the left side of the frame, above the grip panel. The single-stack magazine holds nine rounds. The sights are fixed, with the rear being mounted on the slide and the front on the removable silencer body. This set-up is hardly conductive to long-range accuracy, but it is in any case a strictly short-range weapon. The Type 64 pistol is similar except for a silencer body of larger size and weight.

Manual safety Lever, located at the left side of the frame, above the grip panel; push it up for 'safe'; pull down for 'fire'; a red dot will appear on the frame.

TYPE 77
Trigger type: Single action
Calibre/ammunition used: 7.62 × 17 Type 64
Weight: 500g (17.5oz) empty
Length: 148mm (5.8in)
Barrel length: n/a
Magazine capacity: 7 rounds

The Type 77 pistol is unusual for military weapons of the time, being very compact and firing 7.62mm ammunition of marginal power initially developed for a Type 64 silenced pistol, described above. Intended as a self-defence sidearm for military and police officers, the Type 77 pistol had been developed by 1978 and was officially adopted in 1981.

The Type 77 is a simple, blowback-operated pistol with a fixed barrel. The trigger is of the single-action type, striker fired. To combine the simple single-action trigger with high first-shot readiness and adequate safety when carrying and handling the gun, the Type 77 incorporates an unusual (for modern pistols) Chylewsky-type single-hand cocking device. This device consists of a separate front part of the trigger guard, which can be squeezed by the index finger. The pull on the front of the trigger guard results in the retraction of the slide, allowing for manual reloading using only one firing hand. To avoid damage when firing, the front part of the trigger guard is automatically locked in the forward position once the trigger is pulled. The pistol is intended to be carried with an empty chamber and the safety on. A manual safety is located at the left side of the frame, above the pistol grip. This could be considered as a 'low-tech' replacement for a somewhat more complicated double-action trigger with necessary automatic safeties.

Type 77 pistol.

Type 77B
Trigger type: Single action
Calibre/ammunition used: 9 × 19mm Para
Weight: 1,000g (35oz) empty
Length: 190mm (7.4in)
Barrel length: 127mm (4.95in)
Magazine capacity: 9 rounds

The Type 77B pistol was developed at the Chinese state factories as an enlarged and more powerful version of the Type 77 pistol described above. It appears that the Type 77B is a purely commercial offering and it has not been used by any Chinese military or police force. Apparently developed in the early 1980s, this pistol looks like an outdated design from the start, having a single-action trigger and a relatively small-capacity magazine, when compared with contemporary Western designs. It is also quite heavy and bulky. The Chylewsky device, inherited from the much smaller Type 77, is of little value, as the bigger and more powerful 9mm ammunition requires a stronger return spring and a longer travel of the slide to operate reliably. One must have extremely long and strong fingers to operate this device single-handedly. Probably recognizing the lack of sense in the Chylewsky cocking device, Chinese designers also produced two versions of the Type 77B with a solid trigger guard, the NP20 with a nine-round single-stack magazine and the NP24 with a double-stack fifteen-round magazine.

9mm Type 77B pistol.

The Type 77B is a delayed-blowback-operated, fixed barrel weapon. It uses a Destree gas-retarded system with the gas cylinder located below the stationary barrel. Hot gases are bled from the barrel via the gas port, where they act on the gas piston, attached to the slide, resisting the opening of the slide until the pressure drops down to a safe level. The trigger is of the single-action type, with an exposed hammer and a manual safety on the left side of the frame. The front sight is integral to the slide; the rear sight is of the adjustable 'sport' type.

Manual safety A lever, located at the left side of the frame, above the grip panel; pull it down for 'safe'

This is how the Chylewsky device works. It may look useful on the photo, but to operate it on a large 9mm pistol, one must have hands of King-Kong size and strength.

The NP20 pistol is similar to the Type 77B, except for the lack of the useless Chylewsky device, and so it has a solid trigger guard.

(position marked with 'S' on the grip panel); push up for 'fire' (position marked with 'F' and red dot on the grip panel).

Type 80 Machine Pistol

Trigger type: Single action
Calibre/ammunition used: 7.62 × 25
Weight: 1,160g (40.6oz) empty
Length: 300mm (11.7in)
Barrel length: 140mm (5.5in)
Magazine capacity: 10 or 20 rounds

During the turbulent 1930s, China was among the major users of the Mauser C96 pistols, in various self-loading and selective-fired versions. China also bought significant numbers of Spanish Astra 900-series pistols, which were clones of the C96, as well as building C96 copies at Shangsei Arsenal, in the original 7.63 × 25mm Mauser loading as well as in .45 ACP. It seems that these pistols had a significant impact on the thinking of Chinese small arms experts, as in the early 1980s state factories began to produce Type 80 machine pistols, which were improved and modified Mauser C96 clones. Interestingly enough, the origins of the Type 80 lie in the PLA requirements for a lightweight sub-machine gun (weighing no more than 2kg [4.4lb]), issued in 1969. However, the chaos of the Cultural Revolution delayed the development of many weapons, this one included. Apparently, these pistols were to be used as a personal defence weapons for certain elements of the PLA and, possibly, the police.

The Type 80 pistol retained the basic layout and design of the Mauser C96 *Schnellfeuer* (also known as Mauser Model 712), but with plenty of differences, both external and internal. The key external differences are sloped forward, removable box magazines with a shorter magazine well, a pistol grip of better shape and a four-position safety/fire selector switch on the left side of the frame. Instead of a wooden shoulder stock/holster, typical for C96 pistols, the Type 80 is issued with a large leather holster and a separate detachable shoulder stock, folding and made from stamped metal. Alternatively, a combination leather and steel holster/shoulder stock is also available for this pistol.

The Type 80 machine pistol is a short-recoil-operated, locked-breech weapon. A separate bolt of rectangular cross-section recoils within a large barrel extension/slide. The bolt is locked to the barrel extension by a vertically swinging locking piece located underneath it. The barrel is screwed into the barrel

Type 80 machine pistol. (*Courtesy of* Small Arms *magazine [China]*)

extension and pinned in place. The rear part of the barrel extension has cutaway sides to expose the cocking serrations formed on the rear part of the bolt. The trigger is of the single-action type and differs from the Mauser C96 system in having a main spring in a channel bored inside the hammer, much as in the Tokarev TT-33/Type 51 pistol. A single rotary lever at the left side of the frame controls the safety and fire modes. The rear part of the grip is cut to accept a detachable shoulder stock. The sights are fixed, also unlike most Mauser C96 pistols. The detachable box magazines are of double-stack type and hold ten or twenty rounds.

TYPE 84

Trigger type: Single action
Calibre/ammunition used: 7.62 × 17 Type 64 or Type 84
Weight: 380g (13.3oz) empty
Length: 121mm (4.7in)
Barrel length: 56mm (2.2in)
Magazine capacity: 6 rounds

The Type 84 pistol may not look like a true combat sidearm, being more in a class of a concealed-carry self-defence pistol. However, it was designed as both a defensive and offensive weapon for onboard aircraft security operatives. The key requirements for such a weapon are minimal penetration against aircraft structures and hull and compact size for ease of concealment. To achieve minimum penetration Chinese experts developed a special, frangible loading for a 7.62 × 17 Type 64 cartridge; this loading is designated the 7.62 Type 84 and it will not penetrate the aircraft hull when fired at 2m (6.6ft) distance. The maximum effective distance against human targets with Type 84 ammunition is claimed to be 15m (49ft).

The Type 84 is a simple blowback weapon with a fixed barrel. The trigger is of the single-action type, striker fired. A manual safety is located at the left side of the frame, behind the grip.

Type 84 pistol.

QSZ-92/TYPE 92

Trigger type: Double/single action
Calibre/ammunition used: 5.8 × 21mm or 9 × 19mm
Weight: 760g (26.6oz) empty
Length: 190mm (7.4in)
Barrel length: 115mm (4.5in)
Magazine capacity: 15 (9mm) or 20 (5.8mm) rounds

The QSZ-92 pistol is the most recent Chinese development in the field of handguns. It appears to be well thought out and of indigenous design. The development of the new pistol started in 1994 and the pistol has been in limited use by the PAP (People's Armed Police)

5.8mm QSZ-92 pistol, military version.

9mm QSZ-92 pistol, police version.

and PLA since 2001. It is produced in two versions, one chambered for 5.8 × 21 DAP92-5.8 ammunition of domestic origin and another chambered for a variation of the famous 9 × 19 Luger ammunition, designated as DAP92-9.These versions are almost identical in external appearance but have some differences in the action. A 9mm version of this pistol is offered for export sales through the NORINCO Corporation; the 5.8mm version is believed to be manufactured only for domestic use, at least for now.

The 9mm QSZ-92 is a short-recoil-operated, locked-breech pistol. A rotating barrel is used to lock and to unlock the slide in the initial stages of recoil. The 5.8mm version does not use rigid locking, instead relying on a delayed blowback action with a rotating, but non-recoiling, barrel (in the same way as in Searle system used in the 1907–15 era American Savage pistols). The frame is of composite type, with the removable top part made of steel and the bottom part (grip and trigger guard) made of polymer. The upper part of the frame contains integral slide rails, as well as the entire trigger unit, and can be easily removed from the grip for maintenance or repair. There is a double-action trigger with an exposed hammer and ambidextrous safety levers.

Externally, the 5.8mm and 9mm pistols are almost identical, except for the markings on the slide (QSZ-92-5.8 and QSZ-92-9, respectively) and for the shape of the trigger guard. On 'military issue' 5.8mm pistols the trigger guard is of rounded shape, while on 'police issue' 9mm pistols it is of a more angular shape. The magazines are also different, with the 9mm magazine holding fifteen rounds and the 5.8mm magazine holding twenty rounds, both in a double-column formation. Sights are fixed, with the rear sight dovetailed into the frame.

Manual safety Levers located at either side of the frame, above the grips; push up for 'safe'; pull down for 'fire'; a red dot will appear on the frame

Croatia

Croatia claimed its independence from the former Socialist Yugoslavia in 1991. The following years saw continuous war with neighboring ex-Yugoslavian countries, as well as civil war. As every war requires more and more guns, Croatia had a strong need for domestically produced arms, and among the first to appear was the PHP MV-9 pistol, a 9mm cross-breed of Beretta 92 and Walther P38 features. This pistol was of questionable reliability and in 1995 it was replaced in production by the more modern HS-95 (*Hrvatski Samokres 95*, or Croatian Pistol model of 1995). Actually this was a slightly restyled CZ-99 pistol of Yugoslavian design, which, in turn, was more or less a cross-breed between SIG-Sauer P226 and Walther P88 pistols, both of German origin. However, the Croatian Army required even more modern sidearms and by the late 1990s a design team at IM Metal factory, led by

The 9mm PHP was a strange-looking offspring of a WWII-era Walther P38 pistol; made in small numbers and of uneven quality, this pistol soon gave way to more modern designs.

Marco Vukovic, developed an entirely new sidearm designated HS2000. Production of the HS2000 started in 1999, with the first pistols being delivered to the Croatian Army. In 2000 these pistols appeared in the USA, initially under their original name. However, in 2001 the Springfield Armory Company of Genseo, Illinois, became the sole importer of HS2000 pistols in USA. Springfield offers a slightly improved version of HS2000 as the Springfield XD (eXtreme Duty pistol) and by now these pistols (still made in Croatia) are available in a variety of sizes, calibres and finishes.

9mm HS2000 as issued to the Croatian Army.

It must be noted that, while the HS2000 is a standard sidearm of the Croatian Army, the local police mainly use various imported pistols, including the FN High Power and CZ-75.

HS2000

Data for Standard HS2000 Model
Trigger type: Single action
Calibre/ammunition used: 9 × 19mm Para, also .357 SIG and .40 S&W
Weight: 650g (22.8oz) empty
Length: 180mm (7in)
Barrel length: 102mm (4in)
Magazine capacity: 15 (9mm) or 12 (.357, .40) rounds

The HS2000 is short-recoil-operated, locked-breech pistol with a polymer frame. The barrel locks to the slide with a single large lug which engages the ejection port. Unlocking and locking is controlled by a cam-shaped lug which interacts with a steel insert pinned to the frame. If the insert becomes worn or damaged, any competent gunsmith can easily replace it. The dual return spring arrangement is of the captive type; when assembled, the front of a spring guide rod projects slightly forward from the front of the slide; Springfield advertising literature calls this a 'muzzle protector', but the original text of patent WO02059539, published in August 01, 2002 and granted to the Croatian design team, describes the reason for this feature as being to 'enable firing the bullet when the pistol is pushed towards someone's back'. This is 'necessary' because the HS2000, as almost any other modern pistol, has a disconnector safety that does not permit the release of the striker if the slide is not fully closed and locked to the barrel, and such situation can be easily encountered in the circumstances described in the quote.

The trigger is single-action, also described in the advertising literature as an 'Ultra Safety Assurance' system. The pistol is striker-fired, with the firing-pin cocking indicator made in the form of a pin which protrudes from the rear of the slide when the gun is cocked. There are no manual safeties on the HS2000, but several automatic safeties are present, such as a firing-pin block, a trigger safety and a grip safety. The sights are fixed, with the rear being dovetailed to the slide. Current production pistols also have an integral accessory rail on the frame, below the barrel; early production pistols had no such feature.

The double-stack magazines are made of steel and hold fifteen rounds of 9mm ammunition. Magazine release buttons are duplicated on either side of the frame, at the base of the trigger guard. Military issue

ABOVE: *Latest model of 9mm HS2000, as sold under the Springfield XD9 trademark.* (image: Springfield)

ABOVE: *Latest model of HS2000 in .40 calibre and with a longer barrel, as sold under the Springfield XD40 Tactical trademark.* (Courtesy of Springfield Armory)

RIGHT: *HS2000, partially disassembled.*

pistols have a black finish, while commercial pistols may have either black or a 'military green' frame with a black or polished steel slide. Commercial models are also available in compact and tactical versions, with barrel lengths of 76mm (3in) or 127mm (5in) respectively.

Manual safety None.

Field stripping procedure 1) remove the magazine by pressing the magazine release button; 2) check that the chamber is empty; 3) retract the slide and lock it open with slide-stop lever; 4) rotate upwards the disassembly lever (located at the left side of the frame, above the trigger guard); 5) pull the slide back to release it from the slide stop, then slowly let it go forwards and into the battery; 6) point the gun in a safe direction and pull the trigger to release the firing pin from the sear; 7) push the slide forwards and out of the frame; 8) remove the return spring assembly from below the barrel; 9) remove the barrel from the slide.

Assemble in reverse order.

Czech Republic/Former Czechoslovakia

With the end of World War II, the liberated Czechoslovakia fell into the zone of Soviet interest and soon joined the Warsaw Pact. During the early post-war period the Czechoslovak Army and police retained sidearms of pre-war design, both of early and contemporary manufacture. The major pistol of the period was the 9 × 17 Vz.27 (Vz. Stands for *Vzor*, or 'Model of'), also known as the CZ-27. These were made at various state-owned factories, all known under the name of *Ceská Zbrojovka* (Czech Arms factory, or CZ in short), but located in the various cities such as Brno, Uhersky Brod (CZ-UB) or Strakonice.

In 1948 the Czechoslovakian authorities started two pistol development programmes. One was for a full-size military pistol, chambered for 9 × 19 Luger ammunition. Another was for a smaller police pistol in 7.65 × 17. From the start, the major contender for a military pistol was the design from Jan Kratochvil, who worked at CZ-Strakonice. This was a double-action, locked-breech pistol with an unusual roller-locking system, borrowed from the German MG-42 machine gun. Under pressure from the USSR Czechoslovakia standardized the 7.65 × 25 round of Soviet origin as its pistol and sub-machine gun ammunition and the new pistol was redesigned for the 'new' cartridge by Jiří Čermák, who worked at CZ-Uhersky Brod (CZ-UB). During the redesign this pistol also lost its double-action capability and in 1952 it was adopted by the Czechoslovak Army as 7.62mm Vz.52. Of course, there were other contenders, such as CZ471 and later CZ481 by Frantisek Myska and Jiří Čermák. The CZ471 was more or less an enlarged pre-war CZ-38 pistol with its typical DAO trigger, adapted for 9 × 19 ammunition with the introduction of the Browning-type locking breech action. The CZ481 was a further development of the same design, with a more conventional double-action trigger and manual safety. It also had a double-stack, high-capacity magazine.

The Vz.52 pistol remained in service until the mid-1980s, when it was declared obsolete and replaced by the 9 × 18 Vz.82 pistol. Most of the Vz.52 pistols were

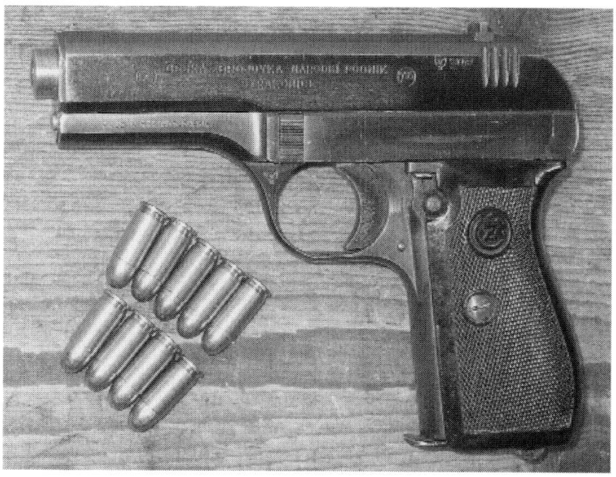

Vz.27 (CZ-27) pistol of post-war manufacture.

9mm Prototype CZ471 pistol of 1947.

then refurbished and sold as surplus, either in the standard 7.62 × 25 or in 'historically original' 9 × 19 chamberings. It must be noted that Vz.52 pistols were rather unpopular among Czechoslovak military users. One of key reasons for that attitude was the standard '7.62 Pi' ammunition which was loaded quite hot for use in the M25 sub-machine guns. Because of this powerful ammunition, Vz.52 pistols showed significant wear, developing cracks in the frame and slide in fewer than 10,000 rounds (and sometimes in just 3,000 rounds or so). Significant felt recoil was another reason, plus the poorly designed decocker (a heritage of the prototype DA pistols of 1948) sometimes caused accidental discharges when applied, if parts were worn. However, it must be noted that with standard pressure 7.62 × 25mm or 9 × 19 ammunition most Vz.52 pistols are reasonably durable and have acceptable recoil. Some 'experts' often claim that the Vz.52 pistols are stronger than the Soviet Tokarev TT pistols of the same calibre, but this is not true. While the Vz.52 has a *potentially* stronger locking system, its design has several weak points in the frame and slide, in the most stressed areas.

During the same period (late 1940s) another pistol was developed. The small, pocket-sized Vz.50 was intended for police use and chambered for the marginally powerful 7.65 × 17 Browning ammunition. Apparently influenced by the Walther PP in design, the Vz.50 had several differences in safety and disassembly. In the early 1970s the basic design went through minor facelift and the 'new' design received the designation of Vz.70. Many Vz.50 pistols were refurbished to Vz.70 standard, with different finish, grip panels and a slightly different safety lever. New Vz.70 pistols were also manufactured between 1970–83. After the replacement in service of the older Vz.50 and Vz.70 by more modern pistols, most were sold as surplus.

During the early 1980s the USSR forced its allies under the Warsaw Pact Treaty to adopt the 9 × 18 Makarov as the new ammunition for pistols and submachine guns. With this decision, the CZ factory in the city of Uhersky Brod developed and manufactured the Vz.82 pistol, which in several respects was superior to the already established Makarov PM pistol of Soviet origin. The Vz.82 has better overall ergonomics and a bigger magazine capacity (twelve rounds versus eight), but it is also more complicated and probably more expensive to make. It is also somewhat heavy for a modern pistol of average power, but the weight does provide the benefit of reducing the felt recoil. The Vz.82 remained the standard military pistol for the Czechoslovak military and police until the country fell apart into the independent Czech and Slovak republics during the early 1990s. It is still the standard military pistol in both republics; however, in Czech police use it is being replaced by the more modern 9 × 19 CZ 75 P-01 pistol, officially adopted in 2001.

Prototype 9mm CZ 75 pistol.

The mid-1970s saw the development of the most successful example of the CZ pistols, the famous CZ 75. Developed at the Ceská Zbrojovka factory in Uhersky Brod (CZ-UB) by two Koucky brothers, Josef and Franticek, the CZ 75 was first displayed at a military exhibition in Madrid, Spain, in 1976. This all-steel, high-capacity military-type 9mm pistol with excellent ergonomics gained much attraction. Developed for export, it has been widely sold on world markets; it was also widely copied and cloned, with the most famous clones being made in Italy by Tanfoglio and in Switzerland by ITM and, later, Sphinx. The Tanfoglio Company is also responsible for a number of other CZ 75 clones, made from parts supplied from Italy, such as the American Springfield P9, the Israeli IMI Jericho, Turkish/Italian Bernardelli 2000 and several others. To underline the instant success of the CZ 75, we shall quote the famous American shooter and gun writer Jeff Cooper, who, in a brief write-up in one of the 1977 *Guns & Ammo* issues, described the CZ 75 as 'possibly the finest 9mm combat pistol in the world'. And one must not forget that Jeff Cooper always considered the American .45 M1911 pistol as the ultimate combat pistol. With the fall of the Iron Curtain, the CZ 75, manufactured in many variants and sizes, found even wider acceptance by police and civilian users. In 2001, a semi-compact version of the CZ 75, designated CZ 75 P-01, was officially adopted as the new sidearm for Czech police.

The .40 S&W calibre CZ-40 pistol was initially developed as a joint effort between CZ-UB and Colt (USA) to produce a modern double-action pistol with M1911-style and ergonomics. Originally sold in the USA as the Colt Z-40, it later appeared under the original CZ name, but did not last for long, being overshadowed by the original CZ 75 pistols of same calibre.

The 9mm CZ 75 P-01 pistol, new sidearm of the Czech police.

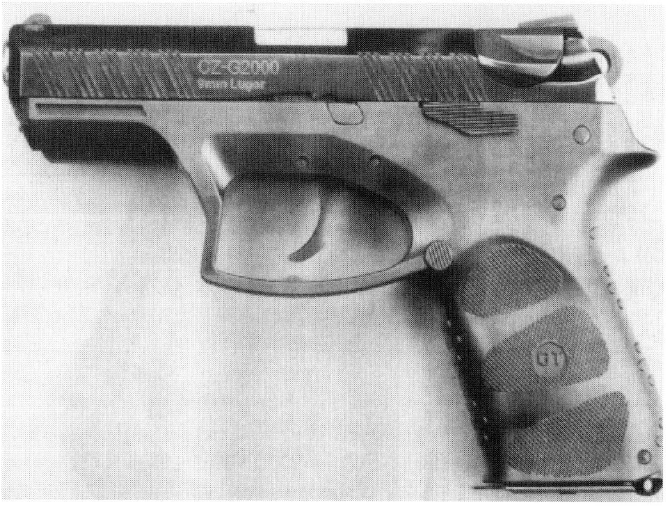

The CZ-G 2000 pistol briefly appeared in the 1999–2001 timeframe, made by one of the CZ factories and marketed by the Czech trading company, Arms Moravia. *Advertised as 'designed for Czech police', this 9mm double-action pistol with a polymer frame quickly disappeared from the market without a trace.*

Vz. 50, Vz. 70 (CZ 50, CZ 70)

Trigger type: Double action
Calibre/ammunition used: 7.65 × 17mm Browning
Weight: 710g (24.9oz) empty
Length: 167mm (6.5in)
Barrel length: 96mm (3.7in)
Magazine capacity: 8 rounds

The Vz.50 is a compact pistol of all-steel construction, firing the relatively weak 7.65mm ammunition. It has a simple blowback action with a fixed barrel. The trigger is double-action, with an exposed hammer and a frame-mounted decocking safety lever. The sights are fixed. Magazines are single stack. The magazine release button is located relatively high on the left side of the frame, behind the trigger. Vz.70 pistols are similar, with minor differences in the finish, shape of the safety lever and grips.

Manual safety Lever at the left side of the frame, behind the trigger guard. To set on safe: pull the lever down; if the hammer was cocked, it will be automatically brought down to rest. To disengage safety: push the lever up; a red dot will appear on the frame.

Field stripping procedure 1) remove the magazine by pressing the magazine release button; 2) check that the chamber is empty; 3) push the disassembly button, located at the right side of the frame, above the front of the trigger guard; 4) while holding the disassembly

7.65mm Vz. 50 pistol, left side.

7.65mm Vz. 50 pistol, right side.

7.65mm Vz. 70 pistol.

Vz. 70 pistol, partially disassembled; grip panels are removed and not shown.

button, retract the slide all the way back, raise the rear part of the slide off the rails, then slowly release it forwards and off the barrel; 5) remove the return spring from the barrel.

Reassemble in reverse order.

Vz. 52 (CZ 52)

Trigger type: Single action
Calibre/ammunition used: 7.62 × 25mm (also 9 × 19mm in export/surplus version)
Weight: 950g (33.3oz) empty
Length: 209mm (8.2in)
Barrel length: 120mm (4.7in)
Magazine capacity: 8 rounds

Introduced into Czechoslovak Army service in 1952, this pistol was declared obsolete in 1982 and most retired pistols were sold as surplus after 1987. The Vz.52 was not a popular weapon amongst its users, but mostly because of over-powered standard issue ammunition intended for sub-machine guns and loaded to higher pressures. With the standard, 'pistol grade' 7.62 × 25 or 9mm ammunition it was more manageable and may show good accuracy and an adequate life cycle.

The Vz.52 pistol is a short-recoil-operated, locked-breech pistol of somewhat unusual construction. The barrel recoils linearly, without any tilting, which increases the accuracy potential (at least, in theory). Two rollers, located below the breech part of the barrel, are used to interlock the slide and barrel in the early stages of recoil. The rollers are allowed to move laterally towards the axis of the gun and this movement is controlled by cam-shaped cuts in the frame. When the barrel and slide are in battery, the rollers are extended

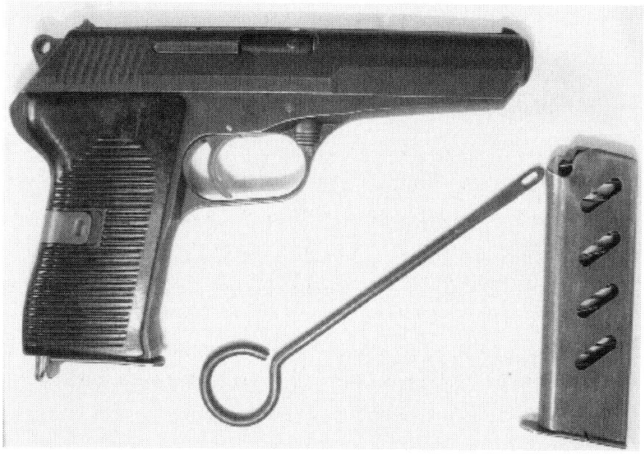

7.62mm Vz. 52 pistol, right side, with spare magazine and cleaning rod.

7.62mm Vz. 52 pistol, left side.

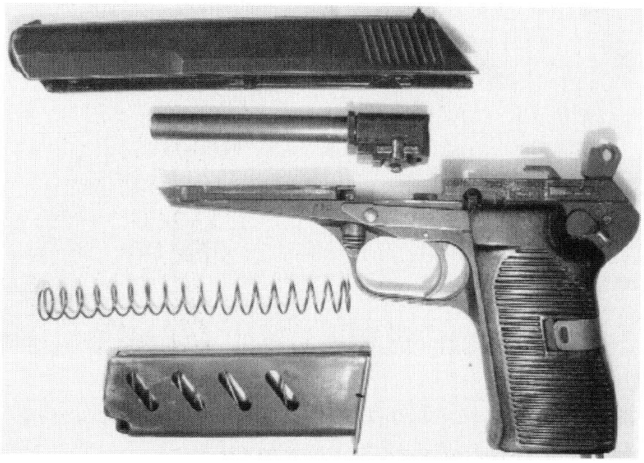

ABOVE: Slide and barrel assembly of the Vz.52 pistol, bottom view. Locking rollers are clearly visible under the breech of the barrel.

LEFT: Vz.52, partially disassembled.

from the barrel sideways and engaged in the respective cuts in the slide. Upon recoil, the rollers are withdrawn from the slide cuts following the cam-shaped tracks in the frame, thus unlocking the slide from the barrel and allowing it to resume the reloading cycle. Another unusual feature is that the return spring is located around the barrel and its rear end is resting against a special part, which is integral to the barrel unit but does not move upon recoil, being locked to the frame. The trigger is of the single-action type, with an external hammer and a frame-mounted, decocking safety.

Yet another unusual feature (at least for a full-size service gun) is that while it has a slide stop, which holds the slide open when the last cartridge has been fired, the slide stop does not have a manual release lever. The only sure way to release the slide from being held open is to remove the empty magazine, pull the slide all the way back, then release it and let it slam home. One could try to push the slide-stop lever (located above the trigger guard at the left side of the frame) with a finger, but this may or may not work, especially with wet or oiled hands. The grip panels are made from bakelite or plastic and held in place by a U-shaped spring clip. The sights are fixed, with the rear being dovetailed to the slide. Magazines are of the single stack type and the magazine release is located at the heel of the grip.

Manual safety Lever at the left side of the frame, behind the grip panel. To set on safe: push the lever up.

To disengage safety: pull the lever down; a red dot will appear on the frame. The safety blocks the action regardless of the hammer position, cocked or not. To decock the hammer: push the safety lever up; ensure an empty chamber to avoid accidental discharges if parts in the safety mechanism are worn.

Field stripping procedure 1) remove the magazine by pressing the magazine release button; 2) check that the chamber is empty; 3) push the safety lever up to decock the hammer; 4) pull down the serrated takedown knobs (located at the top of the frame, above the front parts of the trigger guard); 5) holding the knobs down, push the slide slightly forwards and remove up and out of the frame; 6) while holding the slide upside-down, use the front tip of the magazine bottom plate (or a flat screwdriver) to move the roller cam (located below the barrel breech) forward, so the rollers will be retracted from the slide; 7) tilt the rear part of the barrel slightly up and pull it out of the slide; 8) remove the return spring from the barrel.

Reassemble in reverse order.

Vz. 82/CZ 83

Trigger type: Double/single action
Calibre/ammunition used: 9 × 18mm PM, also 9 × 17mm Browning, 7.65 × 17mm Browning
Weight: 800g (28oz) empty
Length: 172mm (6.7in)
Barrel length: 96mm (3.7in)
Magazine capacity: 12 (9mm) or 15 (7.65mm) rounds

The Vz.82 pistol, chambered for the Soviet 9 × 18 PM (Makarov) cartridge, entered production at the CZ-UB factory in 1983. In the same year, CZ-UB also started the manufacture of a commercial (export) version of the Vz.82, designated CZ 83 and chambered for 9 × 17 Browning or 7.65 × 17 Browning ammunition. In 1992 the management at CZ-UB decided to merge the two almost identical models in a cost-saving measure, and to make the CZ 83 in all three calibres and in a variety of finishes, for both domestic military and police users, and for commercial sales. The CZ 83 is still being made at the CZ-UB factory.

The Vz.82/CZ 83 is a blowback-operated pistol of all-steel construction. The barrel has polygonal rifling to decrease wear and fires specially designed steel-cored 9 × 18 Vz.82 ammunition. The trigger is double-action, with an exposed hammer and an ambidextrous frame-mounted safety. Surprisingly, there's no decocking mechanism, but the pistol has an automatic firing-pin safety that is especially useful when the gun is carried with the hammer down on a loaded chamber. The sights are fixed, with the front

The 9 × 18 Vz.82 pistol, standard military and police sidearm of the former Czechoslovakia.

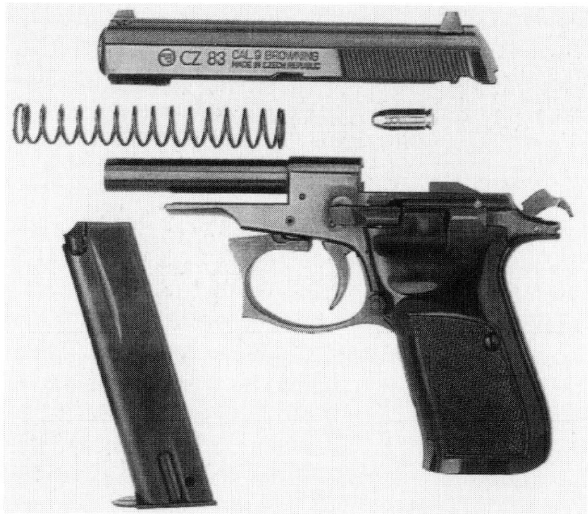

The 9 × 17 Browning CZ 83 (commercial version of the Vz.82), partially disassembled.

sight pinned and the rear sight dovetailed to the slide. Double-stack magazines hold twelve 9mm or fifteen 7.65mm rounds.

Manual safety Levers at both sides of the frame, above the grip panels. To set on safe: push the lever up. To disengage safety: pull the lever down; a red dot will appear on the frame. The safety can be engaged only if the hammer is cocked.

Field stripping procedure 1) remove the magazine by pressing the magazine release button; 2) check that the chamber is empty; 3) pull the trigger guard down and cant it slightly to the side to lock it open; 4) retract the slide all the way back, then lift its rear end and ease the slide forwards and off the barrel; 5) remove the return spring from the barrel.

Reassemble in reverse order.

CZ 75

Data for Current Production CZ 75B Pistols
Trigger type: Double/single action
Calibre/ammunition used: 9 × 19mm Para, .40 S&W
Weight: 1,000g (35oz) empty
Length: 206mm (8in)
Barrel length: 102mm (4in)
Magazine capacity: 16 (9mm) or 12 (.40) rounds

CZ 75 pistols were among the first 'Wonder Nines' (high-capacity, double-action 9mm pistols) which had begun to appear on the market by the mid-1970s. The basic design went through several modifications and by now is widely accepted as an excellent combat sidearm. It is manufactured in several versions that differ in size and safety arrangements. There are also a few specialized military or civilian-only versions, such as the CZ 75 Automatic, the CZ 75 Cadet (a .22LR conversion kit for inexpensive training) and the CZ 75SA (single-action version for sport shooting).

CZ 75 pistols are short-recoil-operated, locked-breech weapons that use Browning-style locking. The barrel locks to the slide with two lugs above and ahead of the chamber area; on recoil, the unlocking and locking movement of the barrel is controlled by a cam-shaped cut in the lug, located below the chamber. The pistol is normally of all-steel construction, but several

9mm CZ 75 pistol, original model, left side.

9mm CZ 75 pistol, original model, right side.

ABOVE: 9mm CZ 75B pistol of current manufacture.

RIGHT: 9mm CZ 75B pistol, partially disassembled.

9mm CZ 75 SP01, a very recently manufactured full-size variant of the smaller CZ 75 P01; sometimes referred to as the 'Tactical' model; it has a full-length frame with an integral Picatinny rail for mounting accessories like laser sights or lights.

CZ 75B Automatic pistol, a selective-fired version with the spare magazine being used as a front grip.
(Courtesy of CZ-UB)

compact models are produced with aluminium or polymer frames. The trigger is double-action, with an exposed hammer and a frame-mounted manual safety on basic models. Since 1994, all versions of the CZ 75 have been equipped with an automatic firing-pin block safety; these versions are marked with a 'B' suffix after the model number, for example CZ 75B or CZ 75BD.

There are two other versions of the trigger/safety arrangement available in the CZ 75 line – 'D' and 'DAO' versions, respectively with a decocker instead of a safety or with the DAO trigger and no safety at all. In size there are standard (full-size) pistols, semi-compact models with the standard frame but a shortened barrel and slide, and compact models, which have a shortened grip frame in addition to the shorter barrel and slide.

Among more specialized versions, the most interesting is the CZ 75 Automatic, a selective-fire version of the basic gun. This pistol may have an elongated barrel with compensator cuts and the manual safety has

CZ 75BD, with the decocker instead of the manual safety. (Courtesy of CZ-UB)

CZ 75DAO, double-action-only version with no manual safeties at all. (Courtesy of CZ-UB)

three instead of two positions – for safe, semi-automatic and fully-automatic fire modes. Another difference from the standard pistols is the addition of a spare magazine holder below the front part of the frame. The spare magazine is inserted into the holder upside down and is used as a front grip for improved control during automatic fire. Unlike most other automatic pistols, there is no provision for shoulder stock attachments. The most recent versions of the CZ 75, such as the semi-compact CZ 75 P-01 or a CZ 75 Tactical, may also have an integral accessory rail under the front part of the frame. Sights are fixed, with the rear being dovetailed to the frame. Magazines are of the double-stack type, and hold fifteen (early) or sixteen (current production) rounds for full-size and semi-compact models and thirteen rounds for compact models. Compact models will also accept full-size magazines.

CZ 75B Compact.
(Courtesy of CZ-UB)

Right view of CZ-85 pistol shows controls (safety and slide release), set for left-hand shooter.

Manual safety Standard model: lever at the left side of the frame, above the grip panels. To set on safe: push the lever up. To disengage safety: pull the lever down; a red dot will appear on the frame. The safety can be engaged only if the hammer is cocked.

Decocker ('D') model: decocking lever at the left side of the frame. To decock the hammer, push the lever down and it will automatically drop the hammer and then return to the 'fire' position.

DAO models: no manual safety.

Field stripping procedure 1) remove the magazine by pressing the magazine release button; 2) check that the chamber is empty; 3) put your left thumb through the trigger guard and with the other fingers grasp the top of the front part of the slide firmly. Retract the slide slightly until the two vertical notches on the rear part of the frame and slide are aligned with each other; 4) while holding the slide in position as described above, push the slide stop pin inwards from the right side of the frame, then pull the slide stop out of the frame from the left side; 5) push the slide and barrel assembly forwards and out of the frame; 6) remove the return spring from below the barrel; 7) pull the barrel out of the slide.

Reassemble in reverse order.

CZ 85

Trigger type: Double/single action
Calibre/ammunition used: 9 × 19mm Para
Weight: 1,000g (35oz) empty
Length: 206mm (8in)
Barrel length: 102mm (4in)
Magazine capacity: 16 rounds

The CZ-85 is an ambidextrous version of the CZ 75 pistol; it is entirely similar to contemporary CZ 75 pistols, except that the manual safety and slide-stop levers are located on both sides of the frame. Standard versions of CZ 85 are fitted with fixed sights; so-called 'Combat' versions are fitted with adjustable rear sights.

CZ 100/CZ 110

Trigger type: Double-action only (CZ 100) or double/single action (CZ 110)
Calibre/ammunition used: 9 × 19mm Para, .40 S&W
Weight: 665g (23.3oz) empty
Length: 180mm (7in)
Barrel length: 98mm (3.8in)
Magazine capacity: 13 (9mm) or 10 (.40) rounds

The CZ 100 and CZ 110 pistols are similar in almost every respect except for the trigger system – while the

CZ 100, a double-action-only pistol, with adjustable rear sight.

CZ 110, with fixed sights.

CZ 100 is a DAO pistol with a relatively heavy trigger pull, the CZ 110 is a conventional DA/SA pistol which can be fired in either mode.

CZ 100 and CZ 110 pistols are locked-breech, short-recoil-operated weapons with polymer frames. The barrel locks to the slide with a single lug above the chamber area, which engages the ejection port in the slide; on recoil, the unlocking and locking movement of the barrel is controlled by the cam-shaped cut in the lug, located below the chamber. This lug interacts with the steel insert set into the polymer frame. Pistols are striker-fired, with automatic firing-pin block safety. The CZ 100 has a DAO trigger, in which the firing-pin spring must be compressed by the pull of the trigger for every shot. The CZ 110 has a conventional DA/SA trigger, with a manual decocking feature. Another optional safety feature available on CZ 100/110 pistols is a loaded chamber indicator of a slightly unusual system. If installed, this indicator has a shape of a small pin, which protrudes from the left side of the slide just behind and below the ejection port once the round is chambered. Sights are fixed or adjustable, with the rear being dovetailed to the frame. Magazines are double stack.

On the top of the slide, just behind an ejection port, there's another unusual device in the form of a flat latch, dovetailed to the slide. In the CZ manuals it is called the 'barrel stop' and is intended for single-hand cycling of the slide. To do this, one must rest this 'barrel stop' against a hard surface such as the corner of the table or a belt buckle and then push the pistol forward. The 'barrel stop' will hold the slide against the movement and the action will be cycled without recourse to the non-firing hand – at least, so says the theory. While this part may slightly interfere with the fast-draw techniques, it cannot be removed easily, as it also holds in place the firing-pin block and its removal will compromise the safety of the pistol in question.

Manual safety CZ 100: none; CZ 110: decocking button at the rear of the frame, just below the slide. To decock the firing pin, push the button and it will automatically release the firing pin to rest and then return to the 'fire' position.

Field stripping procedure 1) remove the magazine by pressing the magazine release button; 2 check that the chamber is empty; 3) put your left thumb through the trigger guard and with the other fingers grasp the top of the front part of the slide firmly. Retract the slide slightly back until the two vertical notches on the rear part of the frame and slide are aligned with each other; 4) while holding the slide in position, described above, push the connecting pin (located in the frame above the trigger guard) inwards from the right side of the frame, then pull it out of the frame completely from the left side; 5) return the slide to its original position; 6) while holding the pistol in the usual 'shooting' manner, point it in a safe direction, pull the slide backward for about 5–10mm (0.2–0.4in) and pull the trigger; 7) push the slide and barrel assembly forwards and out of the frame; 8) remove the return spring from below the barrel; 9) pull the barrel out of the slide.

Reassemble in reverse order.

FRANCE

At the end of World War II, the French Army had a mixed stock of pistols, which included domestically produced Model 1935 (in original M1935A or simplified M1935S versions), captured Walthers and some US M1911A1 pistols. The police were armed with an even wider array of various pocket pistols made by French companies (such as various MAB and Unique brand pistols in 7.65 × 17 and 9 × 17 Browning), as well as some foreign pistols, mostly imported from Belgium and made by FN (such as the 9 × 17 FN Browning M1922).

Recognizing the need for a more modern and powerful military pistol (which would fire NATO-standard 9 × 19 ammunition also suitable for sub-machine guns), in late 1946 the French Army issued a requirement for a new 9mm military pistol. Several designs were tested and in 1950 the Army adopted a design from the state-owned MAS arsenal based in St Etienne (*Manufacture d'Armes de Saint-Etienne*). This pistol, known as the Mle.1950 or M1950 was broadly based on the pre-war M1935 pistol, strengthened to use the more powerful 9 × 19 ammunition which also was intended for MAT-49 sub-machine guns. These pistols were made by two state-owned armouries, MAS and MAC (*Manufacture d'Armes de Chatellerault*) between 1956 and 1978. During their lifetime, many pistols were factory refurbished and returned for service. In general, the troops liked these guns for their strength and durability; accuracy was less than ideal, but still acceptable for a military pistol. When production of the M1950 pistols ceased, the French Army obtained small numbers of commercially available MAB P15 pistols, also in 9 × 19. These pistols were never adopted for general Army service, but saw some use in the hands of French Naval commando units, military police and marines. The long-barrelled version of the P15 was used in very small numbers by the French Army shooting teams under the designation of PAP F-1.

After prolonged trials held in the mid-1980s, the French *Gendarmerie* decided to replace its aged, single-action and 'low capacity' M1950 pistols with the more modern Beretta 92G. Designated PA MAS G1, these pistols were manufactured under Italian licence by the GIAT military industrial corporation. After yet another trial, in 1992 these pistols were adopted by the French Air Force and in 1999 by the French Army and Navy. It has to be said that these pistols have experienced plenty of problems with quality control and the durability of parts. For certain periods of time, PA MAS G1 pistols were withdrawn from active duty and temporarily replaced by the older M1950 pistols. French users also encountered the same slide rupture problems that had plagued US Marine Corps users during the previous decade. Special operations units of the French Army and Navy prefer to use the more robust and dependable SIG-Sauer P226 and HK USP (Universal Self-Loading) pistols of German manufacture.

During the post-war decades, French law-enforcement agencies used various pistols of domestic manufacture, chambered mostly for 7.65 × 17 and 9 × 19 Browning ammunition, such as the MAB Model D

The military issue Mle.1935S pistol, chambered for the indigenous 7.65 × 20 Longue ammunition, has seen some police use in the post-war period. (Courtesy of Cherry's Fine Guns)

The smaller and less powerful MAB Modelé D pistol, chambered for 7.65 × 17SR Browning ammunition, also was a popular police weapon during the early post-war decades.

9mm MAB Model R pistol, developed during the late 1940s for military trials; this design later evolved into the more famous MAB P15.

or Unique RR51. The French company *Manufacture des Armes de Rhine* (Manurhin) manufactured large numbers of Walther PP, PPK and P38 pistols under German licence and many of these pistols also were used to arm various units of the *Gendarmerie*. Manurhin-marked pistols were also adopted by the West Berlin police, as under international treaties they were not allowed to carry German-made weapons. Other weapons used by French law-enforcement agencies were older 7.65 × 20 M1935 pistols and several types of revolvers, made in France by Manurhin and in the USA by Colt, Ruger and Smith & Wesson. Special CRS (*Compagnie Républicaine de Sécurité*) units of the Ministry of Internal Affairs were issued with 9 × 19 M1950 pistols.

Recognizing the general dissatisfaction with the PA MAS G1 pistols, in 2002 the French Government ordered 150,000 new SIG-Sauer SIGPro 2022 pistols (described below), which will be used to rearm *Gendarmerie* and police units.

9mm PAP F-1 target pistol, with a long barrel and adjustable sight, made by MAB.

9mm PA MAS G1, a licence-built copy of the Italian Beretta 92G pistol.

Left side view of MAS Mle.1950 pistol.

Right side view of MAS Mle.1950 pistol.

Modele 1950 (M1950)

Trigger type: Single action
Calibre/ammunition used: 9 × 19mm Para
Weight: 860g (30.1oz) empty
Length: 192mm (7.5in)
Barrel length: 112mm (4.4in)
Magazine capacity: 9 rounds

The M1950 pistol was manufactured by the MAC (Chatellerault) factory between 1956–63 and by the MAS (St Etienne) factory between 1961–78.

The Model 1950 (also designated as Mle.1950 or M1950) is short-recoil-operated, locked-breech weapon of all-steel construction. It uses Browning-type locking with the barrel engaging the slide via two lugs made on the top of the barrel, just ahead of the chamber. Unlocking and locking is controlled by the swinging link, which connects the barrel to the frame. The trigger is single-action, with an exposed hammer and a manual safety mounted on the slide. The hammer/sear module is made as a single, easily removable unit, following the 1935-vintage Charles Petter

Mle.1950 pistol disassembled. Note the Petter-type separate hammer unit.

Right view of MAB P15 pistol, with magazine removed.

Left view of MAB P15 pistol.

patents. Magazines are of single-stack variety and the pistol is fitted with an automatic magazine safety; the sights are fixed.

Manual safety lever at the left side of the slide. To put on the safety, push the lever up; to set to 'fire', pull the lever down.

Field stripping procedure 1) remove the magazine by pressing the magazine release button; 2) check that the chamber is empty; 3) pull the slide slightly back and hold it there; 4) while holding the slide slightly retracted, push in the slide stop pin from the right, then pull the slide stop out of the frame to the left; 5) release the slide carefully and then push it forwards and out of the frame; 6) put the slide upside down and remove the return spring with its guide; 7) remove the barrel from the slide; 8) pull the hammer unit up and out of the frame.

Reassemble in reverse order.

MAB P15/PA-15

Trigger type: Single action
Calibre/ammunition used: 9 × 19mm Para
Weight: 1,070g (41.7oz) empty
Length: 203mm (7.9in)
Barrel length: 117mm (4.6in)
Magazine capacity: 15 rounds

The MAB P15 pistol (also known as the PA-15, for Pistolet Automatique 15) was developed by the *Manufacture d'Armes Automatiques de Bayonne* (MAB) during the early 1970s based on the earlier 9 × 19 MAB Modele R Para pistol, which was originally created for the French Army trials of the late 1940s. The P15 saw minor use by the French military and was also widely exported for commercial sales. Production ceased in the late 1980s and while there were rumours of this pistol being made in the former Yugoslavia, it seems that this venture never went past talks or experimenting. The P15 was a solid, accurate pistol with a large magazine capacity and low felt recoil. Probably the only 'drawback' of this system was its single-action trigger, a slightly old-fashioned set-up by the standards of the time.

The MAB P15 is a delayed-blowback pistol with a rotary barrel. It uses a variant of the Savage-Searle system, developed in the USA in 1905–7. The breech part of the barrel has two lugs, one at the top and one at the bottom. The lower lug is engaged in the perpendicular cut made in the frame insert and allows the barrel to rotate but not to recoil. The upper lug follows a curved track cut in the inner surface of the slide. When the gun is fired, the pressure of the propellant gas forces the cartridge case back against the slide. The recoil of the slide, via the cam track, forces the barrel to rotate to the right at the initial stages of the shot. When the bullet is still in the barrel and pressure is high, the rotation of the barrel and, subsequently, the opening of the slide, are opposed by the inertia of the barrel plus the torque created by the bullet following the rifling. As soon as the bullet leaves the barrel, the torque force disappears and the slide forces the barrel to rotate all the way through the cam arc. There's a straight cut in the slide which then allows it to go back without any further resistance from the barrel. The trigger is single-action, with an external hammer and a manual, frame-mounted safety. There is also an automatic magazine safety. Magazines are of double-stack design and hold fifteen rounds. Sights are fixed, with the rear being dovetailed to the slide.

Manual safety A lever at the left side of the frame, above the grip panel. To put on the safety, push the lever up; to set to 'fire', pull the lever down.

Field stripping procedure 1) remove the magazine by pressing the magazine release button; 2) check that the chamber is empty; 3) cock the hammer; 4) pull the slide rearwards for about 6mm (0.2in), to align the rear end of a slide stop with the disassembly notch in the slide; 5) push the slide stop pin inwards from the right and then pull it out completely from the left; 6) pull the slide forwards and out of the frame; 7) turn the slide upside down and push the base of the return spring guide slightly forward, then pull it up and out of the slide; 8) move the barrel forwards until it completes its rotation in the slide (about 8mm [0.3in]), then pull it out of the slide.

Reassemble in reverse order.

Germany

At the end of World War II Germany lay in ruins, separated into two occupation zones (one subdivided into US, British and French zones), which later turned into two countries. Allied bombing raids had destroyed most of the arms factories and most of the surviving machinery was brought out of the country by the Allied countries as reparations. Many designers and engineers either emigrated to other countries in the search for a better life (such as former Mauser employee Ludwig Vorgrimler, who, among others, moved to France and then to Spain), or were forcefully taken abroad to work for the victors (such as Hugo Schmeisser who was captured by Soviet troops and spent several years in the Soviet Union, transferring his valuable knowledge in manufacturing technologies to Soviet engineers, or Werner Von Braun who worked for the Americans, although on weapons much larger than pistols). Most of these later returned to Germany and lived either in the FRG (Federal Republic of Germany, or West Germany) or in the GDR (German Democratic Republic, or East Germany).

Arms development in the GDR was very limited, and while there were a few experiments in the field of handguns, no significant results ever appeared. The NVA (Army of the GDR) and VoPo (People's Police) were initially armed with what remained from wartime stocks, such as Luger P08 and Walther P38 and PP pistols. Many of these pistols were refurbished before being reissued; apparently, some Walther PP copies were also produced in the GDR as new guns. Most of these weapons were later replaced in service and sold for export as surplus. Soviet Makarov PM pistols (designated Pistole M) were made in the GDR and used

The Walther P38 (in this photo – made in 1944) was in some respects a quite modern design even by post-war standards. Not surprisingly, it was adopted by the West German Army and, in slightly modified form, served until the mid-1990s.

The Walther PP, in this example made in France by Manurhin under Walther's licence, was a very popular police weapon in Western Europe during the first two or three post-war decades. Several countries also issued these pistols to certain military personnel.

The Walther PPK, a smaller derivative of the PP, also saw significant use by the plain-clothes personnel of many West European police and security organizations. In some cases these pistols also were issued to military officers (especially aircraft pilots and crew members) as personal defence weapons.

Prototype Walther P88 pistol.

as standard military and police sidearms from the late 1950s. These pistols were produced to a reasonably high standard of fit and finish.

West Germany started its armed forces with the creation of the police and the border guards (*Bundesgrenzschutz*). These forces were originally armed either with pistols of pre-war or wartime manufacture, or with newly purchased foreign pistols. The police used various Walther and Sauer pistols, as well as imported FN M1922 and GP-35 pistols. The border guard also used imported 9mm Spanish Astra 600/43 pistols (designated in German service the P3) and Swiss SIG P210-2 pistols (P2). The reinstatement of a true Army (the *Bundeswehr*) and membership of NATO resulted in a requirement for domestic production of military-type weapons, including pistols. Since the new *Bundeswehr* was formed with many *Wehrmacht* (WWII Army) veterans, it was logical to adopt a weapon that was already familiar to troops, especially if the weapon in question, the Walther P38, was of a modern and battle-proven design.

The Carl Walther Company, which developed this pistol during the mid- to late-1930s, quickly resumed its operations and by 1953 had built a new factory in the city of Ulm. Walther also licensed production of its key pistols (PP, PPK and P38) to the French company Manurhin; most of the Walther pistols manufactured during the first post-war decades were actually made in France and then shipped to Germany, where they were assembled, proof-fired and properly marked. A slightly modified P38 pistol, with an aluminium instead of steel frame, became the P1, which was adopted by the *Bundeswehr* and served as the standard military sidearm until the mid-1990s. During the following decades Walther also produced a number of newer designs, which included the P5 and P88 pistols. However, by the late 1980s the company had lost most of its position in the market and in 1993 it was bought by the UMAREX Group. New ownership brought the company a new lease of life and, with the introduction of a modern and commercially successful Walther P99 pistol in 1996, the company re-established its strong position on the handgun market.

Another major pre-war arms maker, J.P. Sauer & Sohn of Suhl, was also re-established in 1951, the year of its two-hundredth anniversary. Originally making shotguns and rifles, the 'new' Sauer expanded greatly in the late 1970s, when it was bought out by the Swiss SIG Holdings AG Company to produce the newest SIG pistol, the P220 and its smaller offspring, the P225, for world markets. This was necessary because of strict limitations imposed on arms exports by Swiss law. The

Two generations of German-made SIG-Sauer combat pistols – bottom, the older P225; top, the more modern P228. Both were and still are widely used as military, police and civilian weapons.

pistol, known as the SIG-Sauer P220, swiftly became very popular; it was adopted by the Swiss and Japanese armies and widely sold for export and it is still manufactured and used as a police sidearm in many countries, including the USA. It also spawned a whole family of pistols based on the same design, which found their way to several armies and a wide spectrum of military, Special Operations and law-enforcement agencies. Despite the fact that most production of the SIG-Sauer pistols took place in Germany and even in the USA, the description of all SIG-Sauer pistols can be found in the chapter on Switzerland below, because most of these guns were designed there. Today, SIG-Sauer pistols represent the top class of combat pistols, widely used by military and police forces, as well as by civilian shooters in many countries of the world. Recent valuable contracts with both French and American governments (SIGPro 2022 pistols for the French *Gendarmerie* and P226 and P229 pistols for the US Department of Homeland Security) have effectively positioned the international SIGARMS Company (of which the Sauer factory is a core member) as a leading supplier of military and police small arms.

Still another well-known manufacturer of small arms, the Mauser Werke, was completely dismantled under Allied control in 1945. However, it had been reborn by 1954 and during the 1960s Mauser reintroduced the pre-war HSc and Luger Parabellum pistols;

The prototype 9mm Mauser HSp pistol was submitted to the German police pistol trials but failed and so never went into production. This was a locked-breech pistol with a double-action trigger and probably the last original pistol design to carry the venerable 'Mauser' name.

the latter were built on ex-Swiss machinery and were similar to Swiss P06/29 Lugers in appearance and design. Both of these weapons were made for commercial sale. During the mid- to late-1970s Mauser attempted to develop a 9mm police pistol, known as the HSp, but apparently failed and subsequently lost interest in this field, switching towards larger calibre weapons, such as automatic cannon. However, during the 1980s and 1990s the Mauser trademark was used on certain non-German weapons that were sold in Europe; the most notable of those being the Mauser 80SA and Mauser 90DA pistols, both made in Hungary by FEG.

The fourth and probably most successful German manufacturer of military small arms was, unlike the others, an entirely new name. The Heckler & Koch Company, which was founded in 1949 on the remains of the Mauser factory in the city of Oberndorf by three former Mauser employees (Edmund Heckler, Theodor Koch and Alex Seidel), built its first pistol in 1968. This was the relatively compact HK 4 pistol, available in four chamberings; almost half of the entire production of this pistol went to various German police agencies, including the West Berlin police.

Their next pistols were more along the lines of true combat sidearms – the HK P7 and P9 appeared during the late 1960s–70s, the first as an entrant for trials for a new police pistol and the second as a military pistol. Both pistols found many users among German police and other law-enforcement agencies; they were also widely exported. However, of these two the P7 found much more popularity, as it remained in production up until 2005, while the P9 had been discontinued by the late 1970s.

In the early 1970s HK also produced a very novel military-type pistol, the VP70. This DAO, high-capacity pistol was extremely simple and inexpensive to make and was the world's first pistol to have a polymer frame. However, it did not succeed, and HK's next success (in the field of handguns) did not occur until the mid-1990s, with the introduction of the USP line of pistols. These are widely used by law-enforcement organizations worldwide and a variant of the USP (designated P8) was also adopted by the *Bundeswehr* in 1995 as its new standard sidearm. An offspring of the USP family, the large Mk23 mod.0 pistol, has been adopted by the US SOCOM as an Offensive Handgun Weapons System (OHWS). For a period of time during the 1990s the British Royal Ordnance Company owned HK, but now HK is again German-owned and is one of the largest and most important manufacturers of small arms worldwide.

Despite having settled early on for a basic military pistol, the Walther P1, for many years West Germany had no general policy regarding police pistols. Each police department was armed with whatever weapons they found suitable and affordable, most weapons being pocket-type pistols firing mild cartridges such as the 7.65 × 17SR or 9 × 17 Browning. There were some exceptions, in the form of 9 × 19 pistols like the SIG P210 used by the Border Guard or the FN GP-35/High Power, the Astra 600/43 and some Star B guns used by various police organizations. It was the infamous Munich massacre of 1972, when a group of Palestine terrorists killed a number of Israeli hostages during the Olympic Games, that forced the West German authorities to rethink the role of the police pistol and its effectiveness. The search for a new police sidearm started with debate about its calibre and ammunition, which are key to the combat effectiveness of any firearm. The major contenders were the already proven and familiar 9 × 19 Parabellum and the new 9 × 18 Police, which was more or less a reincarnation of the experimental 9 × 18 Ultra cartridge developed by the Gustaw Genschow Company for the *Luftwaffe* beforeWorld War II. During these debates several companies quickly produced pistols chambered for the 9 × 18 Police, such as the SIG-Sauer P230, Walther PP Super or Benelli 76. However, by 1974 officials had settled on the 9 × 19 Parabellum as the future ammunition for the German police.

The following basic requirements for future police pistols were issued in 1974:

Calibre	9 × 19 Parabellum
Trigger and safety	Double action, suitable for safe carrying with the hammer lowered on the loaded chamber, ready for first shot in DA mode
Magazine capacity	No fewer than 8 rounds
Length	Less than or equal to 180mm (7in)
Height	Less than or equal to 130mm (5in)
Width	Less than or equal to 34mm (1.3in)
Weight	No more than 1kg (35oz) loaded

To fill the gap until the proposed trials could result in new pistols, the German authorities ordered a new and improved design based on the military P1 pistol from its manufacturer, Carl Walther Company. The quickly developed Walther P4 pistol was more or less the P1 but with the barrel shortened by 25mm (1in) and with a modified safety arrangement which discarded the manual safety for a decocking lever combined with an automated firing-pin safety. Most of the P4 pistols went to the *Bundesgrenzschutz* (Border Guard), serving until the early 1980s, when they were declared obsolete and sold as surplus. In the meantime, all four major German arms makers had submitted their designs for the police trials. Heckler & Koch submitted its PSP (*Polizei Selbstlade Pistole* – police self-loading pistol, later renamed P7), an unusual and elegant design, as well as the P9S (a double-action version of the P9). Mauser submitted its HSp, which proved unsatisfactory. Sauer initially submitted the P220 and later replaced it with the smaller P225. And, finally, Walther submitted its P5, a further development of the P4 concept. After the trials, the German authorities officially approved the Walther P5, SIG-Sauer P6 (P225) and HK P7 (PSP) for police use. Police departments of every federal *Land* (region) were allowed to choose any pistol on the 'approved' list. The list was subsequently extended to include more modern weapons and German police forces are now also using 9mm pistols like the SIG-Sauer P228, Walther P99, Heckler & Koch P10 (HK USP Compact) and HK P2000. However, many older 9mm pistols, such as the P6 and P7, are still in use.

The HK 4 pistol was the first venture of Heckler & Koch into the field of handguns. Used by the West German police, this pistol was somewhat unique in that it could be easily converted to fire any of four calibres: .22LR (5.6 × 16R), 6.35 × 16SR, 7.65 × 17SR and 9 × 17 Browning. It was a simple blowback weapon with double-action trigger. (Courtesy of Cherry's Fine Guns)

The HK UCP (Ultimate Combat Pistol) is the most recent development of the Heckler & Koch Company. Apparently developed at the request of the Bundeswehr, *it first appeared in 2003 in prototype form. Chambered for HK's proprietary 4.6 × 30 cartridge originally developed for the HK MP7 PDW sub-machine gun, this pistol appears to be a direct rival to the already established 5.7 × 28 FN Five-seveN pistol of Belgian origin.*

Not much information about this pistol has emerged, but it is believed that it will have a twenty-round magazine and a modular DA or DAO trigger with various safety options.

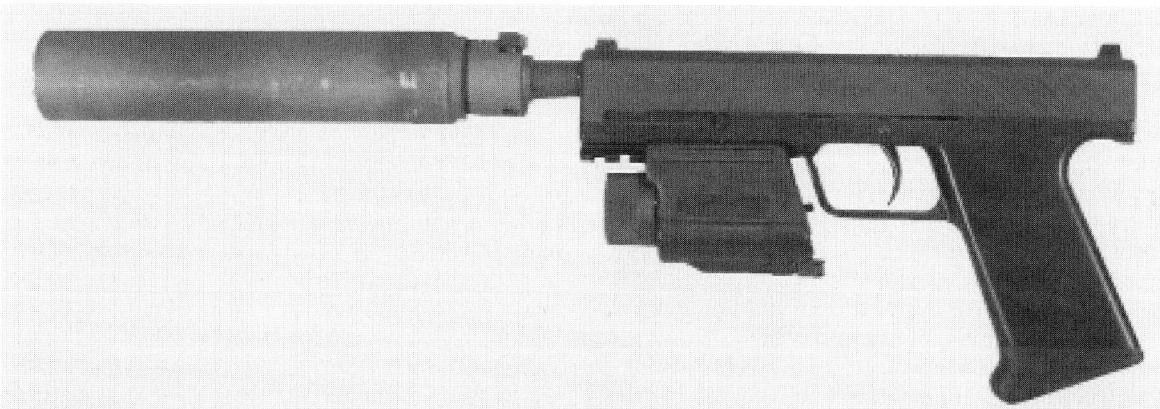

Heckler & Koch HK 4
Trigger type: Double/single action
Calibre/ammunition used: .22LR (5.6 × 15 rimfire), 6.35 × 16SR, 7.65 × 17SR and 9 × 17
Weight: 480g (16.8oz), less magazine
Length: 157mm (6.1in)
Barrel length: 85mm (3.3in)
Magazine capacity: 7 rounds (9mm) or 8 rounds (other calibres)

The HK 4 was the brainchild of Alex Seidel, one of the co-founders and masterminds of Heckler & Koch and a former Mauser employee. During his years at Mauser, Seidel participated in the development of several pistols, including the famous Mauser HSc. In 1952 Seidel, now at HK, developed a prototype .22LR calibre pistol based on the Mauser HSc, but legal problems with Mauser precluded its production. In 1964, Seidel started the development of a multi-calibre pocket-size pistol for civilian and police use. This pistol, designated HK 4 (Heckler & Koch, four calibres), had some features based on the Mauser HSc, but in severely modified form. Mass-production of this pistol started in 1968 and continued until 1984.

The HK 4 pistol was one of the first handguns to employ the idea of modularity. Each pistol can be chambered for any of the four basic calibres, simply by replacing the barrel, return spring and magazine. Many HK 4 pistols were sold with calibre change kits, which included three additional barrels with springs and appropriate magazines.

HK 4 pistols were simple blowback pistols with fixed barrels. To function properly with all four calibres, the slide was somewhat lighter than is necessary for the 9 × 17 ammunition, so the 9mm barrel had radial grooves in the chamber to slow down the blowback action during the high-pressure periods of each shot. Additionally, a polymer recoil buffer was installed in the frame below the barrel. The frames of the HK 4 pistols were made from aluminum alloy; the slides formed from sheet steel. The double-action trigger and, especially, the partially exposed hammer were somewhat similar in design to the Mauser HSc pistol. A manual safety was located at the left side of the slide, which, when engaged, moved the rear part of the firing pin out of the reach of the hammer and locked the firing pin against any movement towards the cartridge in the chamber. Additional automatic safeties included a magazine safety and a disconnector safety. The return spring was located around the barrel, which was easily detachable from the frame and the barrel lock lever was located inside the trigger guard, at the front.

HK 4 pistol.

Same pistol, right side; a 9mm Kurz (9 × 17) barrel is installed, which is marked by the '9 K' stamp visible on the barrel through the ejection port.

To accommodate either centrefire or rimfire ammunition, the HK 4 featured a removable breech plate in the slide, with two channels for the firing pin – one for centrefire ammunition and another for rimfire ammunition. To convert between those types of ammunition, one must disassemble the pistol, remove the breech plate from the slide and set it so the side marked with 'R' will face the barrel breech for rimfire ammunition, or the side marked with 'Z' must face the barrel breech for centrefire ammunition. Early production pistols also had different extractors for each calibre, but later

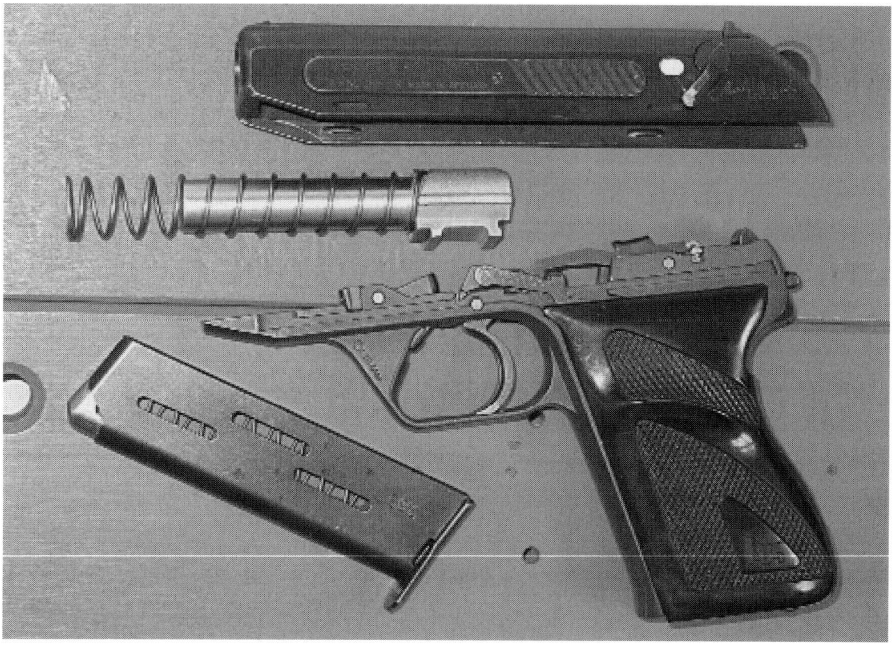

HK 4 pistol, partially disassembled.

in production these were replaced with a single 'universal' extractor. The HK 4 has an internal slide stop mechanism, which locks the slide open after the last shot from the magazine has been fired. The slide stop is automatically disengaged when a new magazine is inserted. To close the slide on the empty chamber, the slide stop is disengaged by the trigger pull. Magazines are single stack and for each calibre there is a specific magazine, properly marked. The magazine release is located at the base of the grip, near the magazine baseplate. Sights are fixed; the rear sight is dovetailed to the slide.

Manual safety A lever at the left side of the slide. To put the safety on, pull the lever down; to set to 'fire', push the lever up; a red dot will appear on the slide. Note that the safety does not lock the operation of the trigger and hammer.

Field stripping procedure 1) remove the magazine by pressing the magazine release; 2) check that the chamber is empty; 3) make sure that the safety is engaged; 4) press the barrel latch down (located inside the trigger guard at the front of the trigger) and hold it; 5) push the slide forwards as far as it will go, then raise it up from the frame; 6) to remove the barrel from the slide, turn the slide upside-down and push the barrel forwards until its rear part can be raised from the slide, then pull it out of the slide completely. (For conversion between calibres, please refer to Owner's manual.)

Reassemble in reverse order.

Disassembled HK 4 pistol with barrels in all four calibres, from .22LR (5.6mm rimfire) top to 9 × 17 (bottom).

Heckler & Koch P9S

Trigger type: Double/single action
Calibre/ammunition used: 9 × 19mm Para, also .45 ACP
Weight: 880g (30.8oz) empty
Length: 192mm (7.5in)
Barrel length: 102mm (4in)
Magazine capacity: 9 rounds (9mm) or 7 rounds (.45)

9mm HK P9S pistol.

Development of a new military-type pistol was started at Heckler & Koch in 1965, at the request of some foreign customers who desired to have an inexpensive yet efficient 9mm pistol for their armies, to go along with HK-made 7.62mm G3 rifles. The first prototype of the Pistole 9 appeared in 1966 and by late 1968 the pistol was ready for production. Actual production started in the following year, but proceeded at a very slow pace, because the new, more modern version was already available. This new version, the P9S (*Pistole 9 mit Selbstspanner* – Pistol 9 with self-cocking, that is, with double-action trigger) was developed between 1967–70. Only a few (about 500) of the original P9 with single-action triggers were made. P9S pistols were made in two basic calibres, the 9 × 19 for the European market and the .45 ACP for the USA. Pistols were also made for the civilian market, equipped with adjustable sport-type sights and, sometimes, with extended barrels and sport-type grips. P9S pistols were used by several West German police organizations, including the famous GSG9 counter-terror unit. P9S pistols of .45 calibre were also used by US Navy SEAL units during the 1970s and early 1980s, sometimes fitted with silencers.

Two versions of the P9S: top, late production; bottom, relatively early production pistols. The late production pistol is distinguishable by its more angular trigger guard. This particular example is also fitted with adjustable target sights.

The HK P9S is a delayed (retarded) blowback pistol, with a fixed barrel. It uses a roller-delayed blowback system, developed during the war at the Mauser factory and later refined in Spain by CETME and in Germany by HK. In this system, there is a separate bolt, or breechblock, made from two parts, front and rear. Between these parts, a pair of rollers is interposed. When the slide is in battery, the front part of the bolt is resting against the barrel breech and the base of the cartridge case. The rear part of the bolt, pressed forwards by the return spring, holds the rollers extended sideways from the bolt, into the recesses in the barrel extensions. Once the pistol is fired, the pressure in the barrel forces the cartridge case back against the breech bolt. This pressure causes the front part of the bolt to start the recoil, which, in turn, forces the rollers inwards as they are leaving the recesses. This inward movement of the rollers is resisted by the inclined planes, made on the rear part of the bolt, so it starts to recoil faster than the forward part of the bolt. Once the rollers are completely out of the recesses, both parts of the breech bolt recoil together, along with the external slide, to which the rear part of the bolt is securely pinned. To ensure reliable extraction, the chamber of the barrel is fluted. The barrel has polygonal rifling for improved accuracy and service life.

The slide and frame of the P9S are made from stamped steel; parts of the grip are made from plastic. The trigger system is of the double-action type and rather original in design. The P9S has a hammer, completely enclosed within the slide, with a hammer-cocking indicator in the form of a pin, which extends backwards from the frame once the hammer is cocked. A manual safety lever is fitted to the left side of the slide, and it blocks the firing pin when applied. There is an additional multifunctional lever at the left side of the frame; this is used to cock or decock the concealed hammer and also as a slide stop release. Military P9S pistols are fitted with fixed sights, while some civilian pistols are equipped with target-type adjustable sights. Magazines are made from steel and hold cartridges in a single stack.

Manual safety A lever at the left side of the slide. To put on the safety, push the lever down; to set to 'fire', push the lever up a red dot will appear on the slide. To safely decock the pistol, put the safety on, then point it in a safe direction, pushing on the decocking lever; while holding the lever down, pull and release the trigger and then slowly release the decocking lever.

Field stripping procedure 1) remove the magazine by pressing the magazine release button; 2) check that the chamber is empty; 3) push the barrel lock lever, which is located inside the trigger guard, forward; 4) while holding the barrel lock pressed, push the slide forwards as far as it will go, then pull it up and off the frame; 5) turn the slide upside down and push the barrel forwards until it can be lifted up and out of the slide; 6) to remove the breechblock, use the barrel extension or a flat screwdriver to push the breechblock lock located on the right (when looking down at the slide turned upside-down); the breechblock will spring forwards from its original position.

Reassemble in reverse order.

Heckler & Koch VP70

Trigger type: Double-action only
Calibre/ammunition used: 9 × 19mm Para
Weight: 820g (28.7oz) empty
Length: 204mm (7.95in)
Barrel length: 118mm (4.6in)
Magazine capacity: 18 rounds

The VP70 pistol was another brainchild of Alexander Seidel. During the latter stages of the war he participated in the development of a *VolksPistole* (VP), or people's pistol. In 1965 Seidel decided to reintroduce the idea of an inexpensive, extremely simple yet effective people's pistol using the latest technology. Development of the new pistol took five years, with the first prototypes appearing by 1970 and production starting by 1973. This pistol was widely tested around the world but found very few military customers, mostly because of its very stiff trigger and the blow-

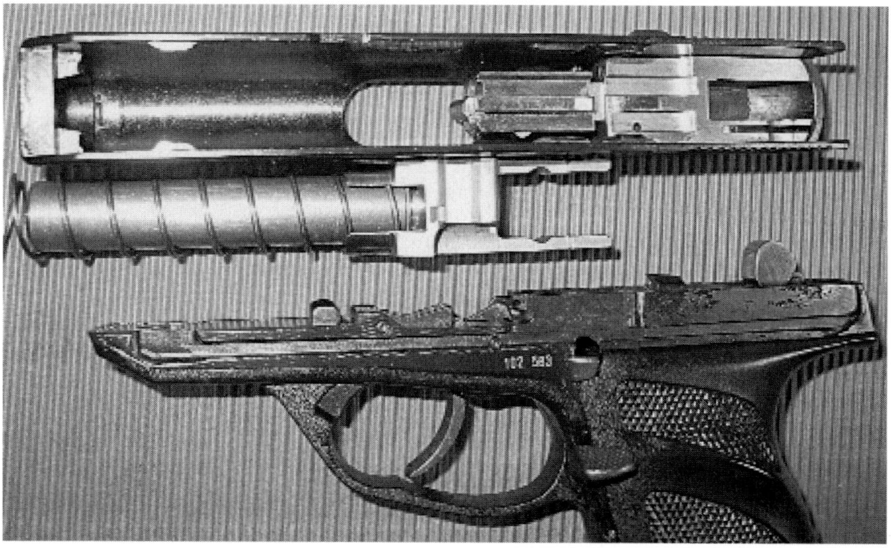

HK P9S, partially disassembled. Clearly seen on the photo are the barrel extension with cuts for rollers and the separate breechblock inserted into the stamped slide, with rollers projecting from its sides.

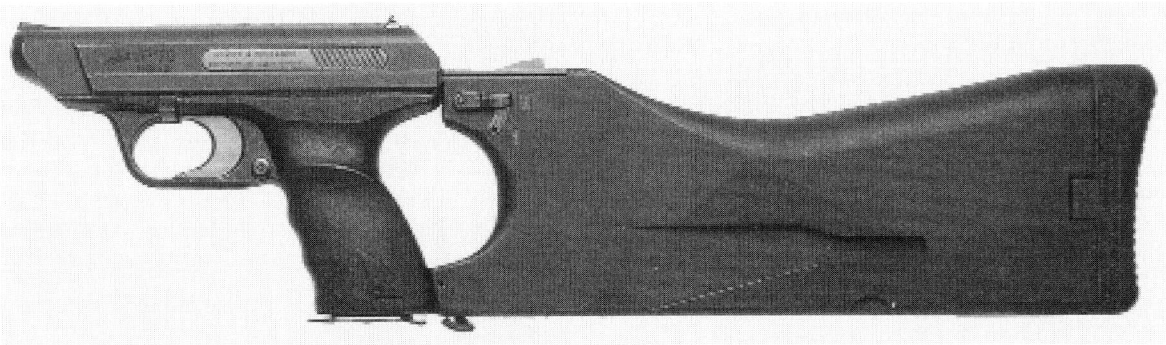

ABOVE: 9mm HK VP70M pistol, with shoulder stock/holster attached. Note the fire mode selector built into the holster near the pistol slide.

ABOVE: 9mm VP70Z pistol, the civilian version of the VP70.

ABOVE RIGHT: Firing the VP70M. (Courtesy Heckler & Koch)

RIGHT: VP70Z, partially disassembled.

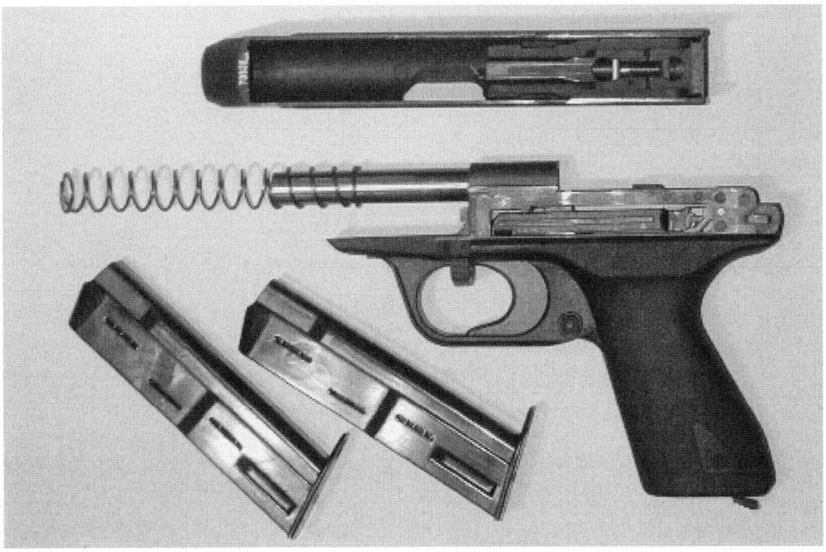

back operating system which, when combined with a fairly powerful round like the 9 × 19, was looked at rather suspiciously by most experts. A civilian version of this pistol was produced but also found few buyers. Only about 3,000 VP70s were made in total.

When it appeared, the VP70 was a rather revolutionary design, well ahead of its time. First of all, it had a lightweight polymer frame, which was inexpensive to make and maintain and resistant to the elements. Second, it had a double-action-only trigger with a very simple design; the trigger pull, however, was very stiff and compromised the accuracy. Third, it had a large-capacity, double-stack magazine. And fourth, in a military version, known as the VP70M (as opposed to the civilian VP70Z), it was equipped with a plastic holster that could be attached to the pistol to serve as a shoulder stock. The holster also held a special mechanism which was connected to the pistol's sear and allowed it to fire either single shots or three-round bursts, depending on the position of the fire mode selector, also built into the holster.

The VP70 is a simple blowback pistol with a fixed barrel. To decrease the recoil, there is a special recoil buffer built into the frame, below the rear part of the barrel. The DAO trigger is linked to a striker instead of a hammer. The frame is made from polymer and the slide is made from sheet steel by stamping and forming, with a machined breechblock insert. Sights are fixed, with contrast inserts.

Manual safety A cross-bolt button at the base of the trigger guard, which locks the trigger. To turn the safety on, push the button from the left side. To turn the safety off, push the button from the right side; a red ring will appear on the left side of the button.

Field stripping procedure 1) remove the magazine by pressing the magazine release button; 2) check that the chamber is empty; 3) pull down the slide lock, which is located within the trigger guard; 4) pull the slide all the way back, then raise its rear part out of the frame rails; 5) ease the slide forwards and out of the frame and barrel; 6) remove the return spring from the barrel.

Reassemble in reverse order.

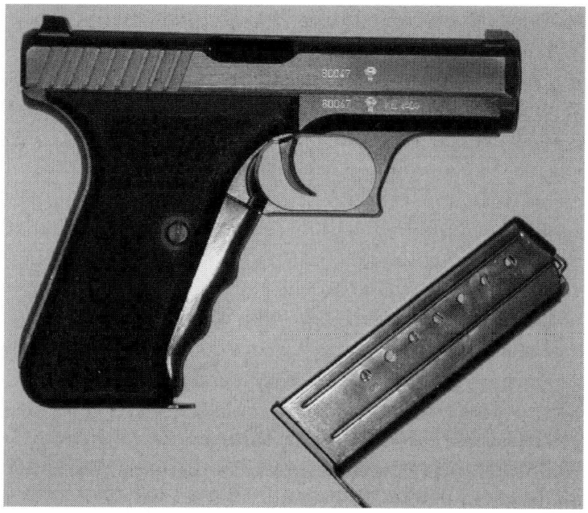

9mm HK P7 pistol.

Heckler & Koch PSP, P7, P7M8, P7M13, P7M10

Heckler & Koch started development of a new semi-automatic pistol for the police in 1976. The first prototype of the 9mm PSP (*Polizei Selbstlade-Pistole* – police self-loading pistol) had appeared by mid-1976. Two key features, the hand-cocking mechanism and the gas-delayed action, were tested on the HK 4 and HK VP70 pistols, respectively, before being incorporated into the new weapon. In 1978 the PSP was

Heckler & Koch PSP, P7, P7M8, P7M13, P7M10

	P7, P7M8	P7M13	P7M10
Trigger type		Single action	
Calibre/ammunition used	9 × 19mm Para	9 × 19mm Para	.40 S&W
Weight	780g (27.3oz)	850g (29.8oz)	1,200g (46.8oz)
Length	171mm (6.7in)	175mm (6.8in)	175mm (6.8in)
Barrel length	105mm (4.1in)	105mm (4.1in)	105mm (4.1in)
Magazine capacity	8 rounds	13 rounds	10 rounds

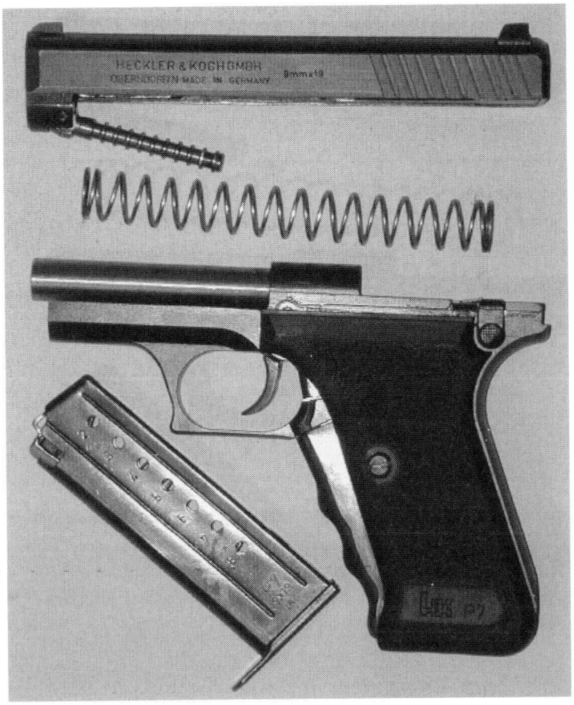

HK P7, partially disassembled. Note the gas piston attached to the slide.

9mm HK P7M8 pistol.

Prototype 9mm pistols, as submitted to the US XM9 trials. Front: the P7A13 with a double-stack magazine; rear: the P7A10 with an extended single-stack magazine. (Courtesy of Heckler & Koch)

officially approved for German police service and received the official designation 'P7'. Before that, about 500 PSP pistols were made for trials and commercial sales. Mass-production of the P7 started in 1977 and lasted until 1981, when it was replaced by the improved P7M8 variant. It was based on the prototype P7A10 pistol, made for the American XM9 trials. In the same year HK also produced a double-stack version of the P7, the P7A13, for the XM9 trials. Two years later, the slightly improved P7A13 entered production for police and commercial sales as the P7M13. In 1987 HK tried to produce a .45 calibre version of the P7 for American markets; this had a single-stack magazine for seven rounds and an artillery-type oil-filled recoil damper, which replaced the gas system. Only seven prototypes were made before the project was abandoned on the grounds of excessive cost and complexity.

The last mass-produced version of the P7 was the P7M10, a double-stack pistol chambered for the new and promising .40 S&W cartridge. This pistol was also made mainly for the US market. In 2005, Heckler & Koch terminated the production of all P7 variations, mostly because of their high price. However, regardless of an expensive price tag and a relatively heavy weight, P7 pistols had a strong following among civilian and police shooters, mostly because of its low felt recoil and excellent accuracy, thanks to a fixed barrel with a low axis. P7-type pistols were used by the German Military Police, as well as by a number of police organizations in Germany, Greece, Malaysia, Mexico and the USA.

9m P7M13s pistol, as manufactured for the Mexican police. It is similar to the standard P7M13 except for an added manual safety, made in the form of a slider above the trigger, on the right side of the frame. (Courtesy of Heckler & Koch)

.40 S&W calibre P7M10 pistol. Because of the heavier recoil of the .40 calibre round, the slide is necessarily bigger and heavier than on the 9mm pistols.

Prototype .45 calibre P7M7 pistol.
(Courtesy of Heckler & Koch)

HK P7 pistols use a gas-retarded blowback action, derived from the pre-war design of the American inventor, J. Destree. The opening of the slide in the initial stage of recoil, when the bullet is still in the barrel and pressure is high, is resisted by the pressure of the propellant gas, fed from the barrel into a gas cylinder located in the frame below the barrel. A gas piston is attached to the slide and enters the gas cylinder from the front; the gas port is located next to the chamber. The barrel has polygonal rifling and the chamber is fluted to help extraction of the spent cases. The pistol is of all-steel construction. On the P7M8 and later pistols, the trigger guard is extended forwards and a plastic heatshield is mounted above the trigger.

The gun is striker-fired, single-action and of unusual design. It is fitted with a squeeze lever, built into the front of the grip. The only way to cock the striker for the first shot is to squeeze the grip firmly, thus pushing the cocking lever backward. As long as the lever is pulled, the striker will be held back until the trigger is pressed. After the shot, the striker will be automatically recocked and held there until the next shot or the grip is released. As the grip is released, the striker is also released automatically into the safety position. This way, the pistol can be safely carried with a round in the chamber, yet rapidly brought into action and fired accurately, as the trigger pull is the same for every shot, unlike in the traditional DA pistols. The original PSP and P7M8 pistols had single-stack magazines, while P7M10 and P7M13 pistols had double-stack magazines. The magazine release was at the heel of the grip on the PSP and P7 pistols and at the base of the trigger guard on later models, in the form of ambidextrous levers that had to be pulled down to release the magazine. The slide stop release is operated by the same squeeze-cock lever – once the empty magazine is replaced by the fresh one, the shooter only has to squeeze the grip to close the slide automatically and cock the striker for a rapid first shot. Sights are fixed, both front and rear being dovetailed to the slide.

Manual safety None.

Field stripping procedure 1) remove the magazine by pressing the magazine release button; 2) check that the chamber is empty; 3) press the disassembly button, located at the rear left side of the frame; 4) pull the slide all the way back, raise its rear end from the frame, then slide it forwards and off the frame.

Reassemble in reverse order, observing that the gas piston fits securely into the gas cylinder.

Heckler & Koch USP, P8, P10

Heckler & Koch started the development of a new *Universal Selbstlade Pistole* – Universal Self-loading Pistol, or USP in short, in mid-1989. The concept of the new design was to provide a universal weapon for police and military forces, available in a variety of versions and sizes. The target market for this new weapon was apparently the USA, as the first version of the pistol, which appeared in 1993, was chambered for a new American cartridge, the .40 S&W. A 9mm version appeared a little later and in 1995 the slightly modified 9mm USP pistol was adopted by the *Bundeswehr* as its new service pistol, under the designation P8. The same year, the company also brought out another 'America-special' version of the USP, this time in .45 ACP calibre. In 1996 the first of the compact series of USP pistols entered production and next year it was officially approved for German police use as the 9mm P10. Several versions of the USP were also made specifically for civilian markets, such as the USP Expert, USP Elite and USP Match. There are also several special purpose versions of the USP. These include two original variations fitted with extended threaded barrels to be used with silencers, the .45 calibre USP Tactical and 9mm USP 9 SD, recently joined by the .45 calibre USP 45 Compact Tactical, sometimes preferred by the US military users over the much bulkier and heavier Mark 23 Model 0 pistol of the same calibre.

USP pistols are short-recoil-operated, locked-breech pistols with a polymer frame. The barrel is locked to the slide by a single lug, which engages the ejection port. Unlocking is controlled by a cam-shaped

9mm HK USP pistol.

Heckler & Koch USP, P8, P10

Data for Standard HK USP Pistols

Trigger type		Double/single action	
Calibre/ammunition used	9 × 19mm Para	.40 S&W	.45 ACP
Weight	720g (25.2oz) empty	780g (27.3oz) empty	785g (27.5oz) empty
Length	194mm (7.6in)	194mm (7.6in)	200mm (7.8in)
Barrel length	108mm (4.2in)	108mm (4.2in)	112mm (4.4in)
Magazine capacity	15 rounds	13 rounds	12 rounds

Data for Compact HK USP Pistols

Trigger type		Double/single action	
Calibre/ammunition used	9 × 19mm Para	.357 SIG, .40 S&W	.45 ACP
Weight	645g (22.6oz) empty	695g (24.3oz) empty	710g (24.9oz) empty
Length	173mm (6.8in)	173mm (6.8in)	180mm (7in)
Barrel length	91mm (3.6in)	91mm (3.6in)	96mm (3.7in)
Magazine capacity	13 rounds	12 rounds	8 rounds

extension below the chamber, which interacts with a steel insert set into the frame. Barrels on all USP pistols have polygonal rifling, except the earliest production pistols made in 1993, and all German Army P8s. The captive recoil spring assembly includes an additional recoil spring buffer on full-size pistols and a polymer buffer on compact pistols. The trigger is double-action, with an external hammer, an automatic firing-pin safety and a frame-mounted lever, which may operate differently in different versions of the pistols, as

HK USP.

.45 calibre HK USP Compact.

.45 calibre HK USP Tactical with extended threaded barrel and adjustable sights. (Courtesy of Cherry's Fine Guns)

described below. Each pistol can be easily adjusted by a gunsmith to any of the trigger mode configurations, with the replacement of a few parts that include a cam plate and a safety lever. The sights are dovetailed to the slide. Magazines on most USP pistols are of steel construction, but *Bundeswehr*-owned P8 pistols have semi-translucent polymer magazines. Magazine release levers are ambidextrous and located at the base of the trigger guard.

There are nine basic variations of the USP trigger/safety arrangements, plus one specific to P8 pistols, as follows:

1) Double-action trigger with manual safety on the left side. If the lever is pressed below the 'fire' position, it acts as a decocker, bringing the hammer down and then returning to the fire position.
2) Same as No.1, but with the safety lever on the right side.
3) Double-action trigger with a decock-only lever on the left side.
4) Same as No.3, but with the lever on the right side.
5) Double-action-only trigger with a manual safety lever on the left side.
6) Same as No.5, but the lever on the right side.
7) DAO trigger without any manual safety.
8) P8 pistol, with 'inverted' safety lever, that is, the upper position of the lever is for fire and the lower is for safe. When the safety is applied, the hammer is decocked automatically.
9) Double-action trigger with manual safety on the left side. No safe decocking functionality is provided.
10) Same as No.9, but with the lever on the right side.

HK USP, partially disassembled.

Manual safety Lever on the frame; operation and position (left or right) may differ, see above.

Field stripping procedure 1) remove the magazine by pressing the magazine release button; 2) check that the chamber is empty; 3) retract the slide partially, so that the slide release lever axle is aligned with the disassembly notch on the slide; 4) press on the slide release axis pin from the right side of the frame, then pull the slide release lever completely from the left side; 5) carefully ease the slide forward, then pull it towards the front out of the frame; 6) remove the return spring assembly from below the barrel; 7) pull the barrel out of the slide.

Reassemble in reverse order.

.45 calibre Mark 23 Model 0 (or Mk23 Mod.0 in short) US SOCOM pistol, fitted with a military-issue silencer made by Knight's Armament Co. (USA).

Heckler & Koch Mark 23 Model 0 US SOCOM Pistol

Trigger type: Double/single action
Calibre/ammunition used: .45 ACP
Weight: 1,100g (38.5oz)
Length: 245mm (9.6in)
Barrel length: 150mm (5.9in)
Magazine capacity: 12 rounds

The Mark 23 Model 0 pistol (or Mk23 Mod.0 in short) is the result of the US Special Operations Command (US SOCOM) requirement for an Offensive Handgun Weapon System (OHWS), set out in 1991. Two companies were originally contracted for OHWS, the US-based Colt Company and the German-based (but then British-owned) Heckler & Koch. In 1994, after initial trials, the HK prototype was chosen for the next stage of development and the first production pistols were shipped to US SOCOM in 1996.

The Mk23 Mod.0 is an extremely strong and reliable, as well as quite accurate, weapon, but it is also heavy and bulky. Despite being promoted as the 'state of the art', it is in fact not very popular among the men who have to carry these on real offensive or defensive operations. In recent years, there have been constant rumours that many SOCOM operators prefer the smaller and lighter HK USP Tactical pistols to the theoretically more effective, but also more obtrusive, Mk23s. Nevertheless, the Mark 23 Model 0 pistol is a formidable weapon, which can use almost any .45 ACP ammunition, from heavy subsonic up to +P+ JHP loadings. Standard equipment includes the quick-detachable silencer, made in the USA by Knight's Armament Co., and the laser-aiming module.

Mk23 Mod.0 pistol is broadly based on the HK USP pistol described above. The basic action with the modified Browning-type locking mechanism and the patented recoil-reducing spring set-up is the same, as

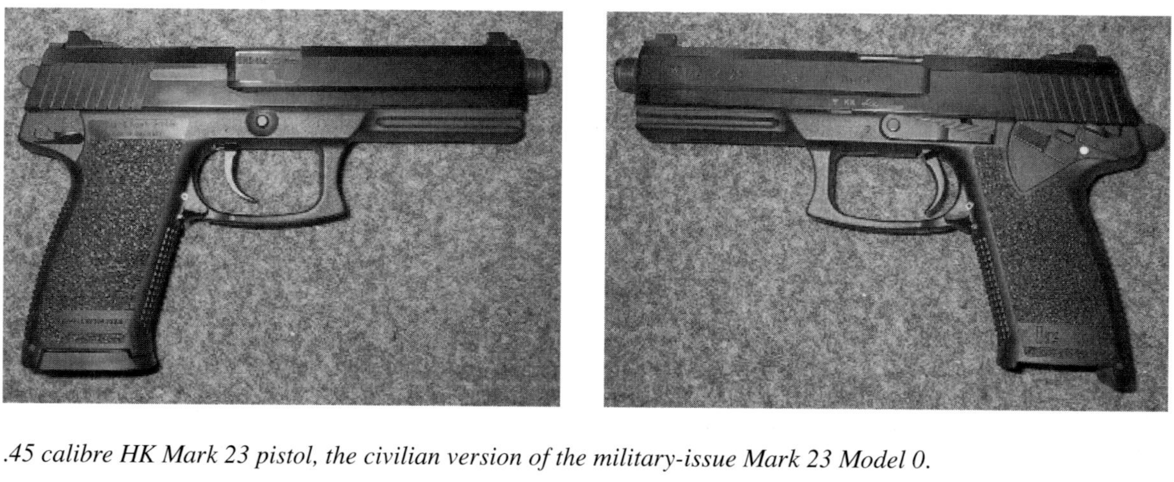

.45 calibre HK Mark 23 pistol, the civilian version of the military-issue Mark 23 Model 0.

well as the general design of polymer frame. However, the barrel is fitted with a polymer O-ring at the muzzle to ensure a tighter and more uniform fit into the slide, thereby improving accuracy. The barrel is threaded at the muzzle to accept a silencer. Safety arrangements also differ from the USP, with separate ambidextrous safety levers and a decocking lever which is located ahead of the safety, on the left side of the frame. Low-profile sights are adjustable and have luminous tritium inserts for improved low-light visibility.

While the original Mk23 Mod.0 pistols were intended only for US Government use, HK later decided to bring these pistols to the civilian market. 'Civilian' pistols are exactly the same as US-contract ones, except for the slide markings and, before the end of 2004, the magazine capacity. Military Mk23 Mod.0 pistols are marked 'Mk23 USSOCOM' on the slide, while civilian pistols are marked 'Mark 23'. Also, before the end of the so-called 'assault weapons ban', civilian Mark 23 pistols were sold in the USA with ten-round magazines, as opposed to standard Mk23 Mod.0 twelve-round magazines.

Manual safety Levers on both sides of the frame, above and behind the grip; the safety is applicable only with the hammer cocked, to provide users with a 'cocked & locked' carry feature. A separate decocking lever allows the hammer to be lowered safely; the decock function works only when the safety is off.

Field stripping procedure 1) remove the magazine by pressing the magazine release button; 2) check that the chamber is empty; 3) retract the slide partially, so that the slide release lever axle is aligned with the disassembly notch on the slide; 4) press on the slide release axis pin from the right side of the frame, then pull the slide release lever completely from the left side; 5) carefully ease the slide forwards, then pull it towards the front out of the frame; 6) remove the return spring assembly from below the barrel; 7) pull the barrel out of the slide.

Reassemble in reverse order.

Heckler & Koch P2000

The HK P2000 is a further development of the HK USP Compact pistol, intended primarily for those police officers and civilian users who may need a compact yet adequately powerful pistol for everyday carry. The P2000 was first announced in 2000 and entered production in 2001 and it is officially approved for police use in Germany. While the P2000 retained many of the design features of the USP, it has some significant improvements in the ergonomics, such as completely ambidextrous handling (with slide release levers located on both sides of the frame) and with adjustable grips with removable backstraps of a different shape, as in the Walther P99. The trigger/safety system is also different and the P2000 is manufactured in no fewer than six different versions, namely the P2000 and P2000 V1 to V5. These versions differ only in the trigger and decocking mechanism set-up.

The P2000 uses the same modified Browning-type, recoil-operated, locked-breech action as the HK USP described above. The frame of the P2000 is made from high-strength polymer and the barrel features polygonal rifling. The trigger is double-action, with an external hammer, and is available in four basic versions, one of which has three sub-versions that differ only in the weight of the trigger pull. Because of the modular design, pistols of the P2000 family can be easily converted to any of the available versions.

Basic P2000 pistols have a two-part hammer, similar in basic concept to the Canadian Para-Ordnance LDA or FN GP-SFS pistols. This means that the hammer can be disconnected from the main spring and lowered while the main spring remains compressed. In this

Heckler & Koch P2000

	P2000	P2000 SK
Trigger type	Double/single action	
Calibre/ammunition used	9 × 19mm Para, 40 S&W	
Weight	620g (21.7oz) empty	600g (21oz) empty
Length	178mm (6.9in)	163mm (6.4in)
Barrel length	92mm (3.6in)	63mm (2.5in)
Magazine capacity	13 (9mm) or 12 (.40) rounds	10 (9mm) or 9 (.40) rounds

HK P2000.

HK P2000 (right side).

View on the rear part of the P2000 slide and frame. The decocking button is clearly visible below and to the left of the hammer.

P2000 SK pistol: a sub-compact version of the P2000.

mode, the DA trigger pull is significantly lower than in conventional DA designs, as the shooter does not have to compress the main spring. The decocking lever is unusually located at the rear of the slide, below and to the left of the short hammer.

HK P2000 V1, V2 and V4 pistols are similar in design to the standard P2000 except that these do not have a decocking lever and the trigger pull is different in weight, being 2.04kg (4.5lb) on the V1, 3.3kg (7.3lb) on the V2 and 2.8kg (6.2lb) on the V4.

The HK P2000 V3 has a conventional DA trigger system, with a hammer that is permanently connected to the main spring and with a rear-mounted decocker. Finally, there is the HK P2000 V5, which is a recent addition to the breed and is a conventional DAO pistol with an external hammer and no decocker.

Manual safety None; on some models a decocking lever is present at the rear of the slide, next to the hammer, as described above.

Field stripping procedure 1) remove the magazine by pressing the magazine release button; 2) check that the chamber is empty; 3) partially retract the slide, so that the slide release lever axle is aligned with the disassembly notch on the slide; 4) press on the slide release axis pin from the right side of the frame, then pull the *left* slide release lever completely from the left side; note that the *right* slide release lever will stay in place; 5) carefully ease the slide forward, then pull it towards the front out of the frame; 6) remove the return spring assembly from below the barrel; 7) pull the barrel out of the slide; 8) to remove the right side slide release lever, rotate it upwards for about 45 degrees, then pull out of the frame.

Reassemble in reverse order.

Heckler & Koch P11 (Underwater Pistol)

Trigger type: Single action, electronic ignition
Calibre/ammunition used: 7.62 × 36 proprietary
Weight: About 1.2kg (42oz)loaded
Length: 200mm (7.8in)
Barrel length: No data
Magazine capacity: Removable barrel cluster, 5 shots/barrels

The HK P11 is one of the least known weapons to emerge from the Heckler & Koch factory. Developed during the early 1970s to arm frogmen (combat divers), this weapon is apparently in use with the navies of Germany, Italy, France, Norway, the UK and the USA. This highly unusual weapon is probably the only serious rival to the Soviet underwater pistol, which appeared in the late 1960s. The P11 and its ammunition are somewhat bulkier than its Soviet counterparts, but, on the other hand, the P11 has one round more capacity and a larger calibre, probably resulting in better lethality.

The HK P11 pistol is mainly unusual in the fact that it is not self-loading. Instead, it uses a quick-detachable cluster of five barrels, each factory-loaded with a powder charge and a long, needle-like, drag-stabilized projectile. The pistol is reloaded by the replacement of an entire barrel cluster. In combat, the empty clusters are discarded; in training, these are collected and later sent back to the factory for reloading. Loaded barrels are protected from salt water by thin diaphragms, which are destroyed by the fired bullet at the instant of firing. Powder charges in each barrel are ignited using electricity, provided by two 9V batteries, stored in a

7.62mm P11 underwater pistol. Right photo shows pierced diaphragms that cover the barrels from the water when unfired. (Courtesy of Heckler & Koch)

sealed container in the grip of the pistol. The effective range underwater depends on the depth, and is about 15m (49ft) near to the surface. Above the water, the effective range is quoted as being around 30m (98ft). While the long and relatively heavy bullets can be lethal at longer ranges, the chances of hitting any target beyond the given range is minimal. This is because the barrels are not rifled and so the bullets are not stabilized in the air, leading to tumbling and yawing in flight through the air.

Walther PP and PPK

	Walther PP	Walther PPK
Trigger type	Double/single action	
Calibre/ ammunition used	7.65 × 17SR Browning, 9 × 17 Browning Short	
Weight, empty	682g (23.9oz)	568g (19.9oz)
Length	173mm (6.8in)	155mm (6.1in)
Barrel length	99mm	86mm (3.4in)
Magazine capacity	8 (7.65mm) or 7 (9mm) rounds	7 (7.65mm) or 6 (9mm) rounds

Walther PP pistols were among the most important developments of the interwar period. Produced between 1929–45 in significant numbers, these pistols, along with the basically similar but smaller PPK, were widely used as police and military guns in Hitler's Germany. After the war, production of the PP and PPK pistols was resumed in France under German licence. Later on, production was returned to the re-established Walther factory in the city of Ulm and these pistols have seen widespread use by civilians and police, as well as for personal defence by many non-infantry officers in several European armies. Very close copies of the Walther PP were manufactured after the war by East Germany, Hungary, Romania, Turkey and the USA. At the present time, Walther PP, PPK/S and PPK pistols are manufactured in the USA by the Smith & Wesson Company under licence from Walther.

The Walther PP is a blowback-operated pistol with a fixed barrel, usually of all-steel construction. A few aluminium-framed PP pistols were built in Germany before the war and stainless-steel versions are manufactured in the USA under Walther's licence since the mid-1980s. The trigger is double-action, with an exposed hammer and a frame-mounted manual safety/ decocker; the lockwork is somewhat complicated in a typical German manner and has many small parts and pins. Sights are fixed, with the rear sight blade dovetailed into the slide. Magazines are single-stack; the magazine release button is usually located at the left side of the frame, just below the slide and in front of the grip panel. However, at least some PP pistols were made after the war with the so-called 'European-style' magazine release, located at the heel of the grip. Walther PP and PPK pistols are fitted with a loaded chamber indicator, made in the form of a small pin that protrudes from the rear of the slide (above the hammer) when a cartridge is in the chamber.

PPK pistols are essentially similar to the larger PP pistols, except for the different design of the grip frame – while on PP pistols the grip backstrap is integral to

9mm Walther PP pistol of post-war manufacture.

9mm Walther PP (left side).

Walther PP, partially disassembled.

the frame and the grip panels are two separate items (left and right), on PPK pistols the grip frame has a rectangular shape of a magazine channel and the backstrap is formed by the single-piece U-shaped grip unit, also usually made of plastic. While most PPK pistols were made with steel frames, in the post-war period Walther also produced some aluminum-framed PPK/L (Lightweight) pistols.

Also, there are Walther PPK/S pistols, which are a cross-breed between PP and PPK, combining the PP frame with the shorter PPK-style barrel and slide. These pistols were designed to avoid limitations imposed by the American Firearms Owners Protection Act of 1968. This act, in particular, limited the minimum size of 'sporting purpose' pistols that are allowed to be imported into the USA and the use of the larger grip frame allowed importing these pistols instead of smaller PPK pistols, banned from importation under this law.

Manual safety Lever at the left side of the slide. Push lever down to turn safety on; push it up to turn safety off, red dot will appear on the slide. When applied, safety also decocks the hammer automatically.

Field stripping procedure 1) remove the magazine by pressing the magazine release button; 2) check that the chamber is empty; 3) pull the trigger guard downwards, then swing it to one side to lock in the open position; 4) pull the slide all the way back, then raise the rear of the slide to the top and off the frame rails; 5) carefully ease the slide forwards and off the barrel; 6) remove the return spring from the barrel.

Reassemble in reverse order.

7.65mm Walther PPK pistol of post-war manufacture.

WALTHER P38, P1 AND P4

	P38, P1	P4
Trigger type	Double/single action	
Calibre/ ammunition used	9 × 19mm Para	
Weight, empty	840g (29.4oz) (steel frame) 770g (27oz) (aluminium frame)	740g (25.9oz)
Length	216mm (8.4in)	197mm (7.7in)
Barrel length	125mm (4.9in)	104mm (4.1in)
Magazine capacity	8 rounds	8 rounds

9mm Walther P1 pistol, a post-war aluminium-framed version of the P38.

The Walther P38 pistol was developed as a military pistol for the German Army (*Wehrmacht*) during the late 1930s. It first appeared in 1938 and small numbers of the original HP (*HeeresPistole* – Army pistol) were bought by Sweden before the Wehrmacht adopted it as the Pistole 38 and took over all production of the guns. During the war, P38 pistols were made by a number of factories, including Walther itself. After the war, most of the ex-Walther machinery ended up in France as war reparations and many of the post-war P38 pistols were actually built in France, by the Manurhin factory.

In 1957 the newly formed *Bundeswehr* adopted a slightly modified P38 pistol with a lighter aluminium frame as the Pistole 1, or P1 in short. Commercial pistols were still manufactured under the original P38 designation. Most of the post-war P38 pistols were made with aluminium frames; only a handful of commercial pistols were made with steel frames. There were a couple of modifications to the P38, intended for police use, which appeared during the 1970s. The first was the P4, the first pistol to conform to new German requirements for police sidearms. Actually, the P4 was the P38 (or P1) with the barrel cut back by 25mm (1in), fitted with an automated firing-pin safety and decock-only lever. Early P4 pistols were actually made using P38 components, including slides, and thus were marked 'P38 IV' instead of the more common 'P4'.

Another, less well-known variation was the P38K (*Kurz*, for 'short' in German), made for some special users such as the KSK (*Kommando Spezial Krafte*) – an elite counter-terror unit. This was a P4 (rather than the original P38), with the barrel cut down at the front of the slide and it was intended for concealed carry. The front sight was mounted on the slide, unlike the full-size model. It must be noted that a similar model was produced in small numbers during WWII for the *Gestapo* and other such organizations. Wartime *Kurz* P38s differed from the post-war version by having a slightly longer barrel with the front sight mounted on the barrel. However, the service life of police derivatives of the P38 was much shorter than of the military P1 – most P4 pistols were declared obsolete and sold as surplus during the 1980s, while the P1 served until 1995. The last commercial P38 pistol was manufactured in 2000.

While P38 pistols were in some aspects revolutionary in design and concept, their post-war P1 versions were less than popular in the *Bundeswehr*, deserving the unoffical description of 'eight warning shots plus one aimed throw'. Also, these pistols showed a typical German obsession for unnecessary over-complication of design – for example, the P38 pistol had eleven springs (mostly of very small size) – about twice that of the older P08 Luger pistol it replaced in service. It also had plenty of small parts and pins that were easy to lose during full disassembly and a firing pin of intricate shape that easily broke.

The Walther P38 is a short-recoil-operated, locked-breech pistol. The barrel is locked to the slide using a wedge-shaped locking piece, which can tilt up and down below the barrel, while following the inclined surfaces on the frame. When the barrel and slide are in battery, the locking piece is in its upper position and its lugs securely lock the slide to the barrel. Upon recoil, the locking piece drops down and out of engagement with the slide, unlocking it and allowing it to recoil and complete the reloading cycle. Because of the short slide, the P38 has two captive return springs, located on either side of the frame and inside the slide. With time, it was found that the aluminium frame developed cracks in the most highly stressed area, where the locking piece and barrel were slamming against it on recoil, so the frames of late production pistols were reinforced with the addition of a hexagonal cross-pin, made from steel. The trigger is double-action, with an exposed hammer and trigger bar (the link between the trigger and sear) unusually located outside of the frame at the right side. The standard safety also acted as a decocker and was located at the left side of the slide. On the P4 pistol, the slide lever retained only the decock function and, once pressed and released, automatically returned to the fire position. P4 pistols were also fitted with a firing-pin block safety. Magazines were single-stack, with the magazine release located at the heel of the grip. The sights were fixed. P38 pistols were also fitted with a loaded chamber indicator in the form of a small pin that projected from the rear of the slide, above the hammer, when a cartridge was loaded in the chamber.

ABOVE: 9mm Walther P4 pistol visually differs from P1 and P38 only by its shorter barrel; however, the safety arrangement is also slightly different.

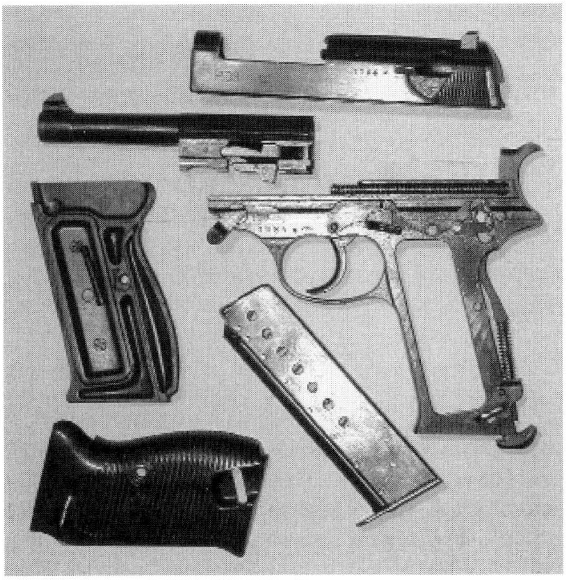

RIGHT: P38, partially disassembled.

9mm P38K pistol (at right) compared to standard P38.

P38K with slide locked open. The barrel-locking piece is visible below the barrel.

Manual safety A lever at the left side of the slide. Push the lever down to turn the safety on; pull it up to turn the safety off; a red letter 'F' will appear on the slide. When applied, the safety also decocks the hammer automatically. Note: on the P4 the lever will return to the fire position once pushed down and released.

Field stripping procedure 1) remove the magazine by pressing the magazine release button; 2) check that the chamber is empty; 3) pull the slide all the way back and lock it there with the slide-stop lever; 4) turn the disassembly lever down and forwards (located at the front of the frame, on the left side); 5) while holding the slide, release the slide-stop lever and carefully ease the slide into battery, then push it forwards and out of the frame; 6) turn the slide upside-down and push forwards the small pin at the base of the barrel; this will unlock the barrel from the slide; 7) pull the barrel out of the slide.

Reassemble in reverse order.

Walther PP Super

Trigger type: Double/single action
Calibre/ammunition used: 9 × 18 Police or 9 × 17 Browning Short
Weight, empty: 780g (27.3oz)
Length: 176mm (6.9in)
Barrel length: 92mm (3.6in)
Magazine capacity: 7 rounds

The Walther PP Super was quickly developed as a possible replacement for the venerable Walther PP as a police pistol. The PP Super retained most of the features of the earlier PP pistols in a slightly larger package that accommodated a reborn 9 × 18 Ultra/Police cartridge. However, the German authorities decided against the 9 × 18 cartridge and the PP Super was quickly dropped from production in favour of the more powerful P5 pistol, chambered for the 9 × 19 round. Despite its short production life, the PP Super did see service with some German police departments in the mid- to late-1970s. Most PP Super pistols were made in the now obsolete 9 × 18 Police chambering; a few also were made in 9 × 17 Browning Short chambering for commercial sales and these pistols are quite rare and valued by collectors.

The PP Super is a blowback-operated pistol with a fixed barrel. It is made entirely of steel, except for the grip panels. The trigger is double-action, with an external hammer and a slide-mounted decocking lever. It is also fitted with an automatic firing-pin block safety, which enables it to be carried with the hammer down on the loaded chamber. Unlike the Walther PP, the PP Super is fitted with an external slide stop release lever. Magazines are of single-stack design and the magazine release button is located at the base of the trigger guard. Interestingly, all PP Super pistols, despite their compact size, were fitted with adjustable rear sights.

Manual safety None; decocking lever at the left side of the slide. Push the lever down to decock the hammer safely and it will return to the fire position once pushed down and released.

Field stripping procedure 1) remove the magazine by pressing the magazine release button; 2) check that the

Walther PP Super pistol.

Walther PP Super and magazine.

PP Super, partially disassembled.

chamber is empty; 3) pull the trigger guard down and cant it slightly to one side to lock it open; 4) pull the slide all the way back, then raise its rear end off the frame rails; 5) ease the slide forwards and off the barrel; 6) remove the return spring from the barrel. Reassemble in reverse order.

P5, partially disassembled.

Walther P5

	P5	P5 Compact
Trigger type	Double/single action	
Calibre/ ammunition used	9 × 19mm Para	9 × 19mm Para
Weight, empty	795g (27.8oz)	750g (26.3oz)
Length	180mm (7in)	170mm (6.6in)
Barrel length	90mm (3.5in)	79mm (3.1in)
Magazine capacity	8 rounds	8 rounds

The Walther P5 pistol was developed at the request of the German police authorities and was approved for service in the late 1970s. It was manufactured until 2000 and served with several police departments in Germany, as well as in the Dutch police, the Portuguese Army and a few other police and military formations in Western Europe. Compact versions of the P5 were produced between 1990 and 2003 and a modified version of the P5 known as the P1A1, which was fitted with cross-bolt manual safety on the slide, was briefly marketed between 1989 and 1992.

From the technical standpoint, the P5 is a further

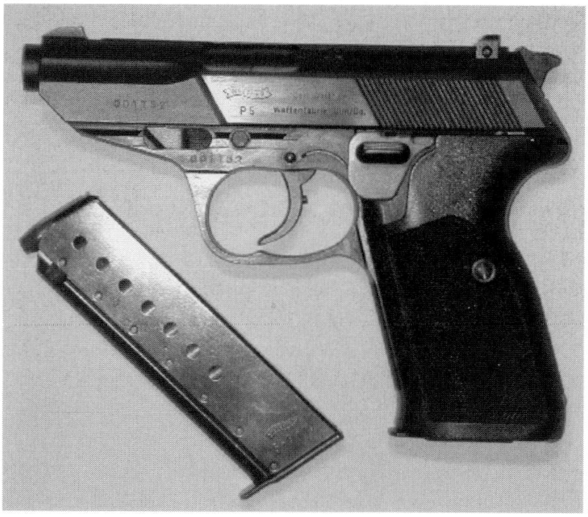

Walther P5 with spare magazine.

Walther P5, right side.

Three versions of the Walther P5, from top to bottom: P5 Lang with long barrel; P5; P5 Compact.

development of the earlier Walther P1/P38 and P4 pistols. It incorporates the same locked-breech, short-recoil system with a vertically tilting lock. However, the slide shape has been considerably revised, as well as the safety arrangements. The P5 has been made in a variety of finishes and in three basic patterns – the 'standard' P5 with a 90mm barrel, the P5 Compact with a 79mm barrel and the P5 Sport, also known as the P5 *Lang* (long), with a 120mm barrel, partially exposed at the front. P5 pistols are neat, easily concealable yet quite accurate weapons of typical German workmanship and intricate design.

For the description of the locking system, please refer to the P38 description above. The trigger of the P5 is double-action, with an external hammer and a frame-mounted decocking lever, which also serves as a slide stop release lever. An additional firing-pin safety block is provided for safe carry with the hammer down on a loaded chamber, as requested by contemporary German police regulations. It is interesting to note that all versions of the P5 are fitted with adjustable rear sights, not only the Sport version. Another less than usual feature of the P5 is the magazine release, located at the heel of the grip.

Manual safety None; decocking lever at the left side of the frame. Push the lever down to decock the hammer automatically; the lever then will return to fire position once released.

Field stripping procedure 1) remove the magazine by pressing the magazine release button; 2) check that the chamber is empty; 3) put the pistol with the barrel vertically down and against the table. Push the pistol down as far as it will go; 4) while holding the pistol against the table, turn the disassembly lever down and forwards (located at the front of the frame, on the left side); 5) while holding the slide, release the slide-stop lever and carefully ease it into the battery, then push it forwards and out of the frame; 6) turn the slide upside down and push forwards the small pin at the base of the barrel; this will unlock the barrel from the slide; 7) pull the barrel out of the slide.

Reassemble in reverse order.

Walther P88.

Walther P88, slide locked open.

Walther P88

	P88	P88 Compact
Trigger type	Double/single action	
Calibre/ ammunition used	9 × 19mm Para	
Weight, empty	900g (31.5oz) empty	822g (28.8oz) empty
Length	187mm (7.3in)	181mm (7.1in)
Barrel length	102mm (4in)	97mm (3.8in)
Magazine capacity	15 rounds	14 rounds

The Walther P88 pistol was developed between 1982–4 in an attempt to produce a modern, military-type pistol. It was boldly submitted to the 1984 American trials for a new 9mm Army pistol, but failed on the grounds of

Walther P88, partially disassembled.

insufficient reliability and corrosion resistance. In 1986 the P88 entered production, which continued until 1994 when it was replaced by the slightly shorter P88 Compact pistol, which, it turn, was dropped from production in around 2000.

Despite its failure in the US XM9 test, the P88 is a decent pistol, well-made and capable of quite good accuracy; what killed it was the expensive price, dictated by the 'old style' manufacturing and high standards of fit and finish.

The P88 is a long step from the earlier Walther military pistols. The older, proprietary Walther locking system was replaced with the more common modified Browning system, in which the barrel locks to the slide by a single lug entering the ejection window and tilts down to unlock upon recoil. Barrel tilting is controlled by a cam-shaped lug located below the chamber, which interacts with the frame upon recoil. The trigger is double action, with an exposed hammer and an automated firing-pin block safety. Original P88 pistols are fitted with ambidextrous decocking levers on either side of the frame, which also doubles as a slide stop release. Later P88 Compact pistols reverted to the more conventional safety/decocking arrangement, with ambidextrous levers located at either side of the slide. All versions of the P88 are fitted with adjustable rear sights. Magazines are of double-stack design; the magazine release button is located at the base of the trigger guard, in the form of an ambidextrous button.

Walther P88 Compact. Note the different position of the safety lever.

Manual safety P88: none; decocking lever at either side of the frame. Push lever down to decock the hammer automatically, the lever then will return to the fire position once released.

P88 Compact: ambidextrous lever at either side of the slide. Pull lever down to turn the safety on; push lever up to set safety off; a red dot will appear on the slide.

Field stripping procedure 1) remove the magazine by pressing the magazine release button; 2) check that the chamber is empty; 3) rotate the disassembly lever down and forwards (located at the left side of the frame, above the trigger); 4) pull the slide forwards and out of the frame; 5) remove the return spring from below the barrel; 6) pull the barrel out of the slide.

Reassemble in reverse order.

Walther P99

Trigger type	Double/single action (also see text)	
Calibre/ ammunition used	9 × 19mm Para	.40 S&W
Weight, empty	630g (22.1oz) empty	655g (22.9oz) empty
Length	180mm (7in)	184mm (7.2in)
Barrel length	102mm (4in)	106mm (4.1in)
Magazine capacity	16 rounds	12 rounds

The P99 pistol was a rather radical step forwards for the Walther company, made under new management. The development of a new pistol started in 1994 and the first production pistols were shown to the public in 1997. The earliest pistols were made in 9mm, with a .40 S&W version appearing in 1999. By 2004, the Carl Walther company had presented the second generation of the P99 pistols. These new guns had even better ergonomics and new model designations – the old P99 was renamed (and slightly redesigned) to P99AS (Anti-Stress), the double-action-only P990 turned into the P99DAO, with only the P99QA (Quick Action) retaining its name. The second generation also included the P99 Compact pistols, apparently made with input from the American company Smith & Wesson, as the basically similar SW99 Compact pistol appeared on the US market a little earlier. Smith & Wesson makes a close copy of the P99 as the SW99, using Walther-made polymer frames and its own slides and barrels of slightly different shape. S&W also makes a

Walther P99QA (Quick Action), first generation.

Walther P99AS (Anti-Stress) pistol, second generation.

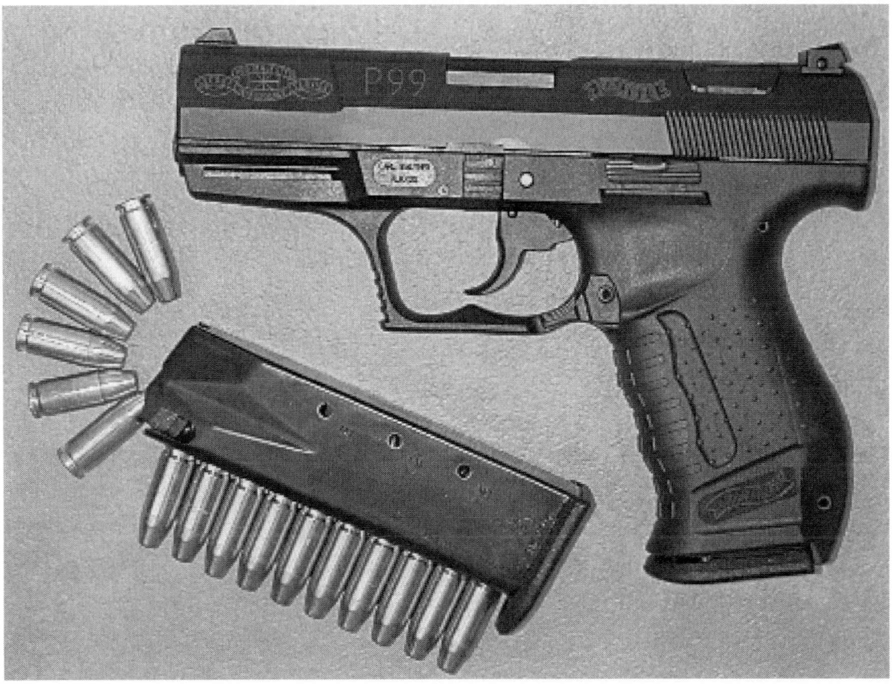

Walther P99 pistol, first generation (pictured is the limited edition 'James Bond – agent 007 commemorative' pistol).

.45-calibre version of this pistol. 9mm P99 pistols are also manufactured under Walther's licence in Poland and are standard police sidearms there.

The Walther P99 is a short-recoil-operated, locked-breech pistol. It uses a modified Browning locking system, with the barrel engaging the slide with a single large lug entering the ejection window. The frame is made from polymer and has interchangeable grip back-straps. Three backstraps of different shape are included with each pistol, so users can adjust the grip to their own preferences. The trigger is striker-fired, with an automated firing-pin block safety, and varies with the model. The standard P99 is double/single action with a decocking button. Earlier P990s and current produc-

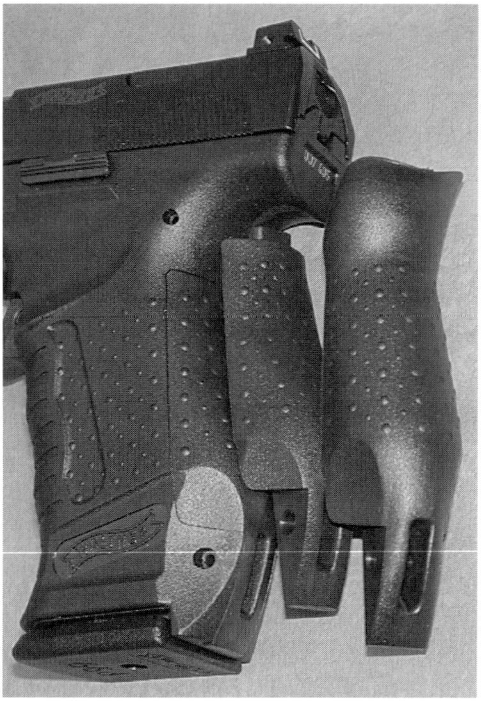

ABOVE: Walther P99DAO pistol, second generation, with optional ambidextrous extended slide release.

LEFT: Set of detachable grip backstraps of different sizes, available for every P99 pistol.

P99, field stripped.

tion P99DAOs are double-action only and the P99DAO also features a slightly different automated safety built into the trigger.

The third variation of the P99 family is the P99QA, which features a partially pre-cocked trigger for a constant trigger pull from the first to the last shot. One feature, unique to the P99 series pistol triggers, is that despite being striker-fired, these pistols do not require the user to press the trigger during disassembly to disconnect the sear from the firing pin. This is done by using the decocking button, built into the top of the slide. This button is relatively large on the standard P99, as it is routinely used to decock the striker, but is much smaller on the P99QA, as it is used only during disassembly to disconnect the firing pin from the sear. The P99DAO does not have such a button, as its firing pin is always disconnected from the sear and at rest, except for the time the trigger is being pressed. The most recent Anti-Stress feature is, in fact, a modification to the trigger system which makes the trigger pull for the first shot in Single Action mode much longer than for subsequent shots. The Anti-Stress feature is activated each time the pistol is reloaded.

Another safety feature is a striker cocking indicator, made in the form of a small pin, which protrudes from the back of the slide when the striker is cocked. First-generation pistols have a slide release only at the left side of the frame; second-generation guns may have optional ambidextrous slide stop release levers which, combined with ambidextrous magazine release levers built into the base of the trigger guard, make these pistols truly left-hand friendly. All variations of the P99 use adjustable rear sights with white inserts. Magazines are of double-stack design.

Manual safety None; a decocking button at the top of the slide, ahead of the rear sight, on the standard P99. Note: the decocking button on the P99QA should not be used as a routine safety measure, as it is intended only for disassembly of the pistol.

Field stripping procedure 1) remove the magazine by pressing the magazine release button; 2) check that the chamber is empty; 3) on the P99, P99AS and P99QA, press the decocking button; 4) press down and hold the disassembly sliders, located at either side of the frame, above the trigger; 5) push the slide forwards and out of the frame; 6) remove the return spring assembly from below the barrel; 7) remove the barrel from the slide.

Reassemble in reverse order.

Hungary

After the end of World War II Hungary fell into the Soviet zone of interest and soon joined the Warsaw Pact. The first post-war military pistol to be adopted by the Hungarian military was the 48M, a straightforward copy of the Soviet Tokarev TT-33, chambered for the Soviet 7.62 × 25 ammunition. In about the same period the state-owned arms factory, FEG (*Fegyver es Gazkeszuelekgyara*), in Budapest began to produce a smaller 7.65 × 17 calibre pistol with the same

The 7.65mm 48M pistol, also known as the Walam 48, was a close copy of the Walther PP.

The 7.62mm 48M pistol, in turn, was an exact copy of the Soviet Tokarev TT; the only visible difference was the Hungarian crest on the grip panels.

designation 48M, also known as the Walam 48. This was more or less an exact copy of the German Walther PP, with only minor differences in details, finish and markings. Most of the Walam pistols were sold for export and went to commercial markets, but it seems that at least a few of these were also used by the Hungarian police. In the late 1950s the FEG factory also produced a relatively large batch of Tokagypt 58 pistols, based on the familiar Tokarev TT design, but chambered for 9 × 19 Luger ammunition and fitted with a manual safety and grips of better shape. These pistols were initially ordered by Egypt but for various reasons the contract felt through and most of the 15,000 Tokagypt 58 pistols made were sold on commercial markets, mostly in West European countries.

During the early 1960s Hungary again followed the Soviet route and adopted 9 × 18 Makarov ammunition as its new standard for pistols and sub-machine guns. FEG developed a new military pistol for this round, known as the PA-63. FEG also produced export versions of the PA-63 in 9 × 17 and 7.65 × 17SR Browning chamberings, which were sold as the AP-63. There was also an even smaller version of the PA-63 in 9 × 18 PM, the SMC-918.

In the early 1980s FEG designers, apparently tired of copying pre-war Walther designs, turned towards more substantial, if not more modern, prototypes and copied the Belgian FN Browning High Power pistol. Initially intended for export purposes, these High Power clones were made in several versions, with different exteriors and dimensions. Later on, FEG improved the basic design with the introduction of a double-action trigger, more or less copied from contemporary US Smith & Wesson automatic pistols. Various versions of these pistols are still available from the reorganized FEG Company as commercial offerings. One of its 9mm double-action pistols, known as the P9R, was adopted by the Hungarian armed forces in around 1996 as the 9mm-ES 96M and should replace the older PA-63 pistols in military service.

In the late 1980s, the Hungarian engineer Robert Voros developed and produced a prototype military pistol of large size and unorthodox shape, known as the 'Danuvia'. This boxy-looking firearm was more or less based on the Israeli UZI sub-machine gun, but fitted with a proprietary helical magazine of quite large capacity – thirty-six rounds. This magazine was located ahead of the trigger guard and parallel to the barrel; the pistol was offered in semi-automatic and selective-fire versions for several years, but, not surprisingly, found no buyers and finally disappeared from the scene in the mid-1990s.

PA-63

Trigger type: Double/single action
Calibre/ammunition used: 9 × 18mm PM
Weight: 595g (20.8oz) empty
Length: 175mm (6.8in)
Barrel length: 100mm (3.9in)
Magazine capacity: 7 rounds

The PA-63 is a blowback pistol with a fixed barrel and an aluminium alloy frame. One specific aspect of the PA-63 design is that the aluminium frame has no finish at all and is white in colour – quite unusual for a

9mm Tokagypt 58 pistol; note the manual safety above the grip panel.

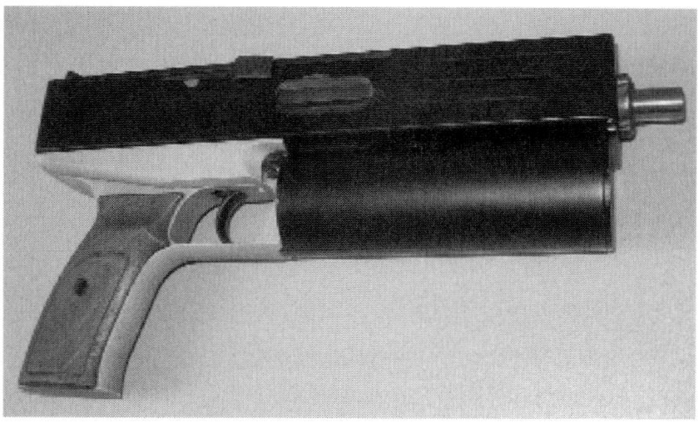

Prototype 9 × 19mm 'Danuvia' assault pistol, designed by Robert Voros; note the large-capacity magazine of cylindrical shape mounted ahead of the trigger guard.

FEG PA-63 pistol; note white aluminum frame.

9mm FEG P9M pistol.

military pistol. The trigger is double action, with an external hammer and slide-mounted manual safety/decocker. The magazine release button is located at the left side of the frame, just above and behind the trigger.

Manual safety A lever at the left side of the slide. To put on the safety, pull the lever down; to set to 'fire', push the lever up; a red dot will appear on the slide. Note: when applied with the hammer cocked, the safety will first bring the hammer down and then lock the action.

Field stripping procedure 1) remove the magazine by pressing the magazine release button; 2) check that the chamber is empty; 3) pull the trigger guard down and cant it sideways slightly to lock it open; 4) pull the slide all the way back, then raise its rear part out of the frame and carefully let the slide to go forwards and over the barrel; 5) remove the return spring from the barrel.

Reassemble in reverse order.

FEG P9M and FP9

Trigger type: Single action
Calibre/ammunition used: 9 × 19mm Para
Weight: 910g (31.9oz) P9M/950g (33.3oz) FP9 empty
Length: 198mm (7.7in)
Barrel length: 118mm (4.6in)
Magazine capacity: 13 rounds

FEG P9M and FP9 pistols are close copies of the Belgian FN Browning High Power pistol; in fact they are so close that some parts are interchangeable,

9mm FEG FP9 pistol.

.45 ACP FEG ACK pistol of current manufacture.

9mm FEG P9R of current manufacture (early production pistols had rounded trigger guards and no finger grooves on the grip).

including magazines. Differences between the FP9 and P9M are only external – the FP9 has a ventilated sight rib, which is more for 'aesthetical' reasons than for anything else. Single-action FEG pistols were widely exported around the world; in Europe they were sold using either their original names, or under the famous German 'Mauser' trademark, as the 'Mauser 80SA'. In the USA, P9M pistols were once sold as the PJK-9HP; at the present time the current American importer, KBI Inc., sells these pistols as 'Charles Daly HP'.

In design and operation, both the FP9 and P9M are exactly the same as the Belgian FN Browning High Power pistols, described above in the section on Belgium.

FEG ACK and P9R

The FEG P9R pistol is a further development of the P9M; the key difference is the introduction of a double-action trigger with a slide-mounted decocking safety, but there are other changes too, including a restyled frame and a magazine capacity increased by one round. Variations of the basic P9R model include the short-barrelled P9RK which retains the same frame and calibre, and the ACK, which is dimensionally similar to the P9RK but is chambered for the bigger .45 ACP ammunition and has a single-stack magazine. The ACK pistol also can be encountered under the designation GKK-45. Some P9R pistols were also exported to Western Europe under the famous German 'Mauser' trademark and were marked as 'Mauser 90DA' and 'Made in Hungary'.

The FEG P9R is a short-recoil-operated, locked-breech pistol of all-steel construction. The locking is of Browning type, with two lugs on the barrel engaging respective cuts inside the slide. Unlocking is controlled by a cam-shaped cut made in the lug below the

FEG ACK and P9R

	ACK	P9R	P9RK
Trigger type		Double/single action	
Calibre/ammunition used	.45 ACP	9 × 19mm Para	
Weight	1080g empty	1,000g (35oz) empty	970g (34oz) empty
Length	197mm (7.7in)	203mm (7.9in)	190mm (7.4in)
Barrel length	105mm (4.1in)	118mm (4.6in)	105mm (4.1in)
Magazine capacity	8 rounds	14 rounds	

chamber of the barrel. The trigger is double action, with an exposed hammer and a manual slide-mounted safety. Magazines are of double-stack design for the 9mm P9R and P9RK and are single-stack for the .45 calibre ACK. Sights are fixed, with the rear being dovetailed to the slide.

Manual safety A lever at the left side of the slide. To put on the safety, pull the lever down; to set to 'fire', push the lever up; a red dot will appear on the slide. Note: when applied with the hammer cocked, the safety will first bring the hammer down and then lock the action.

Field stripping procedure 1) remove the magazine by pressing the magazine release button; 2) check that the chamber is empty; 3) pull the slide all the way back and lock it with the slide-stop lever; 4) while holding the slide back, push the slide stop pin inwards from the right; then pull it out completely to the right; 5) carefully ease the slide into the battery, then pull it further forwards and out of the frame; 6) remove the return spring guide and spring from the slide; 7) pull the barrel out of the slide.

Reassemble in reverse order.

Israel

The state of Israel, which was established in 1948, has fought numerous wars against neighbouring Arab countries which were opposed to the existence of the Jewish state. Interestingly, while Israel is highly militarized and many (if not most) adult citizens are authorized to carry small arms, pistols there have always been of less importance than long arms. Consequently, the TSAHAL (Israeli Defense Forces) never had any specific handgun as a standard issue item, always using everything available. In the first decades, Israel bought large numbers of 9mm FN Browning High Power pistols. In the 1950s or 1960s Israel also purchased significant numbers of 9mm Italian Beretta M951 pistols, probably after examining the same pistols captured from Egyptians. In the 1980s and 1990s, older pistols were gradually replaced with more modern 9mm pistols like Glock (Models 17 and 19) and SIG-Sauer (mostly Model 226). Israel also bought smaller-calibre pistols, like the .22LR Beretta model 71, which were used for training and also issued to female personnel in the Army and police. These pistols, fitted with silencers, were also used by MOSSAD (the Israeli Intelligence service) operatives to execute Palestinian terrorists who had participated in the infamous Munich Olympic Games massacre of 1972.

Not wanting to rely only on imported arms, Israel also manufactured several domestic pistols of more or less original design. Probably the first of these was the ungainly-looking UZI pistol, which was, in fact, a semi-automatic-only copy of the famous Micro-Uzi sub-machine gun. Despite its angular shape and heavy weight, this pistol provided good accuracy and significant firepower thanks to its large magazine; it was chambered for 9mm Luger or .45 ACP ammunition. During the early 1990s, the state-owned Israel Military Industries (IMI) company developed a more conventional combat pistol, the Jericho 941. This was actually an international effort, since the design of the pistol was based on the Italian Tanfoglio system, itself a well-made clone of the Czechoslovak CZ-75 pistol. The original model index, the '941', came from the two

Italian-made Beretta 71 pistol of .22LR (5.6mm rimfire) calibre, fitted with a silencer; similar pistols were used by MOSSAD operatives in their revenge missions against Palestinian terrorists.

The 9mm UZI pistol was rather heavy and bulky for the class, but it also was quite accurate and relatively inexpensive, thanks to its stamped construction. It was based on the famous Micro-UZI sub-machine gun.

calibres initially available in this pistol, the ubiquitous 9mm Luger (9 × 19) and the new .41AE, which was developed in 1986 by the American company Action Arms. This cartridge, which was ballistically similar to the more 'modern' .40 S&W, was designed to be used in 9mm pistols with minimal modifications (new barrel and return spring and possibly new magazine), and thus had a rebated rim of the same diameter as the 9mm cartridge. Early Jericho pistols were often shipped with two sets of barrels and springs, one in 9mm and another in .41AE. The .41 was a good cartridge, by no means inferior to the rival .40 S&W, but it lacked one essential feature – strong marketing support. Because of the failure to sell this calibre to the public, the .41AE has been dropped from the Jericho line and replaced by the more successful .40 S&W chambering. Jericho pistols are widely exported from Israel, as well as being used by Israeli private security and police forces.

At about the same time, the small, privately owned company, KSN Industries, began to manufacture two pistols. First was the Kareen, which was more or less a straightforward copy of the FN Browning High Power pistol. KSN produced three subsequent versions of the Kareen pistol, from Mark I to Mark III. Second was the Golan, a licensed copy of the Yugoslavian CZ-99 pistol. It must be noted that the Golan was produced only in very limited numbers, as its production was moved to South Africa, where it was manufactured as the Tressitu TZ-99. Kareen pistols have seen extensive use by Israeli security companies and guard services.

BUL Transmark, another privately owned Israeli company, entered the pistol market in 1993 with its M5 pistol, based on the proven US M1911 pistol design.

Three basic sizes of the Jericho 941 pistol, top to bottom – standard, semi-compact and compact.

.45 calibre BUL M5 pistol with KBI/Charles Daly markings. *(Courtesy of Charles Daly)*

.45 calibre BUL M5 Commander. *(Courtesy of Charles Daly)*

The key improvement of the BUL M5 over the venerable M1911 is the polymer frame. BUL M5 series pistols are widely used as sport guns by various IPSC (International Practical Shooting Confederation) and IDPA (International Defensive Pistol Association) shooters, as well as by private security in Israel. BUL also makes Tanfoglio-based double-action pistols, with polymer and steel frames; the BUL Impact and BUL Storm. BUL pistols are imported into the USA by KBI Inc./Charles Daly.

In the year 2002 the small arms department of IMI (which was recently put into private ownership as the IWI – Israeli Weapon Industries Ltd) introduced an entirely new and quite original pistol, the SP-21 *Barak* (which means 'Lightning' in Hebrew). This pistol was intended to replace the Jericho pistols as well as to compete with the 9mm Glocks, which are very popular among Israeli users in military and law enforcement. So far, we cannot tell if the SP-21 has been adopted by any major government user in Israel; however, it is exported to Europe and the USA, where it is available through Magnum Research Inc., the major distributor of IMI pistols.

The last gun, certainly worthy of note but more due to curiosity than anything else, is the 'Desert Eagle' pistol of Hollywood fame. Designed in the USA in the mid-1980s as a hunting and sporting pistol to fire powerful .357 Magnum revolver ammunition, this pistol has been refined by IMI under contract from Magnum Research Inc. and is manufactured by IMI. This huge and heavy pistol is now available in .357 Magnum, .44 Magnum and the extremely powerful .50AE, which generates a muzzle energy on a par with several assault rifle rounds. A distinctive appearance and uncommon gas-operated action, combined with heavy recoil (in .50AE) and low magazine capacity, are essential properties of the Desert Eagle. While several respectable sources quote this gun as being 'worthy of note to military and special operations use', the author seriously doubts that. After years of research, the author was able to find only two undocumented accounts of the use of the .44 calibre Desert Eagle pistol as a service sidearm. One was a Hollywood-style sheriff somewhere in the south of the USA, the other was a colonel in a special drug enforcement unit in the

US-designed and Israeli-made, this .44 Magnum Desert Eagle pistol is better suited for sports like hunting or silhouette shooting, than for real combat. For its empty weight (about 1.9kg [4.2lb]), anyone can carry two loaded 9mm pistols such as the Glock 17 or the SP-21 Barak.

Thai Army. In both cases, these pistols were bought with their users' private funds and, in the latter case, the Desert Eagle was finally replaced by the issued 5.7mm FN Five-seveN pistol. While the Desert Eagle is certainly a fearsome and deadly weapon, in real life it is far too heavy and bulky to be routinely carried and is not reliable enough for the extreme conditions often encountered in military and special operations. The confusion caused by the constant portrayal of this pistol in various action movies is further increased by the fact that the entirely different pistol (and quite good as a true combat sidearm), the Jericho 941, was imported into the USA under the trade names of 'Uzi Eagle', 'Baby Eagle' and 'Baby Desert Eagle'. While the Jericho certainly looks similar to the Desert Eagle, it is much smaller, lighter, uses a different action and has twice the magazine capacity of its 'bigger brother'.

JERICHO 941

Data for Full-Size Jericho Pistols
Trigger type: Double/single action
Calibre/ammunition used: 9 × 19mm Para, also .41AE (discontinued), .40 S&W, .45 ACP
Weight, empty: 1,000g (35oz) (steel frame); 800g (28oz) (polymer frame)
Length: 192mm (7.5in)
Barrel length: 115mm (4.5in)
Magazine capacity: 16 (9mm), 12 (.40/.41) or 10 (.45) rounds

Jericho 941 are locked-breech, short-recoil-operated pistols, which are available with either a steel or polymer frame. Locking is of the conventional Browning type, with cam-controlled barrel tilting and dual locking lugs located on the barrel just ahead of the chamber area. The slide rails are located inside the frame, as in all pistols patterned after the CZ-75. There are three basic configurations of pistol: standard, semi-compact (standard frame and short slide and barrel) and compact (short slide and barrel, short grip). Polymer-framed pistols were initially available only in semi-compact and compact versions, but recently IMI (now IWI Ltd) also produced full-size Jericho pistols with polymer frames. Triggers are double-action, with an exposed hammer. Typical safety arrangements may include either a frame-mounted safety (left side of frame only) which may lock the hammer, cocked or not, or a slide-mounted ambidextrous safety which decocks the hammer automatically when engaged. Barrels could have either traditional or polygonal rifling, depending on a customer's preference. Sights are fixed, both front and rear sights being dovetailed to the slide. Magazines are double-stack; the capacity varies with calibre. Full-size and semi-compact Jericho pistols with polymer frames are equipped with a Picatinny-type accessory rail below the barrel, which is integral to the frame.

Manual safety Versions with frame-mounted lever: push the lever up to engage safety; pull the lever down to switch safety off; a red dot will appear on the frame.

ABOVE: 9mm Jericho 941 pistol with slide-mounted safety/ decocking lever; note that it is misleadingly marked 'Desert Eagle', although it is entirely different mechanically from the original, much larger gas-operated 'Desert Eagle' pistols made by the same company.

LEFT: 9mm Jericho 941 pistol with frame-mounted safety.

Polymer-framed semi-compact Jericho 941 pistol with ported barrel and one rather interesting (and rare) accessory – a detachable shoulder stock, which is of telescopic design and can be adjusted for length of pull. Israeli shooters also use the same stock on their Glock pistols.

Versions with slide-mounted lever: pull the lever down to engage safety; push the lever up to switch the safety off; a red dot will appear on the slide.

Field stripping procedure 1) remove the magazine by pressing the magazine release button; 2) check that the chamber is empty; 3) put your left thumb through the trigger guard and with the other fingers grasp the top of the front part of the slide firmly. Retract the slide slightly back until the two vertical notches on the rear part of the frame and slide are aligned with each other; 4) while holding the slide in position as described above, push the slide stop pin inwards from the right side of the frame, then pull the slide stop out of the frame from the left side; 5) push the slide and barrel assembly forwards and out of the frame; 6) remove the return spring from below the barrel; 7) pull the barrel out of the slide.

Reassemble in reverse order.

Top to bottom: Kareen MkIII, MkII and MkI pistols.

Kareen

Data for the Kareen Mark III
Trigger type: Single action
Calibre/ammunition used: 9 × 19mm Para
Weight, empty: about 900g (31.5oz)
Length: 200mm (7.8in)
Barrel length: 118mm (4.6in)
Magazine capacity: 13 or 15 rounds

Kareen pistols were made by the Israeli company KSN Industries, with the help of (and probably using parts from) the Hungarian FEG Company. Kareen Mark I pistols are almost exact copies of the FN Browning High Power pistols, the Kareen Mark II are, in turn, close copies of the High Power Mark II pistols, with the

Kareen MkIII pistol.

typical ambidextrous safety levers of improved shape. The major difference between Kareen and FN Mark II pistols is the material of the frame – while original FN pistols have steel frames, Kareen Mark II pistols use aluminium frames with a more angular trigger guard. Kareen Mark III pistols, which were briefly manufactured during the late 1990s and early 2000s before the company left the business, also had modified, heavier slides.

In design and function, Kareen pistols are exactly the same as the Belgian FN Browning High Power pistols described above.

SP-21 Barak *pistol.*

Barak SP-21

Trigger type: Double/single action
Calibre/ammunition used: 9 × 19mm Para, also .40 S&W and .45 ACP
Weight, empty: 730g (25.6oz) empty
Length: 188mm (7.3in)
Barrel length: 99mm (3.9in)
Magazine capacity: 15 (9mm) rounds

The relatively new *Barak* pistol combines unusual appearance with good handling and some interesting features. It is a short-recoil-operated, locked-breech weapon which uses modified Browning-type locking. The barrel engages the slide with a single large lug that enters the ejection window; unlocking is controlled by a cam-shaped cut made in the lug below the barrel which engages the axis pin of the slide stop release lever. The pistol has an unusual modular recoil spring assembly with dual springs mounted on captive spring guides to form a single unit; the springs are located below the barrel, at either side of the frame. The slide is of unusual, two-part design, with a round-shaped steel element being enclosed into a polymer 'sleeve' at the rear; this sleeve provides cocking serrations and houses the rear sight and decocking mechanism. The trigger is a traditional double-action type, with an exposed hammer.

An ambidextrous manual safety is located on either side of the frame and locks the lockwork with the hammer either cocked or lowered; to provide the necessary safety, the pistol is fitted with a separate decocking mechanism which is activated by the button located at the top of the 'hump' at the rear of the slide, just ahead of the rear sight. Sights are either of fixed or adjustable type, dovetailed to the slide. The polymer frame is made with an integral accessory rail under the barrel. Magazines are double-stack and non-interchangeable between calibres. The magazine release

SP-21 Barak *with ?3 Tactical light by Insight Technologies installed on the integral rail. On the right image, the arrow points to the decocking button.*
(Courtesy of Magnum Research)

button is located at the base of the trigger guard and can be installed on either side.

Manual safety Ambidextrous levers on the frame: push the lever up to engage the safety; pull the lever down to switch the safety off; a red dot will appear on the frame. Decocking mechanism: push the button in the front of the spring to bring the hammer safely down without pulling the trigger; also, there is an integral key lock inside the trigger guard, just ahead of the trigger.

Field stripping procedure 1) remove the magazine by pressing the magazine release button; 2) check that the chamber is empty; 3) retract the slide slightly, then, while holding the slide partially retracted, push the slide stop pin inwards from the right side of the frame, then pull the slide stop out of the frame from the left side; 4) push the slide and barrel assembly forwards and out of the frame; 5) remove the return spring module from below the barrel; 6) pull the barrel out of the slide.

Reassemble in reverse order.

ITALY

At the end of the war, Italy retained significant arms manufacturing potential. Among the core companies in the field of small arms was the famous Pietro Beretta Company, one of the oldest gunmakers in the world. In the mid-1930s Beretta introduced several blowback-operated pistols of strong design and excellent reliability; these were used by Italian armed forces throughout the war, as the 9mm Corto (9 × 17) M1934 and 7.65 × 17 M1935. However, wartime experience showed that military pistols ought to have more power; also, NATO required all of its members (with the notable exception of the USA, which effectively dictated the rules and exceptions) to have firearms using a standard range of ammunition.

In about 1950, Beretta's chief designer, Tulio Marengoni, developed a prototype 9mm pistol; this was more or less a strengthened and enlarged M1934 pistol chambered for the 9 × 19 Luger round. However, a blowback action was not well suited to that ammunition, so by the end of the same year Marengoni had developed two more prototypes, this time with locked-breech actions. One was equipped with the Browning High Power-type locking, the other with the Walther P38-type locking. Both prototypes were built on lightweight aluminium frames. Eventually, the latter was selected for further development. However, problems over the insufficient strength of the new frames delayed production and the new pistol, designated as M951 (or M1951), entered production in 1953. The first large orders for this pistol came from Egypt in around 1955; later, in the mid-1960s, Egypt purchased a manufacturing licence for these pistols and started to make them for domestic use and export under the name

Italy fought World War II with reliable but somewhat underpowered sidearms, like this 9 × 17 Beretta M1934 (used by the Army) and the similar 7.65 × 17 Beretta M1935 (used by the Navy and Air Force).

The first post-war attempt by Beretta to produce a more powerful military pistol. Made circa 1950, this 9 × 19mm Luger prototype pistol with a simple blowback action was based on the pre-war designs and found no success.

The Egyptian-made 9mm Helwan pistol is a licensed copy of the Beretta M951. It is still used as a police and military sidearm in Egypt.

of 'Helwan'. The Italian Army began to purchase Beretta M951 pistols during the late 1950s. In about 1976, the Iraqi Government purchased the manufacturing licence for the M951 pistol, manufacturing them under the name of 'Tariq'.

During the first post-war decades, Beretta also produced a large number of smaller, pocket-type pistols, suitable for police and civilian use, but by the early 1970s most of these pistols began to look obsolete. In about 1970, the team of Carlo Beretta, Juzeppe Mazetti and Vittorio Valle began the work on a more modern military pistol. This was to be a double-action, high-capacity pistol with an aluminium frame. At an early stage of development, two models were made and tested – one with Browning-type locking and another with Walther-type locking. Apparently, the first prototype was codenamed Model 91 (9mm pistol, first model) and the second Model 92 (9mm pistol, second model). After preliminary research and testing, the Model 92 was selected for further development. First shown to the public in 1975, the Beretta Model 92 was among the first 'Wonder Nine' pistols; it entered production in 1976, to meet an Italian Army contract. In the same year, Beretta produced the first of many variations of the Model 92 theme – the Beretta 92S, with a combined safety/decocking mechanism which permitted the gun to be safely carried with the hammer down on a loaded chamber – a method preferred by the Italian police. Subsequently, the Beretta Model 92 went through numerous changes and upgrades and is now available in several versions and calibres. It eventually won the American XM9 trials in 1985 and is now a standard issue item with the American and Italian armies, as well as with numerous law-enforcement and security services around the world.

At about the same time, Beretta also introduced a line of more compact blowback-operated pistols, targeted at police and civilian markets. The first of these pistols, known as Models 81 (7.65 × 17) and 84 (9 × 17), appeared in 1974. These looked much like a scaled-down Model 92 and gradually evolved along similar lines, except for the position of the safety,

The Iraqi-made 9mm Tariq pistol is another licensed copy of the Beretta 951. Its production started in the mid-1970s and Western experts first observed it during the first Gulf War in 1991. (Courtesy of Bradley W. Taylor)

The Helwan 920 is another, lesser-known licensed copy of a Beretta pistol, also made in Egypt; this time it is a Model 92S derivative. (Courtesy of Bradley W. Taylor)

which still remains on the frame. These pistols, as well as their derivative models, found some users in Italian law enforcement; for example, the Italian Border Guard used Beretta 84BB pistols and *Carabineri* officers used the slimmer Beretta 85F pistols. These pistols also sold well in many other parts of the world.

Another offspring of the Model 92 was the Model 93R, a selective-fire, special purpose pistol which appeared in the late 1970s. This was a specialized item used by VIP protection teams and counter-terrorist units such as the *Carabineri* GIS or NOCS. However, the proliferation of compact sub-machine guns rendered these pistols obsolete and Beretta eventually dropped this large machine pistol from production.

By the early 1990s Beretta pistols were well-established in world markets. However, there was a gap in the product line between a somewhat bulky Beretta 92 series and the smaller and less powerful Beretta 81/84 series of pistols. Beretta saw the niche for a more compact yet powerful pistol which could fit a police and civilian 'everyday carry' niche, especially in the most important US market. In 1994, Beretta accordingly presented a new pistol, known as the 'Cougar'. This gun, designed around the new, fast-selling .40 S&W cartridge of US origin, was a radical departure from Beretta's traditional looks and design solutions. Like previous designs from the same company, this pistol is available in a number of calibres and sizes and has proved very successful; it is on the approved list of many American police departments and is well liked by civilian shooters, who find the Cougar to be more suitable for concealed carry than the larger 92-series pistols.

9mm (9 × 17) Beretta Model 32 was an experimental prototype that served as a basis for the later Model 84. (Courtesy of Bradley W. Taylor)

Near the turn of the twentieth and twenty-first centuries Beretta tried to market its first polymer-framed pistol, the Model 9000. It was more of a pocket self-defence weapon than a service one, although chambered for potent cartridges like the 9 × 19 and .40 S&W. This weapon was less popular than previous designs from the same company, so in 2004 Beretta brought to the market its most recent offering, the

The two brothers: 9 × 19 Beretta 92FS (top) and 9 × 17 Beretta 84FS (bottom).

The 9mm Beretta Cougar 8000L is one of the latest offerings from the famous Italian company; this semi-compact pistol is intended for concealed carry by police officers and civilian users.

The Model 9000 was the first pistol from Beretta to have a polymer frame. It was quite compact and thus better suited for concealed carry.

service-type PX4 Storm pistol. Patterned after the successful CX4 Storm pistol-calibre carbine, this new pistol mated the proven Cougar-type rotating barrel action and double-action trigger with a modular polymer frame of sleek appearance. This weapon was offered simultaneously in Europe and the USA. While initial US sales were restricted to law enforcement only, in Europe it was sold to the general shooting public from the start. It is yet to be seen if this weapon will find its place in the holsters of law-enforcement officers on either continent, but initial reports from civilian shooters seem favourable.

Fratelli Tanfoglio is the second most important pistol-maker in Italy. Over the last decades, Tanfoglio has produced a number of CZ-75 clones of good quality; it has also supplied parts and assemblies to a number of foreign makers who wanted to make such pistols; these included companies like the Springfield

One of the latest offerings from Tanfoglio, model Force Police R, with a polymer frame and integral Picatinny-type accessory rail.

(Courtesy of Tanfoglio)

Early production 9mm Tanfoglio pistol, the TA-90.
(Courtesy of Cherry's Fine Guns)

The 7.65mm (7.65 × 21 Luger) Benelli B80 was a civilian-only version of an otherwise similar 9mm B76; it was an unusual but elegant design that shot well, but found no significant sales.

Armory (USA, P9 pistols), IMI (Israel, Jericho 941 pistols), Sarsilmaz (Turkey, Cilik 2000 pistols) and many others. In the USA Tanfoglio pistols are currently imported by the EAA Corporation under the trademark 'Witness'. Other than various CZ-75 clones, available in police/defence and sport versions, with polymer or steel frames and in various lengths, Tanfoglio also makes compact pocket-type pistols for the civilian market and fine M1911-type pistols for sport shooters.

Another well-respected Italian gunmaker, the Benelli Armi Company (which now belongs to Beretta), is better known for its long arms and fine small-bore Olympic-type pistols. However, in the late 1970s Benelli tried to enter the service pistol market with the rather interesting B76 pistol. This gun featured an unusual, delayed-blowback action with a fixed barrel, which contributed to its excellent accuracy, along with a good trigger and comfortable grip angle. Unfortunately, this pistol had a single-stack magazine with capacity of 'only' eight rounds and also was somewhat expensive. Only about 10,000 of the B76-type pistols were made, including variants chambered for 7.65 × 21 Luger/Parabellum (Model B80) or 9 × 18 Police (Model B82) ammunition. There were rumours that the B76, because of its fixed barrel, was a preferred pistol to be used with silencers by several Italian law-enforcement organizations, but we have no proof of that.

The next Italian gunmaker worthy of note is Vincenzo Bernardelli, which for many years was known for high-quality shotguns and hunting rifles and for pocket-type pistols. In the mid-1980s this company decided to produce a 9mm, service-type pistol, the Bernardelli P018. This was a more or less conventional, double-action, locked-breech pistol, available in a variety of versions, including dedicated sport-type pistols for disciplines like IDPA and IPSC. These pistols found almost no sales among government users; subsequently, the P018 was replaced in production by the similar P.One pistol, also available in a number of versions and calibres. Finally, several Turkish-made versions of Tanfoglio pistols also bear the venerable name of Vincenzo Bernardelli.

Beretta Model 1951/M951

Trigger type: Single action
Calibre/ammunition used: 9 × 19mm Para
Weight, empty: 890g (31.2oz) empty
Length: 194mm (7.6in)
Barrel length: 114mm (4.5in)
Magazine capacity: 8 rounds

The Beretta M951 pistol was manufactured in Italy between 1954 and 1980. The special purpose, selective-fire M951R version was in limited production between 1955 and 1982. Beretta also made the M952 pistol especially for civilian shooters; it was similar to the M951 except that it was chambered for 7.65 × 21 Luger/Parabellum ammunition. The M951R machine

A number of versions of the Bernardelli P.One pistols. (Courtesy of Bernardelli)

pistol differed from the basic M951 in several aspects. First, it had a heavier slide to reduce the cyclic rate of fire; it also had an added folding foregrip under the frame and a fire mode selector, located at the right side of the frame, close to the trigger. Because of its limited magazine capacity (usually ten or eight rounds, sometimes twenty) and a high rate of fire, this gun required a lot of effort to master it.

The Beretta M951 is short-recoil-operated, locked-breech pistol. It uses Walther-type locking with a tilting locking piece located below the barrel. Most pistols are of all-steel construction, except for a few early production guns that have aluminium alloy frames. The trigger is single-action, with an exposed hammer and a manual safety. Magazines are single stack; the magazine release button is set into the left grip panel, close to the bottom of the grip.

Manual safety A cross-bolt-type button is set into the grip. Push it from the right side of the gun to engage the safety; push it from the left side of the gun to turn the safety off. The safety works only when the hammer is cocked.

Field stripping procedure 1) remove the magazine by pressing the magazine release button; 2) check that the chamber is empty; 3) pull the slide back until the notch in the slide is aligned with the disassembly lever on the right side of the frame (above the trigger guard); 4) turn the disassembly lever down about 90 degrees; 5) carefully push the slide forwards and out of the frame; 6) remove the return spring from below the barrel; 7) push the pin that projects rearwards from the base of the barrel, then pull the barrel out of the slide.

Reassemble in reverse order.

9mm Beretta M951pistol. Clearly seen are two buttons set into the grip panel – the upper controls the manual safety, the lower controls the magazine release.

9mm Beretta M951E – a modified version of the basic Model 1951, made for an Egyptian contract. It had a slightly shorter slide, modified grips with a lanyard ring and a different magazine release, located at the bottom of the grip, just behind the magazine well. *(Courtesy of Bradley W. Taylor)*

9mm Beretta M951R automatic pistol, with extended ten-round magazine and forward handgrip. The fire mode selector (semi/auto) is visible behind and above the trigger.

Beretta Cheetah Models 81, 84 and 85

All Cheetah series of pistols are essentially similar except for calibre and magazine capacity; these models include 7.65 × 17SR Browning models 81 (with double-stack magazine) and 82 (with single-stack magazine, no longer made) and 9 × 17 Browning Short models 83 (no longer made), 84 and 85. This family also includes the 9 × 17mm model 86 with a tip-up barrel and single-stack magazine and .22 rimfire model 87 pistol, which is more suitable for training or recreational shooting than for service use. First introduced in 1974, these pistols went through several changes. The first significant change came with the introduction of 'BB' models, such as 84BB or 85BB. These included a firing-pin block automated safety, but otherwise were similar to the basic models. Later on, all production pistols were made to the new 'FS' pattern, apparently inspired by the larger Model 92FS pistols of the same company. These pistols have a slightly recurved grip and a reshaped trigger guard which facilitates a more comfortable two-hand hold. More important was the introduction of the decocking mechanism, activated automatically when the safety is applied. Differences between various models of the same 'generation' include only calibre, magazine capacity and the thickness of the grip – single-stack models such as the M85 have grips about 5mm (0.2in) thinner than double-stack models like the M84.

Cheetah series pistols are blowback operated with a fixed barrel (which is, however, easily removed during disassembly). Frames are made from aluminium alloy. Because of the traditional Beretta open-top slides, the return spring is located under the barrel. The trigger is double-action, with an exposed hammer and an ambidextrous frame-mounted safety. Magazines are either single or double stack, depending on the model. Sights are fixed, with the rear being dovetailed to the slide.

Beretta 84 of early manufacture.

Manual safety Ambidextrous frame-mounted levers on either side of the gun. To set on safe: push the lever up. On models with an FS suffix only, if the hammer is in the cocked position, it will be lowered (decocked) automatically. To disengage the safety: pull the lever down; a red dot will appear on the slide above the lever.

Field stripping procedure 1) remove the magazine by pressing the magazine release button; 2) check that the chamber is empty; 3) pull the slide all the way back and lock it with the slide-stop lever; 4) depress the takedown lever stop, which is the small button located on the left frame above the trigger guard; 5) while holding the button down, rotate the takedown lever (located at the right side of the frame) downwards for about 90 degrees ; 6) pull the slide slightly back to disengage the slide stop, then carefully push it all the way forwards and out of the frame; note that the

Beretta Cheetah Models 81, 84 and 85

	Beretta 81FS	Beretta 84FS	Beretta 85FS
Trigger type		Double/single action	
Calibre/ammunition used	7.65 × 17SR Browning	9 × 17mm Browning	9 × 17mm Browning
Weight, empty	685g (24oz)	660g (23.1oz)	620g (21.7oz)
Length	172mm (6.7in)	172mm (6.7in)	172mm (6.7in)
Barrel length	97mm (3.8in)	97mm (3.8in)	97mm (3.8in)
Magazine capacity	12 rounds	13 rounds	8 rounds

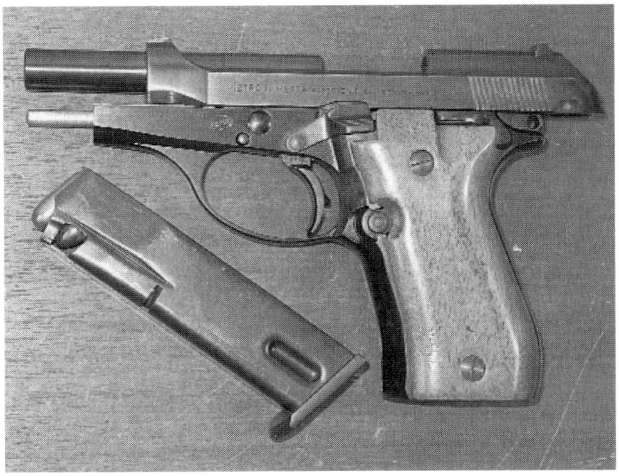

Beretta 84, with slide held back.

Beretta 84, partially disassembled.

barrel will be removed along with the slide and return spring; 7) remove the return spring by pushing its head slightly forwards and then pulling it out of the detent in the barrel; 8) lift up the barrel from the slide.

Reassemble in reverse order.

BERETTA 92 AND VARIANTS

Data for Current Production Model 92FS and 96FS Pistols

	Model 92FS	Model 96FS
Trigger type	Double/single action	
Calibre/ ammunition used	9 × 19mm Para	.40 S&W
Weight, empty	850g (29.8oz)	850g (29.8oz)
Length	217mm (8.5in)	217mm (8.5in)
Barrel length	125mm (4.9in)	125mm (4.9in)
Magazine capacity	15 rounds	11 rounds

Originally designed for the Italian Army and police, the Model 92 pistols earned most of their fame (both good and bad) as the standard sidearm of the US military. First adopted by the Brazilian Army in 1977, this pistol was later adopted in Italy in its Model 92S, 'SB' and finally 'F' modifications. The US military adopted the Model 92SB-F (later renamed as Model 92F) in 1985, as a result of the highly controversial XM9 trials, which are discussed in more detail in the USA section below. In the late 1980s and 1990s, these pistols were also adopted in France. It must be noted that, while

Beretta 84FS of current manufacture. (Courtesy of Beretta)

being entirely adequate as a combat pistol, the Beretta 92 is somewhat bulky for its calibre and magazine capacity, thus less suitable for users with average or smaller hands. Other than the basic 9mm, Beretta also makes these pistols in other calibres, such as .40 S&W (Model 96) and 9 × 21 IMI (Model 98, available for civilian users in certain European countries, including Italy). Beretta also makes a wide variety of models based on the same design; these include not only variations in finishes and sights, but also different trigger types (DA/SA, DA/SA with decock only, DAO, DAO with manual safety).

Beretta 92 pistols are short-recoil-operated, locked-breech weapons with an aluminium frame. The locking

9mm Beretta 92, first production variant, with frame-mounted safety and magazine release on the left side of the grip, at the bottom of the grip panel.

Beretta 92S pistol, the first to have a slide-mounted decocking manual safety. Note that the magazine release remained at the bottom of the grip panel.

system is of the Walther type, with a vertically tilting locking piece located below the breech area of the barrel. The trigger is double-action, with an exposed hammer. Original Model 92 pistols had a frame-mounted safety which was applied only when the hammer was cocked; all subsequent pistols (except for some limited production civilian-only sporting models) either had a slide-mounted safety lever or no safety lever at all. On some pistols, such as the Model 92G adopted in France, the levers do not lock themselves in the lowered position but return to the 'fire' position once released – their function is limited only to safe decocking of the hammer. Some other models, such as the Model 92D, are DAO pistols with no manual safety or decocker. All pistols of current production are fitted with an automatic firing-pin block safety. Magazines are double stack, with the magazine release button located in the base of the trigger guard on all 92-series pistols made since 1981. Sights on service models are of fixed type, with a dovetailed rear blade, usually with high-contrast inserts.

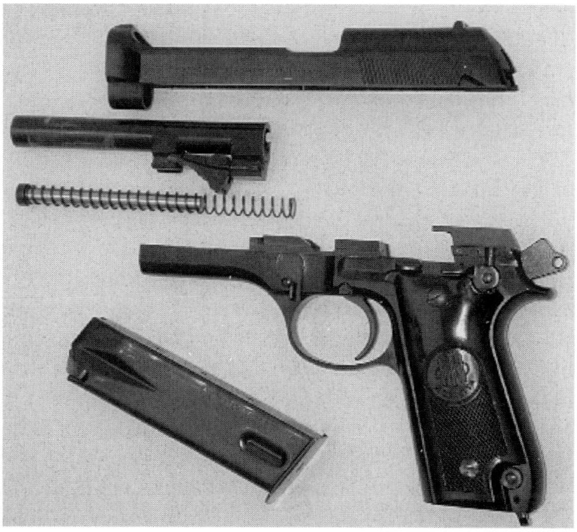

Beretta 92, partially disassembled.

With the introduction of the Model 92FS in the late 1980s, another unusual safety feature was fitted in the form of an enlarged head to the hammer pin. The purpose of this safety is to prevent the rear of the slide from flying back into the firer's face in the case of slide failure. This happened several times during the earlier years of Model 92F service in the US military, apparently because of metallurgical problems, combined with the 'built-in' weak points in the slide where the locking block cuts are made. Recognizing these weak points, the US INS (Immigration and Naturalization Service) requested Beretta to make their Model 96 (.40 S&W calibre version of Model 92) with reinforced slides. This resulted in the appearance of the Model 96 Brigadier pistols and, later on, the same modification was applied to 9mm pistols, available as Model 92 Brigadier. Beretta also produced a number of compact versions of the basic, full-size Model 92 variations. These compact versions had shortened grips, slides and barrels; compact Type M versions also featured single-stack magazines with appropriately thinned grips. At the present time, Beretta no longer makes compact versions of the Model 92; in the firm's product line these were replaced by the entirely different Model 8000 Cougar pistols which are described below.

Below is a brief list of the most important modifications and versions produced in the 92 series over the last thirty years.

Beretta 92S (1976) First modification of the basic Model 92, with a non-ambidextrous slide-mounted safety/decocker and the magazine release button set low in the left grip panel. It was replaced in production by the Model 92SB.

Beretta 92SB (1981) Further evolution of model 92S, initially designated Model 92S-1, later designated 92SB with the introduction of the firing-pin block. The manual safety is ambidextrous, the magazine release relocated to the base of the trigger guard. Discontinued since 1991.

Beretta 92SB-C (1981) Compact version of the model 92SB. Overall length is 197mm (7.7in), barrel length 103mm (4.02in), magazine capacity thirteen rounds (also accepts standard fifteen-round magazines).

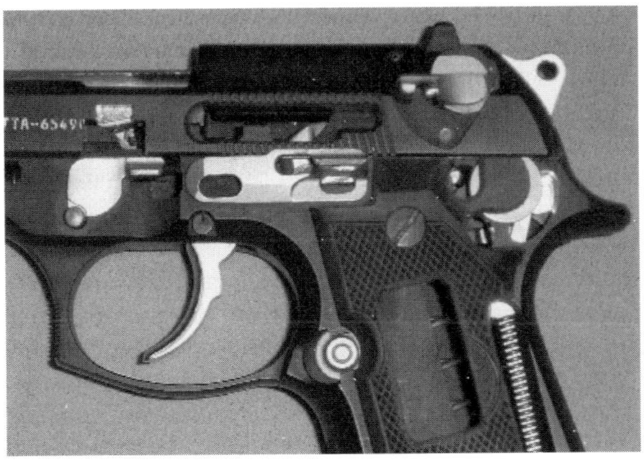

Factory cut-out of a Beretta 92FS pistol; this photo clearly shows the enlarged head of the hammer axis pin (the white disc visible through the cut in the frame, below the hammer head), which is used to prevent the rear part of the slide from flying back into the face of the shooter in the case of slide failure. During recent years this device became more of a redundant safety measure, but it is still built into every modern 92 series pistol. (Courtesy of Bradley W. Taylor)

The Beretta 92S-1 was fitted with an ambidextrous slide-mounted safety/decocker; the magazine release was also moved to a more comfortable spot at the base of the trigger guard, while the left grip panel retained the 'old-style' cut at the bottom. (Courtesy of Bradley W. Taylor)

Beretta 92SB-C-type M (1983) Slimmer version of the model 92SB-C, with a single-stack magazine which holds only eight rounds.

Beretta 92F (1984) Initially designated 92SB-F, later renamed 92F. Evolved from the Model 92SB during the American XM9 trials, with a slightly reshaped grip and trigger guard, also a different finish. The barrel bore and chamber are chrome-plated.

Beretta 92G (1989) The so-called *Gendarmerie* version, created at the request of the *Gendarmerie Nationale de France*; also manufactured under licence in France by GIAT Industries as the PA MAS G1. The pistol is the same as the model 92F except for the operation of the lever, which lost its safety lock function and is used only to decock the hammer safely.

Beretta 92FS (1989) A minor modification of the Model 92F, with an enlarged hammer pin head which prevents the slide from flying back in the case of breakage. Presently, all US M9 pistols are modified to 92FS standard.

Beretta 92FS-C (1989) Compact version of the Model 92FS, similar in dimensions to the earlier model 92SB-C. No longer made.

Beretta 92FS-C-type M (1989) Single-stack version

Commercial Beretta 92FS pistol. It differs from the military Model 92FS/US M9 only by markings and grip panels. (Courtesy of Cherry's Fine Guns)

The Beretta 92D Centurion pistol, with DAO trigger, no manual safeties and a shortened slide and barrel.

The Beretta 92FS Compact Type M is a lesser known variation of the 92 family; it has a smaller and thinner grip that holds a single-stack magazine, as well as a shortened barrel and slide. (Courtesy of Bradley W. Taylor)

The 9mm Beretta 92 Vertec is one of the recent additions to the 92 family. It has a slimmer grip and an integral Picatinny rail on the frame below the barrel. (Courtesy of Cherry's Fine Guns)

of the Model 92FS-C; magazine capacity eight rounds. No longer made.

Beretta 92DS (1990) Double-action-only version of the Model 92FS, with spurless hammer and manual safety.

Beretta 92D (1990) Double-action-only version of the Model 92FS, with spurless hammer and no manual safety levers.

Beretta 92FS Brigadier (1996) Version of the Model 92FS with a reinforced, thickened slide; another change is that the front sight is not integral to the slide, but is dovetailed into it.

Beretta 92FS Centurion (1996) Version of the Model 92FS with a shortened barrel and slide; the frame is the same as on the Model 92FS. Overall length is 197mm (7.7in), barrel length is 103mm (4in), magazine capacity fifteen rounds.

Beretta 92 Vertec (2003) Version of the Model 92FS that addressed constant complaints about the excessive grip width of Model 92 pistols. The backstrap of the grip on Vertec models is made more linear; another change is the addition of an integral Picatinny rail to the frame.

Manual safety (model 92/96FS) Ambidextrous slide-mounted levers on either side of the gun. To set on safe: pull the lever down; if the hammer is in cocked position, it will be lowered (decocked) automatically. To disengage safety: push the lever up; a red dot will appear on the slide.

Field stripping procedure 1) remove the magazine by pressing the magazine release button; 2) check that the chamber is empty; 3) pull the slide all the way back and lock it with the slide-stop lever; 4) depress the takedown lever stop, which is the small button located on the left frame above the trigger guard; 5) while holding the button down, rotate the takedown lever (located at the right side of the frame) downwards about 90 degrees ; 6) pull the slide slightly back to disengage the slide stop, then carefully push it all the way forwards and out of the frame; 7) remove the return spring by pushing its head slightly forwards and then pulling it out of the detent in the barrel; 8) push the locking lever pin (protrudes backwards from below the barrel breech area), then lift up the barrel from the slide.

Reassemble in reverse order.

Beretta 93R

Trigger type: Single-action, selective-fire (single shots and three-round bursts)
Calibre/ammunition used: 9 × 19mm Para
Weight, empty: 1,170g (41oz)
Length: 240mm (9.4in)
Barrel length: 156mm (6.1in)
Magazine capacity: 15 or 20 rounds
Rate of fire, cyclic: 1100 rounds per minute

The Beretta 93R machine pistol is based on the design of the Beretta 92 semi-automatic, but with many modifications. Model 93R ('R' stands for *Raffica* – burst-firing in Italian) inherited the Walther-type locked-breech, short-recoil action of the Model 92, as well as the general appearance and the basic design of the open-top slide. However, both the frame and slide of the Model 93R are of heavier and stronger design, to withstand the recoil forces of burst fire. The trigger is single-action, with an exposed hammer, and separate manual safety and fire mode selector levers are located on the frame, above the left grip. Both the fire mode selector and the safety are mounted on the same axis, with the fire selector lever pointing forwards and the safety lever pointing backwards. The pistol is capable of single shots (fire selector in upper position, marked with a single white dot), or of three-round bursts (selector lever is in the lower position, marked with three dots). The burst-limiting mechanism is located under the right grip panel. To improve control in the burst-firing mode, the pistol is fitted with a hinged front grip

The Beretta 92G 'Special Duty' is another recent version of the 92 line. It combines the G-type decock-only safety levers with a Picatinny-type accessory rail on the frame, below the barrel. (Courtesy of Beretta)

Early model of Beretta 93R, with magazine release set into the bottom of the grip panel. Detachable shoulder stock is shown separately, in folded (storage or carry) position.

Late production Beretta 93R pistol. Front grip is extended, shoulder stock not shown.

and can mount a detachable shoulder stock. The front grip is made of steel and folds up when not in use; to be used, it unfolds down to about 45 degrees. The shooter then puts the thumb of his non-firing hand through the front of the enlarged trigger guard and grasps the foregrip with his other fingers. The detachable shoulder-stock is made of steel and folds up for carrying and storage. Barrels on Model 93R pistols are extended forwards from the slide and are fitted with an integral muzzle flip compensator in the form of several ports on the top, near the muzzle. The standard sights are fixed, the rear sight blade being dovetailed into the slide. Magazines are double-stack, compatible with the Model 92 pistols.

Manual safety A frame-mounted lever with a round, chequered head, mounted above the left grip panel and pointing rearwards. To set on safe: pull the lever down. To disengage safety: push the lever up; a red dot will appear on the frame.

Field stripping procedure 1) remove the magazine by pressing the magazine release button; 2) check that the chamber is empty; 3) pull the slide all the way back and lock it with the slide-stop lever; 4) depress the takedown lever stop, which is the small button located on the left frame above the trigger guard; 5) while holding the button down, rotate the takedown lever (located at the right side of the frame) downwards for about 90 degrees ; 6) pull the slide slightly back to disengage the slide stop, then carefully push it all the way forwards and out of the frame; 7) remove the return spring by pushing its head slightly forwards and then pulling it out of the detent in the barrel; 8) push the locking lever pin (protrudes backwards from below the barrel breech area), then lift up the barrel from the slide.

Reassemble in reverse order.

Beretta Cougar 8000 Series

The Beretta 8000 series pistols, also known under the trademark, Cougar, were in some respects a radical change from previous Beretta designs. Initially developed primarily as police or civilian sidearms, these were built around the new and popular American .40 S&W cartridge. A 9mm version appeared a little later, differing from the .40 calibre version only in the barrel and magazines used. Later on, Beretta also produced a .45 calibre version of the Cougar. Cougar pistols are smaller in size and more comfortable to carry than the Beretta 92/96 series. So far, Cougar pistols have mostly sold to civilian markets, but also have seen some use by various police organizations, including a few US police departments. Despite being NATO-qualified, Cougars have so far not recorded any significant military sales.

Beretta Cougar 8000 Series

Data for standard Cougar pistols

	8000	8000L	8357	8040	8045
Trigger type			Double/single action		
Calibre/ammunition used	9 × 19mm Para	9 × 19mm Para	.357 SIG	.40 S&W	.45 ACP
Weight, empty	925g (32.4oz)	800g (28oz)	920g (32.2oz)	935g (32.7oz)	920g (32.2oz)
Length	180mm (7in)	180mm (7in)	180mm (7in)	180mm (7in)	180mm (7in)
Barrel length	92mm (3.6in)	92mm (3.6in)	92mm (3.6in)	92mm (3.6in)	92mm (3.6in)
Magazine capacity	15 rounds	13 rounds	11 rounds	11 rounds	8 rounds

This pistol is available in four calibres: 9mm, .357 SIG, .40 S&W and .45 ACP. There are also three basic size configurations – Cougar standard, Mini Cougar (with short grip, no longer made; standard magazine capacity was ten rounds for 9mm, eight rounds for .40 and six rounds for .45) and the most recent Cougar L (lightweight) version, with a grip of medium length which houses a shorter magazine (thirteen rounds of 9mm instead of the standard fifteen rounds). The L version also has a lighter barrel and a lighter, bevelled slide. The same lightened slide is also available for standard-size Cougar pistols in the Cougar L Type P versions.

Beretta Cougar pistols are short-recoil, locked-breech weapons which use rotating-barrel locking. The patented recoil-reducing design uses a steel frame insert that houses an unlocking cam path; this insert is built into the recoil spring unit and is allowed to recoil slightly in the frame, decreasing the recoil stress transmitted to the frame upon firing. The barrel has two lugs – one at the top, which locks it to the slide when in the battery, and another below the chamber which controls the locking and unlocking rotation of the barrel during the recoil cycle. Frames are made from aluminium alloy; all other parts are made either from carbon or stainless steel (Cougar Inox).

The trigger is double action, with an exposed hammer. Cougar pistols are available with three types

.40 S&W calibre Beretta 8040 Cougar F pistol.

(Courtesy of Beretta)

.40 S&W calibre Beretta 8040 Cougar D with the DAO trigger, no manual safety and spurless hammer.

(Courtesy of Cherry's Fine Guns)

of trigger – Cougar F pistols have DA/SA triggers with an ambidextrous manual safety/decocker, Cougar G have DA/SA triggers with a decocker only (the lever does not lock the action and returns to the fire position once pressed and released), while Cougar D are double-action-only and have neither manual safety nor decocker. All versions are equipped with an internal firing-pin safety block. Sights are fixed, both front and rear sights being dovetailed to the slide. Magazines are double-stack except for the .45 calibre version, which has single-stack magazines.

Manual safety Model Cougar D: None. Models Cougar F or G: ambidextrous slide-mounted levers on either side of the gun. To set on safe: pull the lever down; if the hammer is in the cocked position, it will be lowered (decocked) automatically. On the G model, the lever will then spring back to the 'fire' position automatically when released. To disengage safety: push the lever up; a red dot will appear on the slide.

Field stripping procedure 1) remove the magazine by pressing the magazine release button; 2) check that the chamber is empty; 3) depress the takedown lever stop, which is the small button located of the left frame above the trigger guard; 4) while holding the button down, rotate the takedown lever (located at the right side of the frame) downwards about 45 degrees; 5) push the slide all the way forwards and out of the frame; 6) remove the return spring from the slide; 7) rotate the barrel counter-clockwise then pull it out of the slide.

Beretta Storm PX4

Trigger type: Double/single action
Calibre/ammunition used: 9 × 19mm Para
Weight, empty: 998g (34.9oz)
Length: 213mm (8.3in)
Barrel length: 122mm (4.8in)
Magazine capacity: 16 rounds

The new PX4 pistol is a further development of the Beretta 8000 Cougar series. The PX4 uses the same recoil-operated, locked-breech system with a rotating barrel. Much like the Cougar, the PX4 has a special frame insert with a top-projecting stud which causes the barrel to rotate on recoil to unlock it from the slide. This insert is mounted on the recoil spring guide rod and is spring-buffered to reduce the peak recoil transferred from the barrel to the frame. The frame is made from high-strength polymer. The slide-stop levers are situated on both sides of the frame and are ergonomically shaped. The modular trigger system of the PX4 Storm is available in four basic models: Type F –conventional double action with manual safety/ decocker; Type D – DAO, no safety or decocker, with a spurless hammer; Type G – conventional double action with manual decocker and no safety; and Type C – with so-called 'constant action' and a spurless hammer (a type of DAO system with less trigger pull weight). In the F and D modifications, safety/decocking levers are located on both sides of the slide.

The grip of the PX4 has removable backstraps; each gun is supplied with three backstraps of different sizes, so the actual grip can be adapted to most shooters. The frame also features a standard Picatinny-type accessory rail under the barrel. Magazines are double-stack and not compatible with previous lines of Beretta pistols in the same calibres. With optional extended capacity magazine baseplates (also available from Beretta), capacity can be further increased to twenty rounds of 9mm or seventeen rounds of .40 S&W. The magazine release button is located at the base of the trigger guard. Sights are fixed, dovetailed into the slide.

Manual safety PX4 Type D or Type C: none; PX4 Type F – ambidextrous levers on the slide; pull levers down to engage the safety; push levers up to disengage safety; a red dot will appear on the slide. When the safety is applied, the hammer is decocked automatically. PX4 Type G – no manual safety; ambidextrous levers on the slide automatically decock

9mm Beretta PX4 Storm pistol.

Left view of Bernardelli PO18 pistol.

Right view of Bernardelli PO18 pistol, with slide locked open.

Bernardelli PO18 partially disassembled.

the hammer once pressed down; when the levers are released, they are returned to the 'fire' position under spring tension.

Field stripping procedure 1) remove the magazine by pressing the magazine release button; 2) check that the chamber is empty; 3) depress inwards the takedown buttons, located at either side of the frame, above the trigger; 4) while holding these buttons pressed, push the slide forwards and out of the frame; 5) turn the slide upside-down and pull the recoil spring guide assembly from the slide; 6) rotate the barrel counter-clockwise then pull it out of the slide.

BERNADELLI P018 AND P.ONE

Trigger type: Double/single action
Calibre/ammunition used: 9 × 19mm Para
Weight, empty: 998g (34.9oz)
Length: 213mm (8.3in)
Barrel length: 122mm (4.8in)
Magazine capacity: 16 rounds

The Bernardelli P018 is a short-recoil-operated, locked-breech pistol which uses a Browning-type locking system. The barrel locks to the slide with two lugs which fit into cuts in the underside of the slide. The P018 is of all-steel construction, with a double-action trigger and an exposed hammer. The safety is mounted on the frame; an ambidextrous safety is optional. Sights are either of fixed or adjustable type, with both front and rear sights being dovetailed into the slide. Magazines are double stack; the magazine release button is located at the base of the trigger guard.

Current production P.One pistols are essentially similar to the P018 except for the wider availability of versions and calibres (9 × 19, 9 × 21 IMI, .40 S&W), the slightly different shape of certain parts and a modified locking system, with the barrel locking into the slide with a single large lug entering the ejection window.

Manual safety A lever at the left side of the frame, above the grip panel (ambidextrous levers optional). Push lever up to engage the safety; pull lever down to turn the safety off. When the safety is engaged, it locks the hammer (either cocked or not) and also blocks the slide movement.

Tanfoglio T95 and T99 Force

	T95	T99 Force
Trigger type	Double/single action	
Calibre/ ammunition used	9 × 19mm Para, .38 Super, .40 S&W, 10mm Auto, .45 ACP	9 × 19mm Para, .40 S&W
Weight, empty	950g (33.3oz)	850g (29.8oz)
Length	205mm (8in)	205mm (8in)
Barrel length	115mm (4.5in)	115mm (4.5in)
Magazine capacity	16 (9mm), 17 (.38), 14 (.40), 11 (10mm) or 10 (.45) rounds	16 (9mm) or 12 (.40) rounds

Polymer-framed Tanfoglio T99 Force pistol with frame-mounted safety; also shown is the optional slide with ambidextrous safety/decocker. (Courtesy of Tanfoglio)

T95-series pistols, made in Italy by the Fratelli Tanfoglio Company, are among the best known and best made clones of the famous Czechoslovak CZ-75 pistols. Made in a wide variety of models, finishes and calibres, these pistols are available either in service-type configurations or as dedicated sport-type versions featuring extended controls, adjustable sights, ported barrels and muzzle brakes, adjustable target-type sights and other 'bells and whistles' typical of high-class sport pistols for disciplines like IDPA or IPSC. Tanfoglio makes two major types of CZ-75 clones – the steel-framed T95 series and the polymer-framed Force series. These pistols are widely exported from Italy and in the USA Tanfoglio pistols are currently imported under the 'Witness' trademark. Apart than complete guns, Tanfoglio also supplies parts and unfinished guns to many manufacturers in the world.

Tanfoglio T95 pistols. Top is the standard (full-size) version with frame-mounted safety; bottom is the compact version with optional slide-mounted safety/decocker. (Courtesy of Tanfoglio)

T95-series pistols are short-recoil-operated, locked-breech weapons with a steel frame. Force series pistols are similar in design, except that these have a polymer frame. Locking is of the most common Browning type, with the barrel engaging the slide with lugs at the top, which fit into cuts made on the underside of the slide. Unlocking is controlled by the cam-shaped cut made in the lug below the chamber area. When the gun is assembled, the slide-stop axis pin passes through this cut and allows the barrel to tip down on recoil. The trigger is double-action, with an exposed hammer. Tanfoglio pistols are available in two basic versions – standard, with a slide-mounted safety/decocker, and Combat, with a frame-mounted safety that allows for 'cocked and locked' carry. Both versions are also fitted with an automatic firing-pin block safety. Magazines are double-stack; the magazine release button is located at the base of the trigger guard. On service-type models the sights are fixed, with the rear sight blade dovetailed into the slide.

Manual safety Combat model: lever at the left side of the frame (ambidextrous levers optional), above the grip panels. To set on safe: push the lever up; to disengage safety: pull the lever down; a red dot will appear on the frame. Standard model: ambidextrous lever on the slide. To set on safe: pull the lever down; to disengage safety: push the lever up; a red dot will appear on the slide. If the safety is applied with the hammer

Polymer-framed Tanfoglio T99 Force Carry Police pistol with Insight Technologies M6 light/laser pointer attached to an integral Picatinny rail. (Courtesy of Tanfoglio)

cocked, it will be automatically dropped down to half-cock position (decocked).

Field stripping procedure 1) remove the magazine by pressing the magazine release button; 2) check that the chamber is empty; 3) put your left thumb through the trigger guard and with the other fingers grasp the top of the front part of the slide firmly. Retract the slide slightly back until the two vertical notches on the rear part of the frame and slide are aligned with each other; 4) while holding the slide in position, described above, push the slide stop pin inwards from the right side of the frame, then pull the slide stop out of the frame from the left side; 5) push the slide and barrel assembly forwards and out of the frame; 6) remove the return spring from below the barrel; 7) pull the barrel out of the slide.

Reassemble in reverse order.

JORDAN

Until very recently, the Army and police forces of the Kingdom of Jordan used imported pistols, mostly of European origin. However, in 2003 the Jordanian King Abdullah II Design and Development Bureau (KADDB), in cooperation with the American designer, Wildey Moore (of the Wildey Firearms Co., designer and maker of the very powerful Wildey pistols), designed a new service-type pistol. This weapon is believed to be in use with Jordanian Special Forces and is offered for export. In the USA, this pistol, known as Viper, is imported by Wildey Firearms.

VIPER

Trigger type: Double/single action
Calibre/ammunition used: 9 × 19mm Para, also .40 S&W and .45 ACP
Weight, empty: about 900g (31.5oz) empty
Length: 219mm (8.5in)
Barrel length: 125mm (4.9in)
Magazine capacity: 15 (9mm), 12 (.40) or 10 (.45) rounds

The Viper pistol has been developed with the main goal to provide users with a modular, multi-calibre pistol

Viper pistols.

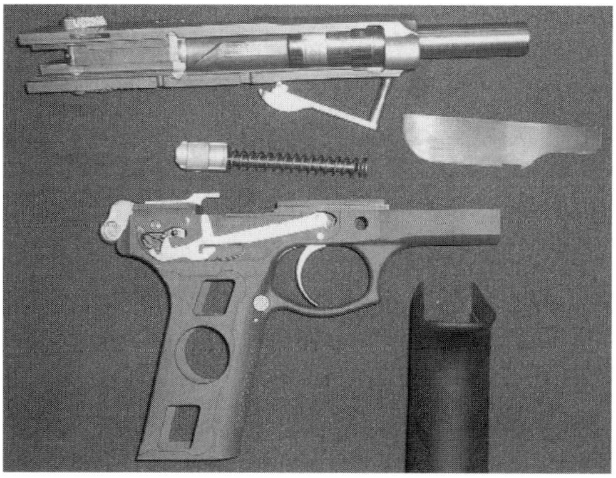

Viper pistol, partially disassembled; note the cam track in the underside of the slide and the removed sideplate, as well as the removable breech face insert (it is of slightly lighter colour than the slide).

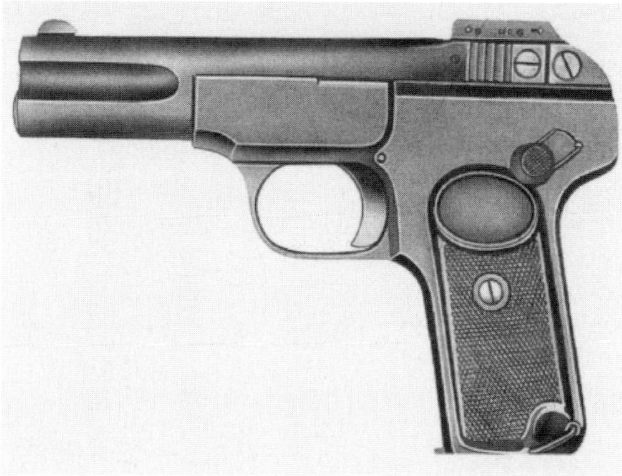

The 7.65mm-type 64 pistol, apparently made for police use, is a copy of a long-gone Browning design.

which is easy to make and maintain and simple to convert between various calibres. To achieve the latter goal, the Viper uses removable breech face inserts in the slide, different for each calibre. To change calibre, the user must partially disassemble the gun, replace the old breech face insert with the appropriate new one, install a new barrel and use an appropriate magazine.

To handle ammunition of different calibres and ballistic properties without any further alterations, the Viper pistol uses a delayed blowback action with a rotary barrel, based on the old Savage/Searle system. The trigger is of conventional double-action type, with an exposed hammer and an ambidextrous, slide-mounted safety. To provide better access to the trigger lockwork, the right side of the frame is cut open and the lockwork is protected from the elements by the removable steel sideplate.

Manual safety Ambidextrous levers on the slide: push the lever up to engage the safety; pull the lever down to switch the safety off; a red dot will appear on the slide.

Korea, North (Democratic People's Republic of Korea)

The North Korean Army has been heavily orientated towards Chinese and Soviet equipment, producing locally made copies or modifications of foreign small arms. The first handguns to be used by the South Korean Army were Tokarev TT pistols of Soviet and Chinese manufacture. In the mid-1960s, North Korea introduced the domestically produced Type 64 pistol, which was a close copy of the long obsolete FN Browning Mle.1900 pistol, chambered for the 7.65 × 17 Browning cartridge. Several years later, the North Korean Army adopted the significantly more powerful Type 68 pistol chambered for 7.62 × 25 Tokarev ammunition. This pistol is a modified Tokarev TT clone.

Type 68

Trigger type: Single action
Calibre/ammunition used: 7.62 × 25mm TT
Weight: 795g (27.8oz) empty
Length: 185mm (7.2in)
Barrel length: 108mm (4.2in)
Magazine capacity: 8 rounds

The Type 68 is a short-recoil-operated, locked-breech pistol. It uses a modified Browning-type locking, with two lugs machined on the top of the barrel engaging notches inside the slide. To lock and unlock during the recoil cycle, the breech end of the barrel is controlled by a cam-shaped cut in the underbarrel lug (instead of the swinging link found in the original TT pistols). The trigger is single-action, with an exposed hammer. There is no manual safety, except for a half-cock notch on the hammer. Single-stack magazines hold eight rounds; they differ slightly from TT magazines by not

Right side view of Type 68 pistol.

having the notch for a magazine catch. The magazine catch is located at the bottom of the grip. Standard TT magazines still can be used in Type 68 pistols, but not vice versa. Sights are fixed, with the rear sight dovetailed into the slide.

Korea, South (Republic of Korea)

In the post-war period, the South Korean military has been strongly orientated towards the USA and was for a long time armed with weapons supplied from the US. By the mid 1980s, the US-made Colt M1911A1 pistols issued to the South Korean Army were showing significant wear. Also, there were constant complaints from the small-stature Korean soldiers about the excessive weight and recoil of the large .45. In 1990, after extensive trials, the South Korean Army adopted a domestically designed 9mm pistol, the K5. Developed and manufactured in Korea by the Daewoo Precision Industries branch of the Daewoo Corporation, the K5 pistol is now the standard sidearm of the armed forces of South Korea. It is also offered for export in several versions and calibres. The basic 9mm K5 is sold as the Daewoo DP-51, its compact variant is sold as DP-51c, and the variant in .40 S&W is sold as the DH-40. Daewoo also makes smaller, pocket-type blowback pistols in 9mm short (9 × 17/.380 ACP) and in 5.6mm rimfire (.22LR).

Daewoo K5/DP-51

Trigger type: Triple action (Double/single action with 'fast action')
Calibre/ammunition used: 9 × 19mm Para
Weight: 800g (28oz) empty
Length: 190mm (7.4in)
Barrel length: 105mm (4.1in)
Magazine capacity: 13 rounds

The Daewoo K5/DP-51 pistol is a short-recoil-operated, locked-breech weapon. It uses a conventional Browning-type locking system, with two lugs on the barrel engaging cuts inside the slide. Locking and unlocking is controlled by a cam-shaped lug at the bottom of the barrel. The frame of the pistol is made from aluminium alloy; all other parts are made from steel. The trigger is of the 'triple-action' type, which

9mm K5 pistol. (Courtesy of Platoon)

9mm K5 pistol. (Courtesy of Platoon)

K5 pistol, partially disassembled. (Courtesy of Platoon)

means that as well as the traditional double- or single-action fire modes, it has a 'fast-action' mode, apparently borrowed from FN's GP 'Fast Action' pistol. In this mode, the hammer is cocked by the cycle of the slide, then is pushed forwards by the pressure on the hammer head. In this position, the main spring remains cocked, so only light pressure on the trigger is necessary to recock the hammer and fire the gun in the 'fast double-action manner'. Or, alternatively, the hammer can be recocked manually to be fired in single-action mode for maximum first-shot accuracy. Sights are fixed, with the rear sight dovetailed to the slide.

Manual safety Ambidextrous levers at both sides of the frame, above and behind the grip panels. To set on safe: push the lever up. To disengage safety: pull the lever down; a red dot will appear on the frame.

Field stripping procedure 1) remove the magazine by pressing the magazine release button; 2) check that the chamber is empty; 3) retract the slide until the notch on the left side of the slide is aligned with the front of the slide-stop lever; 4) push the slide stop pin from the right side, then pull and withdraw the slide stop completely towards the left; 5) ease the slide and barrel assembly forwards and out of the frame; 6) remove the return spring and barrel from the slide.

Assemble in reverse order.

POLAND

During the early post-war period, Poland, which had fallen into the Soviet zone of interest, armed its forces with equipment mostly of Soviet origin. The Polish Army adopted the 7.62mm Tokarev TT as the standard issue sidearm and during the 1950s it began to produce TT pistols at the well-known *Fabryka Brony* arms factory in Radom.

During the late 1950s, following the Soviet movement towards the 9 × 18 Makarov cartridge as a standard issue calibre for pistols, several military officers at the request of the Polish Army designed a compact, pocket-type defensive pistol, chambered for 9 × 18 PM ammunition. After trials which took place between 1961–4, the Polish Army officially adopted the 'Czak' prototype as a next standard sidearm for military and police use. Officially known as the *9mm pistolet wz. 1964* (9mm pistol model of 1964), or P-64 in short, this was a more or less adequate pocket pistol, except for its heavy trigger pull in the double-action mode and somewhat snappy recoil because of the low weight and diminutive grip.

These shortcomings, along with a minimal magazine capacity of six rounds, resulted in a request for a more efficient pistol. In about 1973, Polish Army experts were introduced to the prototype P-70 pistol, which had a precision-cast aluminium frame and a double-stack magazine that held fourteen rounds of 9 ×

18 ammunition. Two years later, a group of Polish designers presented an even more radical firearm, the P-75 prototype, which combined a double-action trigger and a high-capacity magazine with a lightweight polymer frame. However, this pistol turned out to be too radical for the traditionally conservative military experts and in about 1987 two more prototypes appeared, the P-78A and P-78B. These had stamped-steel frames and slides and used single-stack magazines, compatible with Soviet Makarov PM pistols. The key difference between these pistols was the safety arrangement – the P-78A had a decocker only, while the P-78B had a Walther-type manual safety with an automatic decocking function. After several improvements which resulted in modified P-81A and P-81B prototypes, in 1984 the P-81B was officially adopted as the next Polish sidearm for military and police use, receiving the designation of *9mm pistolet wz. 1983*, or P-83 in short. Older P-64 pistols were gradually replaced in service with the new, more potent sidearms; many of the P-64s were refurbished and then sold as surplus to foreign companies, which later offered these pistols to the shooting public.

Following the fall of the Iron Curtain and the disintegration of the Warsaw Pact, Poland turned to the West and eventually joined NATO. This resulted in a need for NATO-standard weapon calibres, pistols included. During 1992–4, the Military Institute of Technical Armaments (WITU – *Wojskowy Instytut Techniczny Uzbrojenia*) developed a new 9mm NATO pistol, known as the WIST-94. Produced at the Prexer factory in the city of Lodz, this pistol won the Army trials and has been the standard sidearm of the Polish Army since 1997. The other major contender at these trials was the

9mm (9 x 18) P-93 pistol.

9mm MAG-95 pistol.

9 × 19mm MAG-95 pistol, developed at the Z.M. Lucznik arms factory. This pistol, as well as the modified MAG-98, was offered for export and commercial sales. The same Z.M. Lucznik factory also offered an upgraded version of the 9 × 18mm P-83 pistol, known as the P-93. This pistol retained most of the P-83 features, but replaced the manual safety/decocker with a decocker-only arrangement, probably borrowed from the P-81A prototype of the early 1980s. Having failed with its own designs, the Z.M. Lucznik factory now makes under licence 9mm Walther P99 pistols of German origin. These pistols are issued to many police units in Poland.

P-64

Trigger type: Double/single action
Calibre/ammunition used: 9 × 18mm Makarov PM
Weight, empty: 635g (22.2oz)
Length: 155mm (6.1in)
Barrel length: 85mm (3.3in)
Magazine capacity: 6 rounds

The P-64 pistol is a simple blowback-operated weapon with a fixed barrel. Heavily influenced by the German Walther PP, the P-64 has a double-action trigger with an exposed hammer and a slide-mounted safety/decocking lever. The P-64 is fitted with a Walther PP-type loaded-chamber indicator, made in the form of a small pin which protrudes from the rear of the slide above the hammer when a cartridge is loaded in the chamber. The P-64 is fitted with a slide stop device, but it has no manual release lever or button – once the slide is locked back open after the last shot has been fired, it

P-64 pistol, left side.

P-64, right side.

must be pulled back by hand and then released to close. Magazines are single stack, with a finger rest; the magazine release is located at the base of the grip. Sights are fixed, the rear sight being dovetailed into the slide.

Manual safety A lever at the left side of the slide. Pull the lever down to turn the safety on; push it up to turn the safety off; a red dot will appear on the slide. When applied, the safety also decocks the hammer automatically.

Field stripping procedure 1) remove the magazine by pressing the magazine release button; 2) check that the chamber is empty; 3) pull the trigger guard downwards, then swing it to one side to lock in the open position; 4) pull the slide all the way back, then raise the rear of the slide to the top and off the frame rails; 5) carefully ease the slide forwards and off the barrel; 6) remove the return spring from the barrel.

Reassemble in reverse order.

P-64, partially disassembled.

P-83.

P-83, partially disassembled.

P-83

Trigger type: Double/single action
Calibre/ammunition used: 9 × 18mm Makarov PM
Weight, empty: 730g (25.6oz)
Length: 165mm (6.4in)
Barrel length: 90mm (3.5in)
Magazine capacity: 8 rounds

The P-83 pistol, also sometimes referred as the 'Wanad' (the codename used during the development and trials period), is a blowback-operated pistol with a fixed barrel. The slide and frame are manufactured from sheet steel by stamping, forming and welding, making this pistol quite inexpensive to manufacture but somewhat rough in appearance. The trigger guard is formed from a separate steel stamping and is permanently welded to the frame. The trigger is double-action, with an exposed hammer and a slide-mounted safety/decocking lever. Magazines are single-stack and are compatible with Soviet Makarov PM magazines. The magazine release is located at the heel of the grip, and the P-83 is fitted with a conventional slide-stop release lever at the left side of the frame. Sights are fixed, the rear sight blade being dovetailed to the slide.

Manual safety A lever at the left side of the slide. Push the lever up to turn the safety on; pull it down to turn the safety off; a red dot will appear on the slide. When applied, the safety also decocks the hammer automatically.

Field stripping procedure 1) remove the magazine by pressing the magazine release button; 2) check that the chamber is empty; 3) pull the disassembly catch downwards; it is located inside the trigger guard, just ahead of the trigger; 4) pull the slide all the way back, then raise the rear of the slide to the top and off the frame rails; 5) carefully ease the slide forwards and off the barrel; 6) remove the return spring from the barrel.

Reassemble in reverse order.

WIST-94

Trigger type: Double-action only
Calibre/ammunition used: 9 × 19mm Para
Weight, empty: 730g (25.6oz)
Length: 190mm (7.4in)
Barrel length: 114mm (4.5in)
Magazine capacity: 16 rounds

The WIST-94 is a short-recoil-operated, locked-breech pistol using the traditional Browning-type action. The frame is made from impact-resistant polymer. The barrel locks to the slide using two lugs, machined on the top of the barrel; these lugs fit into cuts made on the underside of the slide. Barrel locking and unlocking is controlled by a cam-shaped cut made in the lug below the chamber area on the barrel. The trigger is of DAO type, striker fired; the WIST-94 is fitted with an automatic trigger safety. The WIST-94L

9mm WIST-94 pistol.

WIST-94L with integral LAM, built into the 'hump' at the front of the trigger guard. The LAM switch button is built into the left grip panel.

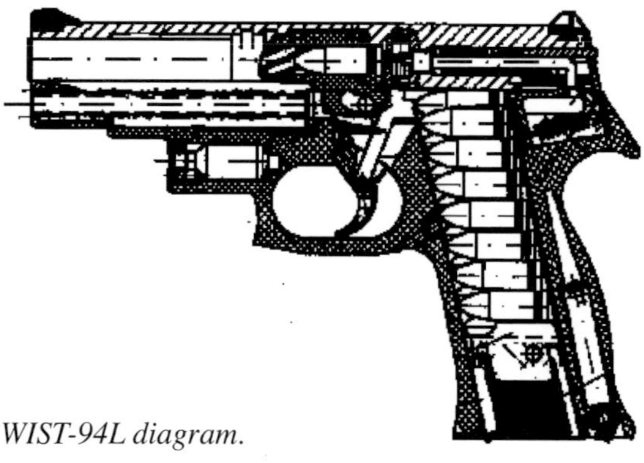

WIST-94L diagram.

differs from the standard version with the addition of the Laser Aiming Module (LAM) that is built into the frame at the front of the trigger.

Manual safety None.

Russia and the Former USSR

The Red Army fought the Great Patriotic War of 1941–5 with both the semi-automatic Tokarev TT pistols and the obsolete Nagant M1895 revolvers. Despite the fact that more potent pistols were designed just before and during the war, the TT remained the mainstream weapon. Wartime experience, which included a close (and sometimes very personal) familiarity with German pistols, resulted in a major change of thinking about the role and necessary features of a military pistol for the Soviet Army. Another factor that played a major role in the development of new requirements for the next military pistol was the realistic prospect of a third world war, with massive nuclear bombing and other such large-scale actions; as a result, pistols played a very minor role in both the strategic and tactical doctrines of the Soviet Army. Furthermore, Tokarev pistols, despite being relatively simple and powerful, showed significant deficiencies, some of them quite serious, such as the lack of positive safety, so almost immediately after the war the GAU (*Glavnoye Artillerijskoe Upravlenie* – General Artillery Department of General Staff) issued a new set of requirements for a military and police pistol.

These requirements asked for a compact, double-action pistol of the Walther PP type. New pistols were to be submitted in three calibres – 7.65 × 17SR Browning (the proposed police round), 9 × 17 Browning and a new 9 × 18. The last of these had been initially developed just prior to the war and refined after the war by the designer Syomin. Apparently, this round was inspired by the German 9 × 18 Ultra, which was designed in the mid-1930s to provide 'acceptable maximum power' in simple, pocket-sized blowback pistols. The key reason for the increase in calibre when compared with the West's 9mm rounds is unknown (the 9 × 18 Soviet has a bullet diameter of 9.2mm, while most Western 9mm rounds have a bullet diameter of 9.02mm). However, with the benefit of hindsight, we can assume that the reasons for a calibre increase were probably the same as for the calibre of Soviet 82mm mortars, which were able to load and fire the slightly smaller German 8.1cm mortar bombs, but not vice versa. Also, while the Soviet Army was ahead of many others in the request for a double-action pistol, it regressed somewhat in adopting an only marginally more powerful round in a weapon that in essence was a pocket-type pistol.

At the same period of time, many other armies, looking for an increase in power, starting to change their 'weak' 7.65 × 17 Browning, 9 × 17 Browning or 7.65 × 20 Longue pistols to the more potent 9 × 19 Luger/Parabellum/NATO weapons. The explanation for this fact, however, is rather simple – while most Western countries relied on full-power rifles (bolt action or semi-automatic) and sub-machine guns as primary individual armament for the infantry, the new Soviet concept had no place for sub-machine guns, as the only primary arm of the infantry was the newly developed assault rifle. Most Western pistols were required to fire 9mm NATO ammunition just to have commonality in ammunition with the standard issue sub-machine guns; Soviet designers had no such requirements, and by the late 1940s 9 × 18 ammunition looked as if it was good enough for a military pistol.

Trials for a new pistol started in 1947. Many designs were submitted and tested, such as pistols by Baryshev, Rakov, Voevodin, Simonov, and Makarov. Some designs were submitted in only one of the desired calibres, some, such as the Makarov design, in two, and a few in all three. At the same time, the Army also tested some larger machine pistols in 9 × 18, which were intended as personal defence weapons for certain officers and NCOs. In 1948, the first trials resulted in the selection of the Makarov pistol in 9 × 18 as the next military sidearm for the Soviet armed forces. However, it took three more years to refine its design, before it was officially adopted in 1951 as the 9mm Pistolet Makarova, or PM in short. In the same year, the Soviet Army also adopted the twenty-shot, selective-fire Stechkin APS pistol in the same calibre. It was chosen over two similar machine pistols, designed by the then-unknown Kalashnikov (the designer of the famous AK assault rifle) and Voevodin (who designed several pistols just before the war).

It must be noted that while the TT was declared obsolete in 1951, it remained in service with the Soviet Army until the early 1970s; in some rural departments of the Soviet *Militia* (police), TT pistols served well into the 1980s.

The Stechkin APS, unlike the Makarov PM, had a relatively short production life. Made only between 1951–5, this gun turned out to be overly expensive and complicated to make (workers called it the 'machinist's worst nightmare'). Its wooden or plastic holster/shoulder stock was bulky and often cumbersome to carry; many complaints from tank and APC crews were caused by the holster snagging on the equipment of armoured vehicles, especially during emergency exits. Most of the APS pistols had been withdrawn from

Tokarev TT-33 pistols, left to right: Soviet-made of 1939 manufacture; Polish-made and Chinese-made, both of post-war manufacture.

ABOVE: Some of the pistols that were tested in the 1947–8 trials. From top to bottom: designs by Rakov, Voevodin and Korovin.

ABOVE: 9mm prototype Makarov pistol of 1947; it had an integral, non-moving trigger guard.

LEFT: A very early production 9mm Makarov pistol, made in 1949. One might expect that such a rare gun must be in a museum collection; however, this particular pistol is still issued on an everyday basis to one of the officers of the Drug Enforcement Department of Saint Petersburg City police. Fifty-six years old, this gun still has all its original parts (many of them handmade), with many tens of thousands of rounds recorded.

ABOVE: 9mm Stechkin APS machine pistol, with holster attached to form a shoulder stock.

RIGHT: Soviet Army officer fires the Stechkin APS pistol, circa 1951.

service and put into storage by the early 1960s, replaced either by Makarov pistols or Kalashnikov AKM assault rifles.

The following decades saw the gradual evolution of Makarov pistols, in continuing attempts to produce new pistols with fewer rejects and at lower cost. Most efforts concentrated on production technologies and by the 1980s many parts that were initially made by machining and hand-fitting became precision-castings requiring minimal, if any, fitting. The early 1960s also saw attempts to make PM pistols lighter by the introduction of an aluminium alloy frame. Apparently, these pistols were less durable than steel-framed ones and development was terminated.

Recognizing the extreme value of various Special Operation Forces (generally known as *Spetsnaz*, after the Russian *Voiska Spetsialnogo Naznachenija* – Special Purpose Troops), the Soviet Army began to request various special purpose weapons. Another important consumer of highly specialized equipment was the KGB (*Komitet Gosudarstvennoj Bezopasnosti* – State Security Committee). Both the Army *Spetsnaz* and the KGB requested silenced weapons, although sometimes of quite different types.

During the early 1950s, the KGB requested a

Prototype Kalashnikov 9mm machine pistol, with holster/shoulder stock attached. Tested against the APS in 1950–51, this blowback-operated, single-action pistol lost the trials and never went any further.

concealed weapon suitable for 'special operations in urban environments'; these operations, known by the Russian euphemism, 'wet deeds', were usually either political murders or involved the self-defence of undercover agents. By the mid-1950s, designer Igor Stechkin (who designed the APS pistol), had developed several such weapons. Some were concealed inside paper cigarette packs, some were disguised as tin cigarette boxes; some fired standard bullets, some fired poison-tipped ones, but all were using ammunition designed using the same concept of 'internal silencing'. Such ammunition has some means to trap powder gases – the key source of gunshot sound – inside a reinforced cartridge case, thus not requiring any form of external sound moderator/silencer. Usually, this was achieved by placing a piston inside the case, between the powder charge and the bullet. When the cartridge was fired, the piston violently pushed the bullet out of the case, then jammed itself in the mouth of the case, capturing the hot powder gases inside. Similar ammunition was later developed in the USA for the specialized S&W 'tunnel revolvers' that were briefly tested in Vietnam, and in Germany for the underwater HK P11 pistol described above.

From 1965 onwards, the State Technical Institute of Precision Machine Building (*TSNII TochMash*) began the development of such silent ammunition with the intention of using it in more conventional weapons for both the KGB and Army *Spetsnaz*. The first production ammunition of this type, known as the SP-2 (*Spetsialnyj Patron 2* – special cartridge no. 2), was produced in about 1970. This cartridge had a blunt-nosed 7.62mm bullet loaded in a bottlenecked case which looked much like a slightly shortened 7.62 × 39 M43 case, containing a two-stage telescoped piston that propelled the bullet and locked the powder gases inside.

In 1972, both the KGB and the Army adopted the 'silenced weapon complex' that included the 7.62mm SP-3 ammunition (similar in design to the SP-2 but loaded with a standard, pointed 7.62mm M43 bullet) and the MSP two-barrelled, derringer-type pistol. This was a rather small weapon, intended for concealed carry and short-range encounters only. Several years later, at the request of the Army, *TSNII TochMash* also developed an S4 pistol of similar design but larger dimensions, which fired the more powerful 7.62 × 63 PZ family of ammunition, also of 'silenced' design. This gun saw action in Afghanistan, in the hands of the Soviet Army's *Spetsnaz* and GRU (Army Intelligence Department) elite strike and recon units.

Of course, these weapons had a very specific niche – other than being only two-shot, slow reloading guns, they, and especially their ammunition, were quite expensive, and (last but not least) were top secret. Because of that, several more conventional silenced weapons were developed for more 'ordinary' users. These included a highly modified PM pistol with a semi-integral silencer, known as the PB (adopted in 1967), and a modified Stechkin APS, known as the APB (adopted 1972). Both weapons saw extensive use during the Soviet invasion of Afghanistan, as well as in many subsequent local conflicts and law-enforcement operations.

The next line of 'special purpose' development included weapons for combat divers. The *TSNII TochMash*, responsible for most *Spetsnaz* developments in small arms, produced the SPP-1 underwater pistol in the late 1960s. Adopted by combat divers of the Soviet Navy, this pistol is still in use, but is no longer secret – it is officially offered for export to foreign government purchasers through the state arms export agency, *RosOboronExport*.

The year 1972 saw the adoption of a somewhat unusual weapon 'complex' that included an entirely new cartridge and a pistol to fire it. Unlike the designs described above, both the pistol and its ammunition were quite standard in nature; the cartridge appeared to be rather unimpressive, firing only a 5.45mm bullet at a moderate velocity of about 300m/s and the pistol was more of a deep-concealment, last-ditch weapon than anything else. The PSM pistol was, in fact, designed at the request of the all-powerful KGB, which required a concealed carry weapon for its plainclothes operatives who were operating 'in country'. Original papers, approved by the Government, requested a 'flat-sized pistol, not thicker than a standard matchbox (17mm [0.7in])'. There were no specifications for calibre and it is not known why *TSNII TochMash* dared to develop an entirely new round when other rounds were already available and had previously been manufactured in the USSR, such as the 6.35 × 16SR and 7.65 × 17SR Brownings. The only real (although doubtful) advantage of the new 5.45 × 18 MPTs round (official designation 7N7) was its deeper penetration, especially against soft body armour at short range. The stopping power of this round is so miserable that some police operatives officially refused to carry this pistol in harm's way, asking for the venerable PM instead. The PSM itself was quite a conventional weapon, but of very thin and flat profile. It was tested against only one other competitor, the BV-025, which was more or less a scaled-down Makarov PM copy, chambered for the same 5.45 × 18 ammunition. The Government offi-

One popular legend about the creation of the PSM says that specific requirements for a 'very flat' pistol were inspired by a hidden desire for a fancy, engraved, 'chiefs only' pistol of distinct appearance. Indeed, a great many PSM pistols were put through an elaborate engraving process, sometimes with gold accents and almost always with political themes. This one is a '30 years of Victory in a Great Patriotic War' commemorative; apparently, it was presented to one of the top-ranking WWII veterans of the Soviet Army in 1975.

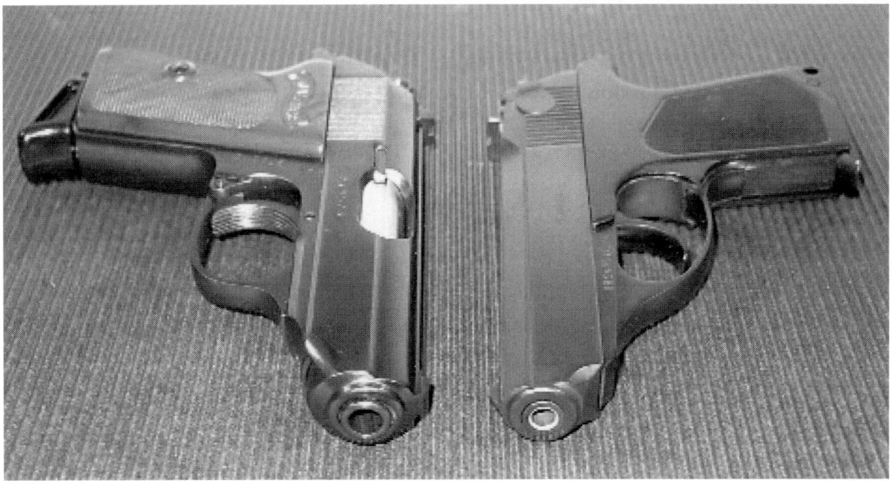

The PSM pistol compared to the Walther PPK, showing how thin it is.

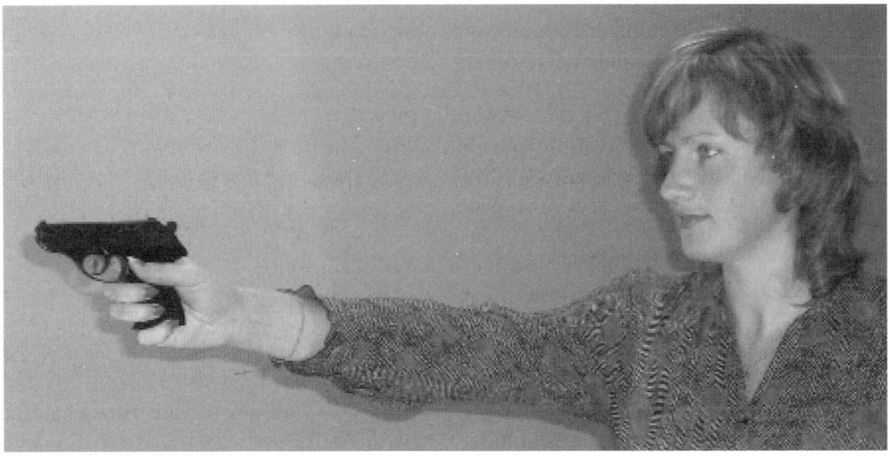

The author's wife, Elena, demonstrates that the PSM pistol looks small even in female hands; in fact, certain Russian female law-enforcement and justice officers are armed with this tiny weapon for personal defence in the line of their duty.

cially approved the PSM for service in 1972 and from the mid-1970s it was issued mostly to top-ranking officials of the military, law-enforcement agencies and the Communist party, as a self-defence weapon (or suicide special, as it turned out during the turbulent first half of the 1990s).

After critical reviews following experience gained in the Afghanistan war and local conflicts in the USSR republics, Soviet military experts finally saw the light and declared that a modern pistol is indeed a very useful weapon and it has much more value as an actual combat sidearm than as a formal badge of office. These conclusions resulted in a research programme, code-named *Grach* (Rook), which was started in the mid-1980s. As a result of this programme, the Soviet Army issued requirements for a new military pistol. It must be noted that these requirements were rather unspecific in several major aspects; for instance, there were no specific tactical roles that the new pistol must fulfil, nor specific requirements for ergonomics or weight limits and so on. Of course, there were stringent requirements for durability and reliability, as well as a requirement for a multi-calibre capability – the Army wanted to use new pistols with either current issue 9 × 18 PM ammunition or with older, but still commonly available, 7.62 × 25mm TT ammunition, as well as with proposed 'high-effectiveness' 9mm ammunition, which was also being developed in the *Grach* programme.

Because of the above factors, the pistols submitted to the trials were quite different in many respects, including different actions and safety arrangements, different materials and so on.

During the early 1990s, the Russian Army (the USSR disintegrated in 1991, leaving all of the newly independent republics and their armies in a rather poor shape, the Russian one included) began to test new pistols.

Among the prototypes tested were the following designs:

- Modular pistol designed in Izhevsk by Zarochintsev; this pistol, known as the *Grach-1*, had a blowback action in 9 × 18 calibre and a delayed blowback action in other, more powerful calibres.
- Locked-breech pistol designed by Yarygin in Izhevsk, which was known as the *Grach-2*.
- Modified Makarov PM pistol, also designed in Izhevsk by Shigapov; this pistol used a high-capacity magazine and fired improved 9 × 18 PMM ammunition. It was tested as the *Grach-3*.
- Multi-calibre, blowback-operated TKB-0220 pistol, designed in Tula by Stechkin and Avraamov.
- Two polymer-framed pistols, designed in Kovrov by Serdjukov – one was locked-breech, chambered for the powerful 9 × 21 RG-052 armour-piercing ammunition, and the other had a delayed blowback action, chambered for the 7.62 × 25 or 9 × 19.

At about the same time (1991–3), recognizing the need for a new, more potent pistol but not seeing any practical results from the ongoing trials, the Russian Government authorized the issue of the modified PMM

9 × 19 6P35 pistol with polymer frame and delayed blowback action, designed at TSNII TochMash by Piotr Serdjukov. This was one of two major contenders during the final stages of the Grach *trials.*

Early version of the Yarygin 9mm pistol, designed at the Izhevsk Mechanical Plant. This was the second major contender in the Grach *trials; its improved version eventually won.*

pistol. However, this pistol was soon limited only to standard 9 × 18 PM ammunition and high-impulse (and high pressure) 9 × 18 PMM ammunition was never widely issued for military or police service. This was because of the obvious dangers of loading the PMM ammunition into older guns, which could result in damage to the gun and injury to the shooter. PMM pistols, with a simplified shape and high-capacity magazines but chambered for the standard 9 × 18 PM ammunition, have been manufactured in Izhevsk since 1994 and are now widely issued to a number of Russian law-enforcement agencies.

The first stages of the trials eliminated several contenders, such as designs by Zarochintsev and Stechkin-Avraamov. Also, by 1993 the Army had decided to go ahead with only one calibre, the newly developed 7N21 9 × 19 cartridge. The 7N21 was dimensionally similar to the world's most popular 9 × 19 Luger/Parabellum cartridge, but loaded to higher (about +P+) pressures and with an armour-piercing bullet of patented design. The interchangeability of the new military cartridge with the 'world's standard' 9mm ammunition obviously increased the export potential of the proposed weapons, but the 1994 agreement between Russian President Yeltsin and US President Clinton, which prohibited the export of Russian-made centrefire pistols to the USA, greatly limited the export market.

The second stage of trials commenced in the mid-1990s and included only two 9 × 19 designs – the Yarygin pistol with a steel frame and manual safety from Izhevsk and the Serdjukov pistol with a polymer frame and no manual safeties from Klimovsk. However, two of the designs rejected before the second stage of trials continued in development under separate requests from the MVD (Internal Affairs Ministry) and the FSB (Federal Security Bureau). The MVD showed great interest in the Stechkin-Avraamov pistol, chambered in 9 × 18, as a possible replacement for aging Stechkin APS pistols borrowed from Army stocks during the late 1980s in a desperate attempt to increase the firepower of local police OMON (*Otryad Militsii Osobogo Naznachenija* – Special Purpose Police Troop) and SOBR (*Spetsialnyj Otryad Bystrogo Reagirovaniya* – Rapid Reaction Force) SWAT-type teams. This pistol, in its production form, is known as Ots-27 *Berdysh*. The FSB, on the other hand, pursued the second Serdjukov design, along with the powerful 9 × 21 RG-052 armour-piercing ammunition. The new pistol received the FSB designation SR-1 (*Spetsialnaya razrabotka* 1 – Special development no. 1), and the improved cartridge received the designation of 9 × 21 SP-10. Later on, this pistol also received the semi-official codename *Vector* and an export-only designation of *Gyurza* (poisonous snake). Both designs had evolved into limited production weapons by the late 1990s, but it seems that the SR-1 was the more successful of two, as it has now been officially adopted for general service and is being issued to various law-enforcement units under MVD, FSB and local police supervision.

OTs-35 was an interesting modification that attempted to adapt standard Makarov PM pistols to high-impulse 9 × 18 PMM ammunition. Developed during the early 1990s, this never went past the prototype stage. Note that a muzzle brake has been pinned to the slide and was used to slow it down on recoil – a necessary measure when using PMM ammunition. However, as the high impulse 9 × 18 PMM ammunition was never issued on a wide scale, the OTs-35 soon vanished into oblivion.

The results of the Army trials were announced in 2000, but it took another three years before the Russian Government officially approved three new 9mm pistols for military and police service. These pistols included *9mm Pistolet Yarygina PYa* (9 × 19), *9mm Pistolet Serdjukova SPS* (9 × 21) and the *9mm Pistolet Gryazeva i Shipunova GSh-18* (9 × 19). While the first two pistols were the latest versions of the *Grach* trials winner and the SR-1, the latter was somewhat different.

The GSh-18 pistol started its life in about 1996 as the P-96, a proposed 9 × 19 police/military pistol, designed at the famous KBP (Instrument Design Bureau, a state-owned development centre known for its outstanding automatic guns and various missile systems). The P-96 was a sort of 'private venture' for KBP, but it failed factory trials and was quickly (and

Drawing of a P96 pistol, which served as a starting point for development of the GSh-18.

extensively) redesigned to became the GSh-18, also known as 'the Russian Glock'. This pistol had not participated in the *Grach* trials, but eventually was approved by the Government and now is in limited use with several Russian law-enforcement agencies.

Of the three newest approved pistols, the GSh-18 seems to be the closest match to a typical modern combat pistol – it is lightweight, relatively comfortable to handle and fire, as well as having significant fire-power with an eighteen-round magazine that will hold 9 × 19 ammunition of any type. The PYa, which was selected by the Russian Army, looks somewhat outdated, with its heavy, all-steel construction and traditional double action with no safe decocking facility, and the SPS is more of a specialized weapon, with its powerful 9 × 21 ammunition and specific handling properties.

While the US export market was (and still is) closed to Russian pistols, there are some other markets in Europe and Canada, for which several Russian gun manufacturers now produce versions of their service pistols (private handgun ownership is prohibited in Russia). The Izhevsk Mechanical Plant, which makes Makarov and Yarygin pistols, now offers several versions of the PM/PMM for commercial export sales; these pistols are available with single- or double-stack magazines, with various sights and finishes and even with a polymer frame. Commercial versions of the PYa include two polymer-framed pistols based on the Yarygin design, the MP-446 Viking and a slightly improved MP-446P Viking-P. The latter, in the author's opinion, could in fact be a better service pistol than the PYa, as it is lighter, has a better grip and includes a decocking function in its safety; but it seems that the Army thinks differently.

Russian service pistols, from top to bottom: Makarov PM, Gryazev-Shipunov GSh-18, Yarygin PYa and Serdjukov SR-1.

The 9mm MP-446 Viking is no more than a polymer-framed export version of the military PYa pistol.

The MP-446P Viking-P is an improved version of the MP-446; it has a removable grip panel that can be had in various sizes; the safety levers can be used to decock the hammer safely if pushed up from the 'safe' position.

Just three of many commercial versions of the Makarov pistol, made in Izhevsk; top to bottom: 9 × 18 MP-448 Skyph *with polymer frame, 9 × 17 IZh-71H with a twelve-round magazine and the 9 × 18 MP-442 with a twelve-round magazine and a side-mounted magazine release.*

TOKAREV TT-33
Trigger type: Single action
Calibre/ammunition used: 7.62 × 25 TT
Weight, empty: 910g (31.9oz)
Length: 195mm (7.6in)
Barrel length: 116mm (4.5in)
Magazine capacity: 8 rounds

The TT pistol got its short name from Tula Tokarev – Tula is the name of the city where this gun was designed and Tokarev is the name of the designer. The official designation of this pistol was *7.62mm Pistolet Tokareva obraztsa 1933 goda* (7.62mm Tokarev pistol of 1933 pattern). This pistol was heavily influenced by two Browning designs – the Belgian-made FN M1903 and the US-made Colt M1911. Actually, Tokarev conducted early tests of his en-block detachable hammer unit using a modified FN M1903 pistol. While the TT was designed in 1929–30 and then modified in 1933, it was manufactured in the USSR until the early 1950s and is still made in countries like China and Yugoslavia. When it was first introduced, the TT was a rather modern pistol, simple to make and use, with good accuracy and reliability. However, its 7.62mm jacketed bullet lacked the stopping power of larger rounds. The lack of any manual safety (except for a hammer half-cock notch) was the major flaw of the design – it was inspired by the Cavalry Marshal Voroshilov, who once failed to switch off the safety on his Browning M1903 during a cavalry charge that took place during the Civil War of 1918–20. With this fearful experience in mind, he told the designers to drop any manual safeties and, not surprisingly, they complied with the 'words of wisdom' from the then leader of the Red Army.

TT pistol, made in 1939.

In the USSR, TT pistols were originally manufactured in the city of Tula, at the Tula Arms Factory (TOZ). From 1942, TT pistols were also made in Izhevsk, at the factory now known as Izhevsk Mechanical Plant. It is believed that these factories produced more than 1,700,000 of these pistols between 1930–53.

The Tokarev TT was a short-recoil-operated, locked-breech pistol with a Browning-type action. The barrel locked to the slide with two lugs, which entered cuts on the underside of the slide; the tipping of the rear of the barrel upon recoil was controlled by a swinging

TT, partially disassembled.

Soviet TT of post-war manufacture.

link that connected the barrel and frame. The trigger was single-action, with an exposed hammer and no manual safety. The hammer, along with its spring and sear, formed a single unit which was easily removed during field disassembly, simplifying the routine maintenance of the pistol. Magazines were single stack, with the magazine release button located at the base of the trigger guard. In 1942, small numbers of modified TT pistols were made with double-stack magazines that held fourteen rounds and with the magazine release located at the base of the grip. Sights were fixed, the rear sight blade being dovetailed to the frame.

Soviet pistols of post-war manufacture were the same as before, except for their better finish and a different pattern of slide serrations.

Manual safety None; the hammer can be put to half-cock but it is not recommended to carry the TT with the chamber loaded, regardless of the position of the hammer.

Field stripping procedure 1) remove the magazine by pressing the magazine release button; 2) check that the chamber is empty; 3) using the tip of the magazine floorplate, or a flat screwdriver, push the slide stop retainer spring (located at the right side of the frame, above the trigger guard) rearwards, then pull it further to unlock the slide-stop axis pin; 4) push the slide-stop axis pin inwards from the right, then pull the slide stop out of the frame from the left; 5) push the slide forwards and out of the frame; 6) remove the return spring with its guide from below the barrel; 7) rotate the barrel bushing for half of the turn, then pull it out of the slide; 8) pull the barrel out of the slide; 9) lift the hammer unit up and out of the frame.

Reassemble in reverse order.

Makarov PM and PMM

	PM	PMM
Trigger type	Double/single action	
Calibre/ ammunition used	9 × 18 PM	9 × 18 PM/PMM*
Weight, empty	730g (25.6oz)	760g (26.6oz)
Length	161mm (6.3in)	165mm (6.4in)
Barrel length	94mm (3.7in)	94mm (3.7in)
Magazine capacity	8 rounds	12 rounds

* pistols chambered for 9 × 18 PMM are no longer made.

The Makarov PM is often rated as one of the best pocket-type pistols. It was inspired by, but not copied from, the German Walther PP and still remains the major service sidearm of Russian law-enforcement and military forces. By modern standards, it is somewhat heavy and only marginally powerful, with a small magazine capacity of just eight rounds. However, it is very simple to use and maintain (but not to make; many parts are of very intricate shape) and is extremely strong and reliable. The modified PMM version has a slide of simplified external shape, as well as a modified frame and grip, shaped to accept a double-stack magazine. Original PMM pistols, which were intended

Makarov PM pistol.

Same gun, slide locked open.

Modified Makarov/Shigapov PMM pistol.

to fire high-impulse 9 × 18 PMM ammunition, had chambers with spiral grooves which helped to slow down the recoil by increasing the friction between the cartridge case and the chamber walls. However, current production PMM pistols have plain chambers and 9 × 18 PMM ammunition is no longer issued because it can damage standard PM pistols. The only factory to make PM and PMM pistols in the USSR and Russia was (and still is) the Izhevsk Mechanical Plant.

The Makarov PM is a blowback-operated weapon with a fixed barrel. It is of all-steel design, with a double-action trigger and an exposed hammer. A slide-mounted safety doubles as a decocker. Magazines are single stack on the standard PM and double stack with single feed on modified PMM pistols. The magazine release is located at the base of the grip. The sights are fixed.

Manual safety A lever at the left side of the slide. Push the lever up to turn the safety on; pull it down to turn the safety off; a red dot will appear on the slide. When applied, the safety blocks the movement of the slide and decocks the hammer automatically.

Field stripping procedure 1) remove the magazine by pressing the magazine release button; 2) check that the chamber is empty; 3) pull the trigger guard downwards, then swing it to one side to lock in the open position; 4) pull the slide all the way back, then raise the rear of the slide to the top and off the frame rails; 5) carefully ease the slide forwards and off the barrel; 6) remove the return spring from the barrel.

Reassemble in reverse order.

PMM, partially disassembled.

PB Silenced

Trigger type: Double/single action
Calibre/ammunition used: 9 × 18 PM
Weight, empty: 970g (34oz)
Length: 310mm (12.1in) assembled; 170mm (6.6in) without detachable part of the silencer
Barrel length: 105mm (4.1in)
Magazine capacity: 8 rounds

The integrally silenced PB (*Pistolet Beschumnyj* – noiseless pistol, official military index 6P9) was introduced in 1967 for use by various *Spetsnaz* units. Manufacture of the PB pistols was stopped in the mid-1980s, but was resumed about fifteen years later due to constant demand for such specialized weapons. It was and still is made at the same factory that makes PM pistols – the Izhevsk Mechanical Plant.

The PB is based on the Makarov PM but has been heavily modified to accept a semi-integral silencer of unusual design. The pistol is designed to fire standard 9 × 18 PM ammunition, and to ensure that the velocity of the bullets remains subsonic regardless of the external conditions (such as temperature), the barrel has small ports that allow some of the propellant gas to escape into the rear part of the silencer. The front part of the silencer can be quickly detached and is mounted

PB pistol in ready to use condition.

PB with the front part of the silencer detached for carry or maintenance.

onto the rear part by means of an interrupted thread lock. The front part of the silencer contains several steel baffles and, while the pistol is intended to be fired with both parts of the silencer in place, in emergency situations it can also be safely fired without the front part of the silencer. The rear part of the silencer is filled with rolled steel mesh, which acts as a heat sink for powder gases. The sound signature in this case will be more significant, of course. The rear part of the silencer, which surrounds the ported barrel, is removed only for maintenance and inspection, and the pistol cannot be fired with the rear part of the silencer removed.

The use of an integral silencer resulted in a shorter slide; because of that, the return spring is located in the grip, under the right-grip panel, and is linked to the slide by a long swinging lever. The trigger unit is similar to the one used in the Makarov pistol, with a

PB pistol in its specially designed holster, along with the detached front part of the silencer, spare magazine and cleaning rod.

double-action trigger, an exposed hammer and a slide-mounted safety/decocker. Magazines are also the same as in the Makarov PM. The button located at the base of the trigger guard, on the left side of the grip, is not a magazine release – it is used to remove the grip panels for disassembly and maintenance. The magazine release is located at the base of the grip. Sights are fixed and equipped with luminous inserts for night shooting. Each PB pistol is issued with a special holster that has a separate pocket for the detached front part of the silencer.

Manual safety A lever at the left side of the slide. Push the lever up to turn the safety on; pull it down to turn the safety off; a red dot will appear on the slide. When applied, the safety blocks the movement of the slide and decocks the hammer automatically.

Stechkin APS

Trigger type: Double/single action, selective fire
Calibre/ammunition used: 9 × 18 PM
Weight, empty: 1,220g (42.7in)
Length: 225mm (8.8in)
Barrel length: 140mm (5.5in)
Magazine capacity: 20 rounds
Rate of fire, cyclic: 600 rounds per minute

The APS pistol could in some respects be considered the first 'Wonder Nine' – the high-capacity, double-action, 9mm calibre pistol. Also, in some respects, it was still influenced by the pre-war machine pistols like the Mauser C96 *Schnellfeuer*, the Astra Modelo F or the Star MM. All considered, when originally issued it failed to fill a predetermined niche, as it was too bulky and heavy for a pistol, but too weak for a carbine or sub-machine gun, still less an assault rifle. Most APS pistols spent years between the early 1960s and late 1980s stored deep in Army warehouse stocks; however, many of them were 'liberated' and recalled for action during the turbulent 1980s and 1990s, when the fall of the USSR caused much commotion, including a rise in crime and various separatist movements. APS pistols are still favoured by many operators of various OMON (SWAT-type) special purpose police units across Russia, mostly because of the large magazine capacity and great accuracy. In most cases, the original holsters/stocks are discarded and more conventional (faster and much lighter) soft holsters are substituted. Automatic fire is only used on very rare occasions, as the heavy weight and moderate-power ammunition allows for very quick and accurate semi-automatic fire.

The Stechkin APS pistol is blowback-operated, with a fixed barrel. The trigger is double action, with an exposed hammer; a three-position safety works as a fire mode selector (safe – semi – auto), and also forces the hammer to decock safely when the gun is set on 'safe'. To provide controllable full automatic fire, the APS is fitted with an inertia-type rate reducer, made in the form of a steel plunger that reciprocates up and down in the grip, just behind the magazine channel. When the pistol is fired in full automatic mode, at

Stechkin APS pistol with the magazine removed.

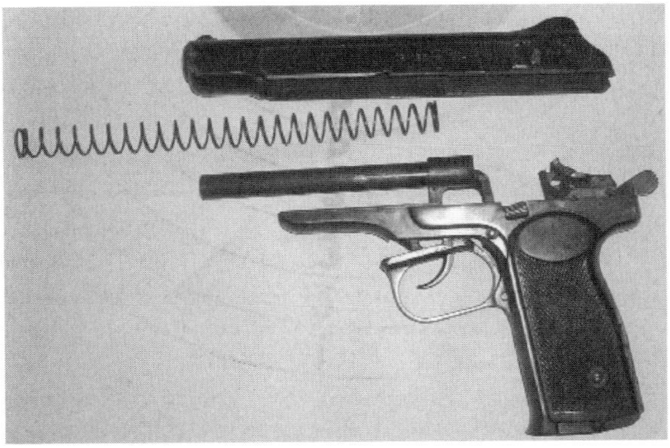

ABOVE: *APS, partially disassembled.*

RIGHT: *Close-up view on the safety/fire selector switch.*

the end of its forward movement the slide hits the plunger and forces it down against its spring; at the same time, the hammer is held cocked by the automatic sear. Once the plunger completes its down-and-up cycle, it strikes the automatic sear and thus releases the hammer to fire the next shot. The magazine is of the double-stack type with a double feed and the magazine release is located at the base of the grip. Sights are of somewhat unusual design, with a fixed front and a range-adjustable rear; rear sight adjustments are made by the rotation of a small drum, which has pre-sets for 25m, 50m, 100m and 200m ranges. The grip is slotted and accepts a wooden (early production) or plastic holster, which then becomes a shoulder stock.

Manual safety A lever at the left side of the slide. Rotate the lever forwards and up to turn the safety on; rotate it down about 60 degrees (position marked with a dot) to turn the safety off, in the single-shot mode; rotate it all the way back to set to the automatic-fire mode. When applied, the safety blocks the movement of the slide and decocks the hammer automatically.

Field stripping procedure 1) remove the magazine by pressing the magazine release button; 2) check that the chamber is empty; 3) pull the trigger guard downwards, then swing it to one side to lock it in the open position; 4) pull the slide all the way back, then raise the rear of the slide to the top and off the frame rails; 5) carefully ease the slide forwards and off the barrel; 6) remove the return spring from the barrel.

Reassemble in reverse order.

APB Silenced

Trigger type: Double/single action, selective fire
Calibre/ammunition used: 9 × 18 PM
Weight, empty: 1,650g (64.4oz) with silencer
Length: 246mm (9.6in); 455mm with silencer attached
Barrel length: 140mm (5.5in)
Magazine capacity: 20 rounds
Rate of fire, cyclic: 600 rounds per minute

All APB pistols (*Avtomaticheskij Pistolet Beschumnyj* – automatic silenced pistol, official index 6P13) were remanufactured from Stechkin APS pistols, described above. This modification was developed at the TSNII TochMash in 1972, with several thousand APS pistols subsequently being refurbished and upgraded to APB and then issued to Spetsnaz units of the Soviet Army and the KGB. The pistol is fairly large and heavy, but it provides users with excellent accuracy both in single shots and in automatic mode. With the silencer in place, the sound of an APB firing is about the same as a typical airgun, or only slightly louder, and of somewhat different tone. According to the author's own experience, the sound of an APB fired with the silencer on is completely masked when other guns are fired at the same time, even pistols like the Makarov. This gives the operator a better chance to escape a firefight without attracting unnecessary attention. This pistol is still in use with Russian Army reconnaissance units and special purpose elements in law enforcement, and is quite popular among its users.

Mechanically, the APB is exactly the same as the Stechkin APS, except for a modified barrel, slide and

APB with silencer and shoulder stock removed; note how the silencer fits into the shoulder stock for storage and carry.

APB pistol, partially disassembled; note the separated barrel jacket tube.

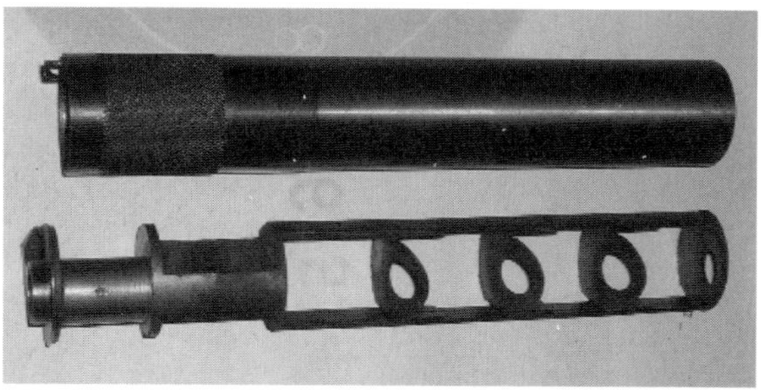

APB silencer, field-stripped. This one probably has never been cleaned inside and out, but still works well.

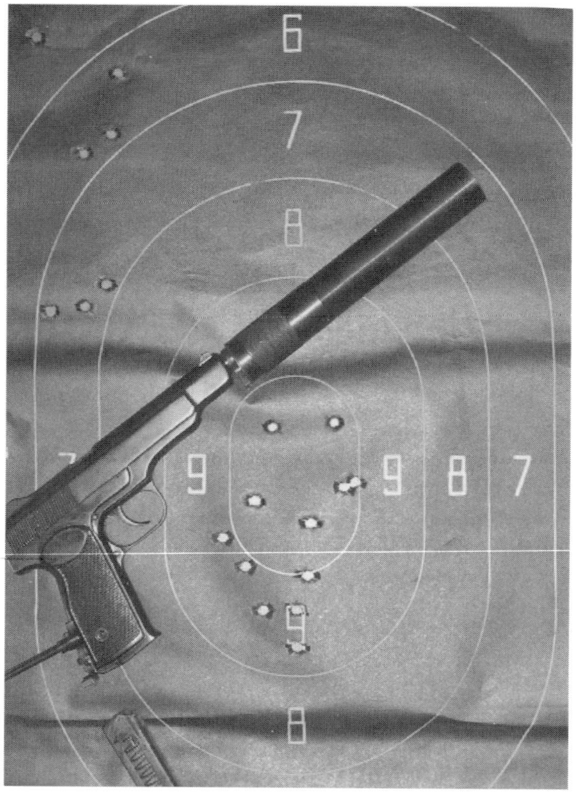

This target has been shot at with an APB pistol set to full automatic mode from 25m (82ft) distance. As you can see, this pistol can show decent accuracy in fully automatic mode (bursts of two to four rounds) and it does much better in single shots.

return spring. The barrel has two sets of ports, one closer to the chamber and another next to the muzzle, which are used to drop the barrel pressure and decrease the muzzle velocity with standard 9 × 18 ammunition below the speed of sound (the standard APS propels the bullet at a supersonic velocity of about 340m/s [1,115ft/s]). To protect the internals of the gun from powder residue and damage from high-pressure powder gases, the barrel is enclosed in a steel jacket tube, which extends forwards from the slide to form a silencer attachment with external threads. This tube is used as a rear expansion chamber, diverting the powder gases into a detachable silencer of all-steel construction.

To accommodate the barrel jacket tube, the return spring is of greater diameter than the same spring of an unaltered Stechkin APS; the slide is also reamed from inside to accommodate the larger spring. The pistol can be safely fired with or without the silencer, but it must not be fired with the barrel jacket uninstalled, as this will damage the weapon. The standard holster/shoulder stock of the Stechkin APS was discarded during the remanufacturing process and replaced with a simple detachable shoulder stock made of steel wire. The stock has clamps to accept the detached silencer for carry and storage. APB pistols are issued with a standard soft holster for a pistol, plus two separate pouches – one for spare magazines and another for the shoulder stock and silencer.

MSP Silent

Trigger type: Single action
Calibre/ammunition used: 7.62 × 38 SP-3
Weight, empty: 530g (18.6oz) empty
Length: 115mm (4.5in)
Barrel length: 66mm (2.6in)
Magazine capacity: 2 rounds in a clip, 1 round per barrel

The MSP (*Malogabaritnyj Spetsialnyj Pistolet* – small special pistol) is one of the very rare, truly silent pistols – when it is fired, the only sound it produces is the quiet 'click' of a firing pin. Yet, this pistol has no silencer, as every specially designed cartridge has an integrated silencing feature in the form of a gas-sealing piston that pushes the bullet and then locks the powder gases within the strengthened case. The pistol is of a very compact and slick design, as it is intended for concealed carry and surprise use. It is of the two-barrel derringer type, with tip-up barrels. For loading and unloading, two rimless cartridges are loaded into a special steel clip and then inserted into the open barrels.

There are no extractors, nor ejectors, as spent cartridge cases or unfired rounds are removed from the barrels by pulling the clip out of the gun manually. This ensures that no cases are left at the scene. The trigger is single action, with concealed hammers which are cocked by a separate lever that protrudes forwards from the trigger guard and must be pulled down to cock the gun. Once the barrels are loaded and closed, the shooter pulls this lever down to cock the hammers and then releases it. Once the pistol is loaded and cocked, the manual safety can be engaged via a lever set into the left grip panel behind the trigger. The crude sights are fixed and this pistol is intended for close range only, as the muzzle velocity is only about 200m/s (656ft/s).

S4M Silent

Trigger type: Single action
Calibre/ammunition used: 7.62 × 63mm PZAM
Weight, empty: n/a
Length: 140mm (5.5in)
Barrel length: 88mm (3.4in)
Magazine capacity: 2 rounds in a clip, 1 round per barrel

The S4M pistol is an enlarged and strengthened derivative of the MSP pistol. It uses silent ammunition of the

MSP pistol.

MSP pistol opened for reloading.

S4M pistol.

same concept, but of much larger size and with greater muzzle velocity and energy (the bullet is the same as in the SP-3 cartridge, borrowed from the standard issue 7.62 × 39 M43 assault rifle cartridge). A longer sight base also ensures slightly better accuracy over extended ranges. This gun saw some use in the hands of elite GRU *Spetsnaz* units in Afghanistan and is still issued to some special purpose elements in Russian military and law-enforcement organizations.

The S4M is a two-barrelled pistol with tip-up barrels. Cartridges are loaded and unloaded in pairs, using a specially designed steel clip. The pistol has neither ejector nor extractor, as the clip is removed manually during reloading. The trigger is single-action, concealed hammers being cocked automatically during reloading when the barrels are closed. A manual safety is located at the left side of the frame. Sights are fixed.

PSS Silent

Trigger type: Double/single action
Calibre/ammunition used: 7.62 × 42 SP-4
Weight, empty: 850g (29.8oz) empty
Length: 170mm (6.6in)
Barrel length: n/a
Magazine capacity: 6 rounds

The PSS pistol (*Pistolet Samozaryadnyj Spetsialnyj* – special self-loading pistol, official index 6P28) was developed between 1979–83 to provide more firepower to the very special users, who had previously carried either MSP or S4M silent pistols. The PSS still relies on specially designed, integrally silenced ammunition, but the SP-4 cartridge has a very short, single-stage piston that does not protrude from the case when fired. Also, the standard pointed M43 bullet has been replaced with a solid steel cylinder with a brass driving band at the front. This bullet is said to be capable of penetrating a standard steel helmet at 25m (82ft), then causing 'enough' damage for a sure kill. This improved ammunition permitted a semi-automatic pistol to be designed.

Because of the unusual ammunition, the overall design of the weapon is also somewhat unusual. The pistol is made almost entirely of steel, except for the grip panels, which are plastic. The action is of more or less the usual blowback type, but the barrel is made from two parts – the front part is rifled and fixed. The rear part, which contains the cartridge chamber, is allowed to recoil inside the frame against its own spring for a short distance. The slide return spring is located above the barrel, around the guide rod. When the pistol is fired, the rear part of the barrel and the slide initially recoil together. This is necessary to use the friction of the bullet, as it leaves the cartridge case and enters the barrel, to push the chamber and slide back for the reloading cycle. After some 6mm (0.2in) of travel the barrel is stopped by the frame and the slide recoils alone; as soon as the barrel stops, it then returns to the battery under the pressure of its own spring, while the slide still moves back. At the very end of the slide recoil cycle, it catches the sliding barrel part to use its mass as a brake, in order to slow down the recoil and decrease the sound of the moving parts hitting the frame in their rearmost position. The extraction and

PSS pistol.

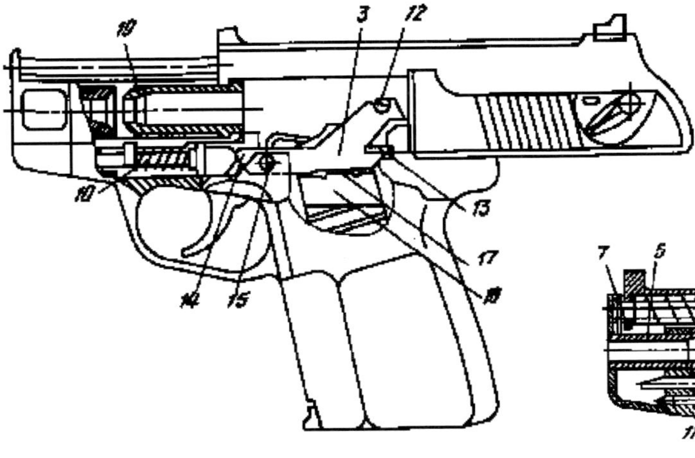

PSS diagram, from patent RU 2 072 076; the pistol is in the maximum slide recoil position. Note how the slide is linked with the sliding chamber.

ejection cycle is standard and, while all this may sound rather complicated, the gun actually works quite well. The trigger is double action, with an exposed hammer and a slide-mounted safety/decocker. The magazine is single stack and the magazine release is located at the base of the grip. Sights are fixed; the rear sight blade is dovetailed to the frame.

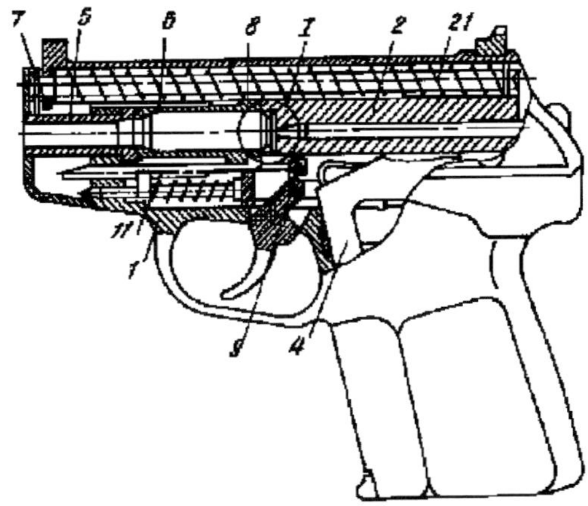

PSS diagram, from patent RU 2 072 076; the pistol is in the ready-to-fire position.

SPP-1M Underwater

Trigger type: Double-action only
Calibre/ammunition used: 4.5 × 40R SPS
Weight, empty: 950g empty
Length: 244mm (9.5in)
Barrel length: 203mm (7.9in)
Magazine capacity: 4 rounds in flat clip, 1 round per barrel

The SPP-1 (*Spestialnyj Podvodnyj Pistolet* – special underwater pistol) was developed during the late 1960s and, in slightly improved SPP-1M form, is still in use by the Russian Navy. Early prototypes were built to fire either 7.62mm rocket-boosted projectiles or more conventional 4.5mm 'needles', using removable barrel clusters of appropriate calibre; however, for various reasons technically successful rocket-boosted ammunition was dropped from further development. The key to the underwater effectiveness of this weapon is specially designed ammunition that consists of a rimmed, bottlenecked cartridge case loaded with a very long steel projectile. The nose of the projectile is cone-shaped, with a small flat tip that creates a cavitation effect around the front of the projectile, providing stabilization when in water. When fired in air, the projectile tumbles in flight and accuracy is poor, although it can still cause severe wounds at typical pistol ranges of engagement. When fired underwater, the effective range degrades as the depth increases – at the 5m (16ft) depth, the effective range is said to be about 17m (56ft), at 20m (66ft) depth it falls down to about 11m (36ft), but, in any case, this is greater than

SPP-1M pistol.

SPP-1M pistol with the barrel cluster tipped open for reloading and the clip with four rounds partially loaded.

SPP-1M striker unit exposed.

the sighting range in typical submerged conditions. SPP-1M pistols are manufactured by the Tula Arms factory.

The SPP-1 pistol itself is a rather simple weapon with four barrels, assembled into a tip-up cluster. The barrels are smoothbore (not rifled), as in water the projectiles are drag-stabilized and in the air it is not possible to stabilize the extra-long bullet by spin, considering the muzzle velocity. The trigger is DAO, striker-fired; a safety lever is located at the left side of the frame and also doubles as a barrel release lever (when pushed up above the 'safe' position). Four cartridges are loaded and ejected simultaneously using flat steel clips. Each pistol is supplied with a set of spare clips, a clip loading tool and a plastic container for spare ammunition (each container hold four rounds in a clip). Sights are fixed and of very crude design.

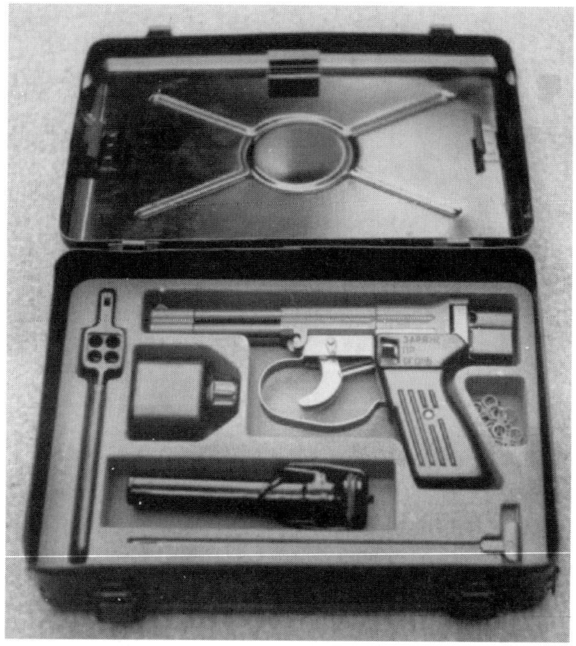

The complete SPP-1M kit with the container for spare ammo, spare clips and clip-loading device.

PSM
Trigger type: Double/single action
Calibre/ammunition used: 5.45 × 18mm
Weight, empty: 460g (16.1oz)
Length: 155mm (6.1in)
Barrel length: 85mm (3.3in)
Magazine capacity: 8 rounds

The PSM pistol (*Pistolet Samozaryadnyj Malogabaritnyj* – self-loading small pistol) is one of the thinnest self-defence guns ever made, as it is only 17mm (0.7in) thick; but it is also one of the least useful defensive weapons, as the pointed, jacketed bullet of small calibre and moderate velocity (about 300 m/s [984ft/s]) can seemingly take forever to disable a target, even in the case of a middle-of-the chest hit. Of course, it is better than bare knuckles, but far too many cases have been recorded of people suffering fatal shots from 5.45 × 18 bullets (usually in the chest area), after which they have continued to fight or run for as much as half an hour before suddenly collapsing and dying from internal bleeding. During the early 1990s this pistol also became a favourite among criminal hitmen, who preferred it for its ease of concealment and good penetration against soft body armour. Some such guns (usually stolen from Army warehouses) were later confiscated by Russian police, often fitted with homemade silencers. Export versions of the PSM, known as IZh-75, were also made in 6.35mm Browning/.25 ACP, which probably was a little more effective as a manstopper than the original 5.45mm load. PSM pistols are manufactured by the Izhevsk Mechanical Plant.

The PSM is a blowback-operated weapon with a fixed barrel. It is of all-steel design, with either aluminium (early) or plastic (modern) grip panels. The trigger is double-action, with an exposed hammer and a slide-mounted safety/ decocker. The safety arrangement is a little unusual in that the lever is made flat with the slide and protrudes rearwards, along with the hammer. This way, the safety does not increase the overall width of the gun and also allows one movement of the thumb to turn the safety off and cock the hammer. The pistol is fitted with a slide stop device, but it has no manual release – to close a slide locked open after the last shot from the magazine, one must pull it rearwards and then release. Magazines are single stack, with a small finger rest and a magazine release located at the heel of the grip. The sights are fixed. Early

PSM pistol of early manufacture, with aluminium grip panels.

PSM pistol, right side.

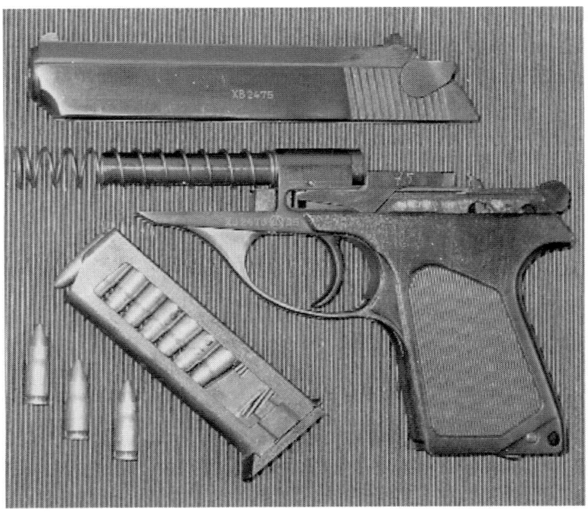

PSM, partially disassembled.

The 5.45mm IZh-75 pistol is an export version of the late-production PSM pistol (with plastic grip panels). Similar guns also were manufactured in 6.35 × 16SR Browning/.25 ACP for European markets.

production pistols had very thin ribbed aluminum grip panels, but these were found to be quite uncomfortable and so were later replaced with plastic wraparound grip panels of better shape.

Manual safety A lever at the left side of the slide. Push the lever up to turn the safety on; pull it down to turn the safety off; a red dot will appear on the slide. When applied, the safety blocks the movement of the slide and decocks the hammer automatically.

Field stripping procedure 1) remove the magazine by pressing the magazine release button; 2) check that the chamber is empty; 3) pull the trigger guard downwards, then swing it to one side to lock in the open position; 4) pull the slide all the way back, then raise the rear of the slide to the top and off the frame rails; 5) carefully ease the slide forwards and off the barrel; 6) remove the return spring from the barrel.

Reassemble in reverse order.

OTs-23 Drotik

Trigger type: Double/single action, selective fired
Calibre/ammunition used: 5.45 × 18
Weight, empty: 850g (29.8oz)
Length: 195mm (7.6in)
Barrel length: 125mm (4.9in)
Magazine capacity: 24 rounds
Rate of fire, cyclic: 1,800 rounds per minute

The OTs-23 *Drotik* (Dart) was developed during the early 1990s in an attempt to provide a type of micro-sub-machine gun for law-enforcement operatives. At the time of its introduction, the 9 × 18 and 9 × 19 armour-piercing ammunition was still in development, but many criminals used body armour, so designer Igor Stechkin proposed the use of small-calibre 5.45 × 18 ammunition, which is known for good penetration against soft body armour. To provide some 'stopping power', Stechkin also proposed to use a burst-fire mode with very high rate of fire. In theory, the key idea was to hit the 'bad guy' with three rounds almost at once and to create more wound shock. However, in real life the tiny 5.45 cartridge lost most of its already diminutive energy quite quickly as the distance increased and even at short ranges and in burst-fire mode (three-round bursts) its effectiveness was rather unspectacular. Very few OTs-23 pistols were produced and the main achievement of this design was to create the developmental basis for more practical weapons such as the OTs-33, which is described below.

OTs-23.

The OTs-23 is a blowback-operated pistol; however, the barrel is allowed to recoil against its own buffer spring and is used as an additional weight to slow down the slide at the end of its rearward movement. To further decrease the muzzle jump, the barrel and slide are ported at the front. The trigger is double action, with an exposed hammer. The safety/fire selector lever is located at the slide and also doubles as a decocker. Magazines are double stack; the magazine release lever is located at the base of the trigger guard. Sights are fixed and the frame has an integral accessory rail below the frame.

Manual safety Levers at both sides of the slide. Rotate the lever to the horizontal position (marked with a white dot) to engage the safety (it will lock the hammer either in cocked or half-cock position). Rotate the lever down to the first position (marked with one red dot) to set to semi-automatic mode; rotate the lever further downwards (to the position marked with three red dots) to set it to burst fire mode. To safely decock the hammer, rotate the lever up from the 'safe' position until the hammer falls to the half-cock position, then release it; the lever will return to the 'safe' position.

OTs-27 *Berdysh*

Trigger type: Double/single action
Calibre/ammunition used: 9 × 18 PM or 9 × 19 Para
Weight, empty: 930g (32.6oz)
Length: 200mm (7.8in)
Barrel length: 125mm (4.9in)
Magazine capacity: 18 rounds

The OTs-27, also known as the PSA (*Pistolet Stechkina – Avraamova* or pistol designed by Stechkin and

OTs-27, early prototype.

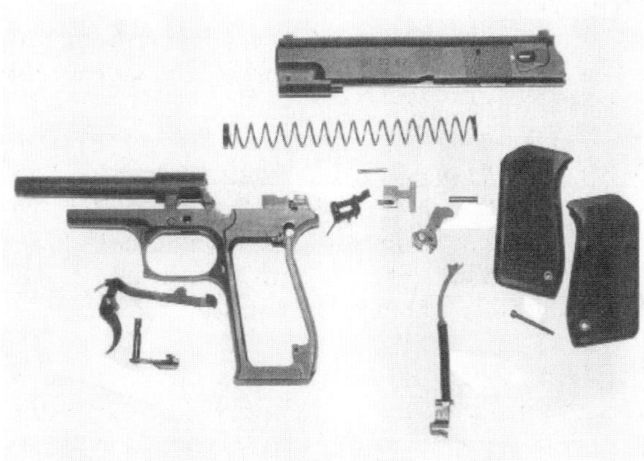

OTs-27, disassembled.

Avraamov) or *Berdysh* (medieval Russian pole-axe), evolved from the TKB-0220 prototype, which failed the early stages of military tests. However, the MVD (Internal Affairs Ministry) saw the TKB-0220 in 9 × 18 as the possible replacement for the popular, but aging Stechkin APS pistols, so financed its further development, which took place at TSKIB SOO (Central Design Bureau for Sporting and Hunting Guns, situated in Tula). When compared with the earlier APS, the PSA is simpler in design and significantly less expensive to make; it also does not have the rarely required fully-automatic fire mode. The PSA pistol is of modular design and calibres can be changed by exchanging the barrel, return spring and magazine; however, only the 9 × 18 PM version saw any use, by a few Russian law-enforcement units.

The OTs-27/PSA pistol is a blowback-operated weapon with a fixed but easily removable barrel. Construction is entirely steel, except for the grip panels. Also it has an additional recoil buffer built into the front of the slide, which takes the shock of the slide hitting the frame in its rearmost position. Barrels are made integral with the feeding ramp which allows for reliable feeding regardless of the calibre used. The trigger is double action, with an exposed hammer and a slide-mounted safety/decocking lever; decocking and safety functions are separated, so the pistol can be safely carried either 'cocked and locked', or with the hammer down on a loaded chamber. The magazines are double stack and not compatible with the older APS pistols, as the magazine release is located differently – at the base of the trigger guard. The sights are fixed, both the front and rear being dovetailed into the slide.

Manual safety A lever at the left side of the slide. Rotate the lever to the horizontal position (marked with a white dot) to engage the safety (it will lock the hammer either in the cocked or half-cock position); rotate the lever down to disengage the safety (position marked with a red dot). To safely decock the hammer, rotate the lever up from the 'safe' position until the hammer falls to the half-cock position, then release it; the lever will return to the 'safe' position.

OTs-33 *Pernatch*

Trigger type: Double/single action, selective fire
Calibre/ammunition used: 9 × 18 PM/PMM
Weight, empty: 1,150g (40.3oz)
Length: 222mm (8.7in)
Barrel length: 150mm (5.9in)
Magazine capacity: 18 or 27 rounds
Rate of fire, cyclic: 900 rounds per minute

The OTs-33 *Pernatch* (*Pernatch* is a medieval Russian pole weapon) was designed at TSKIB SOO in Tula, under the leadership of the Igor Stechkin, in an unsuccessful attempt to produce a more modern and less expensive alternative to the Stechkin APS (from the same designer, but developed about forty years earlier). The OTs-33 is based on the previous OTs-23 pistol described above, but it replaced the three-round burst mode with an unlimited automatic-fire capability and added a detachable shoulder stock of stamped-steel construction. It was made in very limited numbers, as at about the same time (the mid-1990s) several Russian gun manufacturers turned out some more potent 'true' sub-machine guns of compact size, so there was no clear need for a machine pistol. The OTs-33 is much

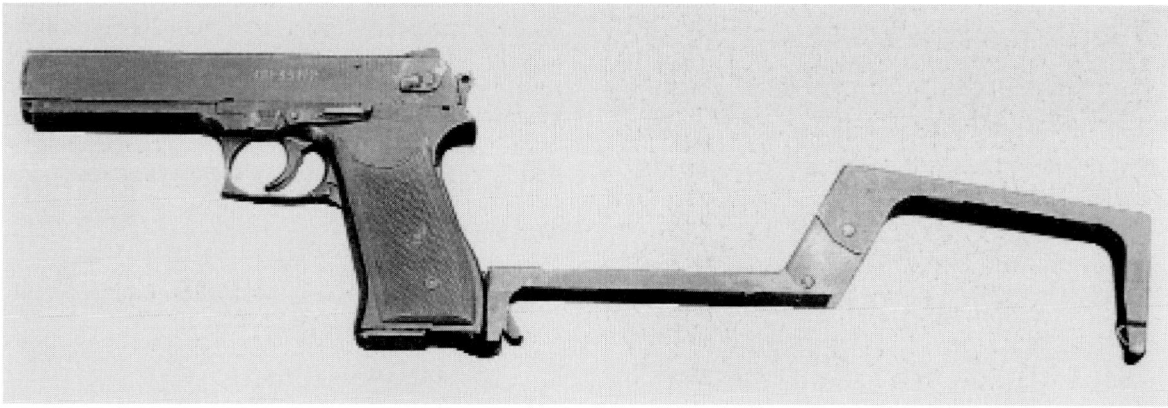

OTs-33 with shoulder stock attached.

simpler and less expensive than the APS, but in automatic fire it has a rate of fire about 30 per cent faster (as it does not have a fire-rate reducer) and is thus significantly less controllable and accurate; in semi-automatic fire, it is essentially similar to the less expensive OTs-27/PSA.

The OTs-33 is a blowback-operated pistol; however, the barrel is allowed to recoil against its own buffer spring and is used as an additional weight to slow down the slide at the end of its rearward movement. The pistol is made of steel, except for the polymer grip panels. The trigger is double action, with an exposed hammer. The safety/fire selector lever is located at the slide and also doubles as a decocker. Magazines are double-stack, of two capacities – standard eighteen-round magazines are a flush-fit with the grip, while extended magazines hold twenty-seven rounds and protrude down from the grip. The backstrap of the grip is slotted to accept the detachable shoulder stock, which is made of steel and folds down when not used. Sights are fixed and the frame has an integral accessory rail below the barrel.

Manual safety Levers at both sides of the slide. Rotate the lever to the horizontal position (marked with a white dot) to engage the safety (it will lock the hammer either in cocked or half-cock position); rotate the lever down to the first position (marked with one red dot) to set to semi-automatic mode; rotate the lever further downwards (to the position marked with three red dots) to set it to full automatic mode. To safely decock the hammer, rotate the lever up from the 'safe' position until the hammer falls to the half-cock position, then release it; the lever will return to the 'safe' position.

P-96S/P-96M

	P-96S	P-96M
Trigger type	Double-action only, pre-cocked	
Calibre/ ammunition used	9 × 17 Browning Short	9 × 18 PM
Weight, empty	450g (15.8oz)	450g (15.8oz)
Length	151mm (5.9in)	151mm (5.9in)
Barrel length	n/a	n/a
Magazine capacity	10 rounds	15 rounds

The P-96S/P-96M pistol was a spin-off from the development of the bigger, but generally unsuccessful, P-96 pistol, designed at KBP in Tula during the mid-1990s. While the full-size P-96 was considered as a possible military weapon, its designers also tried to pursue police and private security markets by simply scaling down the P-96 to create more compact weapons, chambered for the 9 × 18 PM (for the *Militsia*/police market) and the 9 × 17 Browning (for private security companies). These versions were designated P-96M and P-96S, respectively. Despite the fact that neither cartridge was powerful enough to require any form of locked breech, the designers retained the rotary barrel, locked-breech system of the full-size 9 × 19 prototype. The P-96S pistol is rather unpopular among private security personnel in Russia (who are authorized to carry weapons only in 9 × 17 and only under strict Government regulations), as its ergonomics and reliability are inferior to 9 × 17 versions of the Makarov PM pistol. The only positive aspects of the P-96S are its light weight and compact size. The 'police' version of

P-96S pistol chambered for 9 × 17 ammunition.

the same design languished for years, until very recently when it was brought back to the local law-enforcement market. It has yet to be seen if P-96M will prove itself any better than its 'civilian' brother. The P96S and P96M pistols are manufactured by KBP in Tula.

The P-96M/P-96S pistol is a short-recoil-operated, locked-breech weapon with a polymer frame. It uses rotary barrel locking, with a single lug at the top of the barrel which engages the ejection window in the slide. The trigger is DAO, striker-fired; for each shot, the striker is pre-cocked during the initial movement of the slide and then brought to full cock before being released by the pull of the trigger. There are no manual safeties and the trigger is fitted with an automated trigger safety of rather uncomfortable shape. The sights are fixed. Magazines are double stack and hold fifteen rounds in the police version or ten rounds in the private security version. The P-96S ten-round magazines are flush-fit with the grip, while the P-96M fifteen-round magazines are extended and provided with a polymer cover/finger rest at the bottom. The magazine release button is located at the base of the trigger guard.

Manual safety None.

Serdjukov SPS/SR-1 *Vector/Gyurza*

Trigger type: Double/single action
Calibre/ammunition used: 9 × 21 SP-10
Weight, empty: 990g (34.7oz)
Length: 195mm (7.6in)
Barrel length: 120mm (4.7in)
Magazine capacity: 18 rounds

This pistol started its life as the RG055. When this design was selected by the FSB (Federal Security Service of Russia) for further development, it received the designation 'SR-1' and the codename *Vector*. Offered for foreign export sales, in the late 1990s it received the export trade name *Gyurza* (one of the most dangerous poisonous snakes that live in Russia); and finally, when it was approved for general service by the Russian Government in 2003, it received the official name of *9mm Samozaryadnyj Pistolet Serdjukova SPS* (Serdjukov self-loading pistol). As of now, this pistol is made in small batches and issued to FSB and MVD operatives in Chechnya, as well as to selected OMON (SWAT-type) units in Russian law enforcement across the country. These pistols are produced at two factories

SR-1 Vector *pistol.*

Same pistol, right side.

Serdjukov SPS pistol of the most recent manufacture, with the updated shape of the grip and trigger guard, plus the added lanyard ring.

Serdjukov SPS, partially disassembled.

– one is *TSNII TochMash* in the city of Klimovsk and the other is the Mayak factory, located in the city of Kirovsk.

Interestingly, this pistol gets very mixed reviews. It is liked by many users for its serious penetration and effective range – its 9mm SP-10 armour-piercing bullet can penetrate US NIJ Class IIIA body armour at 50m (164ft) and still retain enough power to cause lethal damage. It is also very effective against car bodies and windows. On the other hand, what this pistol seriously lacks is 'human engineering'. It is among the more user-unfriendly creatures emanating from the Russian arms manufacturers, with its trigger and grip safeties of very questionable (and uncomfortable) shape. Trigger arrangement is also somewhat unusual: it is of the traditional double-action type, but has no safe decocking facilities nor manual safety. It also has no manual slide stop release and the internal slide stop is disengaged automatically once a fresh magazine is inserted in the grip. All in all, this pistol requires more training and care during handling than most other combat pistols of modern design, making it in this sense a true successor to the older Tokarev TT.

The SPS pistol is a short-recoil-operated, locked-breech design that uses a modified Walther type of locking, with a vertically tilting locking piece. Unlike most short-recoil-operated designs, the SPS has its slide return spring located around the barrel; the rear end of the spring rests against a special annular spring rest that is connected to the frame and does not move on recoil. Another departure from the Walther system is the use of the locking wedge as a slide accelerator – this is one of several patented features of this unusual design. The frame is of composite design – its top part is made of ordnance-grade steel, while the grip and trigger guard are made from polymer. The trigger is double action, with an exposed hammer and several automatic safeties. These include a trigger safety and a grip safety that locks the sear. There are no manual safeties and yet another strange feature of this gun is that the double-action mode works only if the hammer has been set to the half-cock position; it does not work from the hammer-down position and thus the user has no ability to use the double-action trigger pull for a fast second-strike in the case of a misfire. Magazines are of double-stack, double-feed design; the magazine release is located at the base of the trigger guard. Sights are fixed.

Manual safety None

Yarygin PYa

Trigger type: Double/single action
Calibre/ammunition used: 9 × 19mm 7N21
Weight, empty: 950g (33.3oz)
Length: 198mm (7.7in)
Barrel length: 112mm (4.4in)
Magazine capacity: 17 rounds

The Yarygin PYa pistol, known before its official adoption in 2003 either as the MP-443 (factory designation) or the 6P35 (military index), was advertised as a replacement for the venerable Makarov PM in the holsters of the Russian military. However, because of the very limited funds available, it will take many years before the PYa can replace even half of the many military-issue PM pistols. In 2005, the Russian Ministry of Internal Affairs also announced that it would purchase

Yarygin PYa, right side.

Yarygin PYa, left side, slide locked back.

PYa pistols for issue to various law-enforcement units across the country. The Izhevsk Mechanical Plant manufactures this pistol.

After close examination of this gun, one may suppose that it might have been designed in around 1975 and then shelved for the next twenty or thirty years. Also, it may prompt some questions about human engineering. Thanks to an all-steel construction, this pistol is quite heavy by modern standards. Its shape is somewhat angular, so it can easily cause wear to holsters and other wearable equipment; the grip shape is somewhat angular too. The surprising lack of a safe decocking mechanism, combined with an unusual 'semi-concealed' hammer (which is 'protected' from both sides by rear slide extensions) makes handling of this weapon a little less safe than could be desired. This is especially strange and surprising, as most of the local users are completely used to the integral safety/decocking arrangements of the Makarov PM and Stechkin APS pistols, and also the 'cocked and locked' carry is not authorized in current Russian military manuals. However, this pistol is said to be very durable and reliable – as one may expect from a weapon that has passed the typically exhaustive Russian test and evaluation procedures. It is also reasonably accurate and strong enough to fire on a regular basis the newest Russian 7N21 armour-piercing ammunition, which generates significantly higher pressures than any commercial 9 × 19 Luger/Parabellum round or standard issue 9mm NATO rounds.

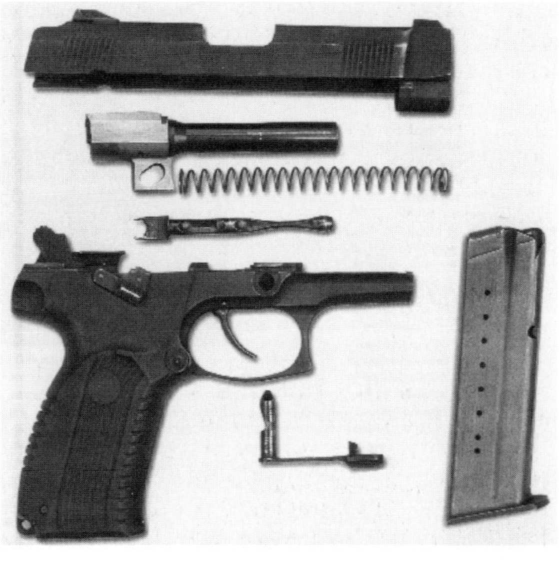

Yarygin PYa, partially disassembled.

The PYa pistol is short-recoil operated and uses the very common modified Browning-style locking system. The barrel locks into the slide with a single lug that engages the enlarged ejection window. Unlocking is controlled by a cam-shaped cut made in the lug under the barrel. The pistol is made from ordnance-grade steel, except for the barrel, which is made from stainless steel, and the polymer grip panels. The trigger is double action, with an exposed hammer. Ambidextrous safety levers are located on the frame. Magazines are double stack and double feed; a reversible magazine release is located at the base of the trigger guard. Sights are fixed, the rear sight being dovetailed into the slide.

Manual safety Ambidextrous levers at both sides of the frame, above the grip panels. To set on safe: push the lever up; to disengage the safety: pull the lever down; a red dot will appear on the frame.

Field stripping procedure 1) remove the magazine by pressing the magazine release button; 2) check that the chamber is empty; 3) push the slide stop pin inwards from the right side of the frame, then pull the slide stop out of the frame from the left side; 4) pull the slide and barrel assembly forwards and out of the frame; 5) remove the return spring from below the barrel; 6) pull the barrel out of the slide.

Reassemble in reverse order.

Gryazev–Shipunov GSh-18

Trigger type: Double-action only, pre-cocked
Calibre/ammunition used: 9 × 19mm 7N21 and 7N31
Weight, empty: 580g (20.3oz)
Length: 183mm (7.1in)
Barrel length: 103mm (4in)
Magazine capacity: 18 rounds

The GSh-18 pistol was designed after the *Grach* trials under the leadership of two men – Gryazev and Shipunov, who lead the famous Instrument Design Bureau (*KB Priborostroenija* – KBP). This pistol is closest to the modern idea of the combat pistol among the three guns officially adopted by the Russian Government in 2003. So far, it is also the lightest and most ergonomic of these. Often compared to the famous Austrian Glock-17 pistol, the GSh-18 is a little lighter and has a grip of better shape (at least, in the author's opinion). However, unlike the Glock, the GSh-18 has yet to establish a proven track record of durability and ruggedness. Originally developed along

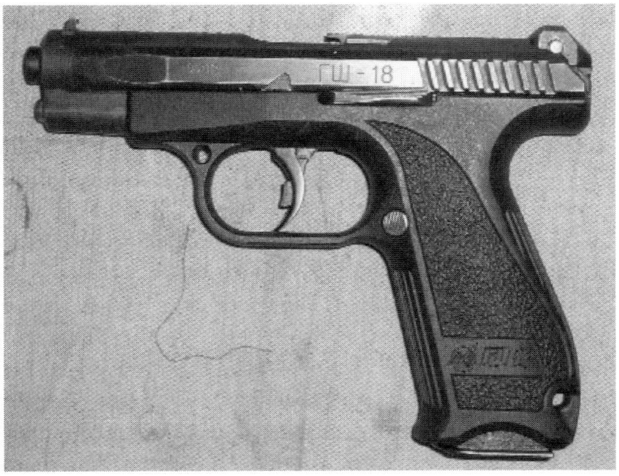

GSh-18 pistol.

GSh-18 pistol, front view.

with the specialized 9 × 19 PBP ammunition (high-pressure, high-velocity load with armour-piercing bullet), the GSh-18 can safely handle all commercial and military 9 × 19 ammunition, including the new standard issue 7N21 AP loads. The 9 × 19 PBP load also evolved into the 7N31, which is intended only for the GSh-18 pistol and the PP-2000 sub-machine gun, which is also designed by KBP. 7N31 ammunition has a lighter and faster AP bullet of similar design, when compared to the 7N21, and it develops higher pressures, way beyond the international industry standards for 9 × 19 ammunition.

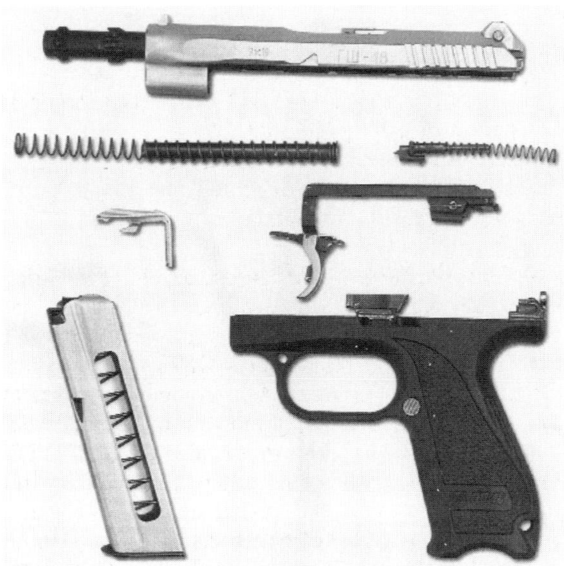

GSh-18 pistol, partially disassembled.

The GSh-18 is a short-recoil-operated, locked-breech pistol with a polymer frame and a stamped-steel slide which has a separate breechblock insert. It uses a rotating barrel with ten radial lugs to lock it to the slide; this provides an extremely strong lock-up, necessary to fire high-pressure loads like the 7N21 and 7N31. The trigger is of double-action-only design, with a pre-cocked striker. The pistol has no manual safeties, and is equipped with an automatic trigger safety, as well as with a firing-pin block safety. The magazines are double stack with double feed; the magazine release button is located at the base of the trigger guard. The small hole made in the front of the trigger guard, which initially puzzled many observers, is used to mount a proprietary laser aiming module. Standard sights are fixed; the front sight is mounted on the slide, while the rear is mounted on the removable breechblock which is set into the rear of the slide.

Manual safety None.

Serbia and the Former Yugoslavia

After the end of the Second World War, socialist Yugoslavia adopted many weapons of Soviet design, including the Tokarev TT; after the late 1950s it was made for the Yugoslavian Army and for export as the M57. This pistol also served as a platform for several modifications which include changes of calibre and size. Of those modifications, the most important were

The 7.62mm M57 pistol was a very close copy of the Soviet Tokarev TT, except for a slightly longer grip, which houses a nine-round magazine.

the M70A, a basic TT rebarrelled to 9 × 19 Luger, and the M88, a shorter-barrelled version of the M70A. For use by police and military non-infantry officers, the major Yugoslavian arms factory, *Crvena Zastava* ('Red Banner'), also produced the smaller 7.65 × 17mm M70 pistol, which was based on a scaled-down Tokarev

The 7.65mm M70 pistol was a smaller derivative of the Soviet TT pistol. Widely used as a military and police sidearm, this pistol retained the Tokarev-type en-block hammer unit with a single-action trigger, but used a simple blowback action because of the lesser power of the cartridge. It also had a manual safety added to the left side of the frame, above the grip.

design, adapted to a simple blowback operation and fitted with a manual safety.

During the mid-1980s, the *Crvena Zastava* factory began development of a more modern pistol, which was intended to replace the obsolete M57 and M70 pistols then in use. Starting with the then-new SIG-Sauer P226 design, Yugoslav designers improved it in some ways and in 1988 the Yugoslavian Ministry of Defence adopted the new pistol, known as the CZ-99, for Army service. It entered production in 1989. However, the start of the bloody civil war and the subsequent disintegration of federal Yugoslavia into several independent republics resulted in a serious decline in small arms development. Licences for the CZ-99 design were subsequently sold to the Israeli company, KSN Industries, which manufactured it briefly as the Golan pistol before passing the documentation package to the South African company, Tressitu Ltd, which made the same gun as the TZ-99.

Despite all of the troubles, the newly formed state of Serbia and Montenegro retained control over the remains of *Crvena Zastava* and after reorganization to *Zastava Oruzje* (Zastava Arms Ltd), this company continued production of both military and civilian arms, including pistols. The most recent developments of *Zastava Oruzje* are the CZ-999 pistol, which first appeared in the late 1990s, and the TOP-XX sporting pistol, which is intended for IPSC and IDPA shooters. It also must be noted that older pistols such as the M57, M70A and M88 are still in production. CZ-99 and CZ-999 pistols are currently imported into the USA by KBI Inc./Charles Daly under the 'Charles Daly ZDA' trademark.

Zastava CZ-99

Trigger type: Double/single action
Calibre/ammunition used: 9 × 19mm Para and .40 S&W
Weight, empty: 860g (30.1oz)
Length: 190mm (7.4in)
Barrel length: 108mm (4.2in)
Magazine capacity: 15 (9mm) or 10 (.40) rounds

The CZ-99 is a short-recoil-operated, locked-breech pistol with an aluminium alloy frame and a SIG-Sauer-type stamped-steel slide with a separate breechblock pinned in place. The lock is of modified Browning type; the barrel engages the slide with a single lug at the top, which enters the enlarged ejection window. Unlocking is controlled by the cam-shaped lug below the barrel that interacts with the steel insert set into the frame. The trigger is double action, with an exposed hammer and automatic firing-pin block. The CZ-99 has

9mm CZ-99 pistol.

no manual safety, but there are ambidextrous levers at either side of the frame which provide both a safe decocking function and a slide stop release function. Zastava Arms also manufactures the CZ-99S pistol with an added manual safety, located on the slide. Magazines are double stack, the magazine release button being located at the base of the trigger guard. Sights are fixed; both front and rear blades are dovetailed into the slide.

Manual safety None; decocking levers at either side of the frame, in the cut of the grip panel.

Field stripping procedure 1) remove the magazine by pressing the magazine release button; 2) check that the chamber is empty; 3) pull the slide all the way back and lock it there with the slide stop; 4) rotate downwards the disassembly lever, which is located at the left side of the frame above the trigger; 5) release the slide and carefully push it forwards and out of the frame; 6) remove the return spring assembly from below the barrel; 7) remove the barrel from the slide.

Reassemble in reverse order.

ZASTAVA CZ-999

Trigger type: Double/single action
Calibre/ammunition used: 9 × 19mm Para
Weight, empty: 828g (29oz)
Length: 190mm (7.4in)
Barrel length: 108mm (4.2in)
Magazine capacity: 15 rounds

The CZ-999 is a further development of the CZ-99 pistol described above. It retains the overall design of the earlier pistol, with a few new features added. The first such feature is a low rounds indicator, in the form of a pin set into the head of the upper right grip screw. Once magazine capacity is down to three rounds or fewer, this pin protrudes outwards from the centre of the upper grip screw in the right grip panel, giving a tactile warning to the shooter. A similar system also may be found on late-production CZ-99 pistols. The second feature, as uncommon as the first, is a trigger mode switch, located at the top left side of the slide. This switch has two positions, pointing rearwards and forwards, marked 'P' and 'R' respectively. These marks refer to two trigger modes, 'Pistol' and 'Revolver'. In Pistol mode, the CZ-999 fires like any other traditional DA/SA pistol, that is, the first shot can

CZ-999 pistol, right side.

CZ-999 pistol, left side.

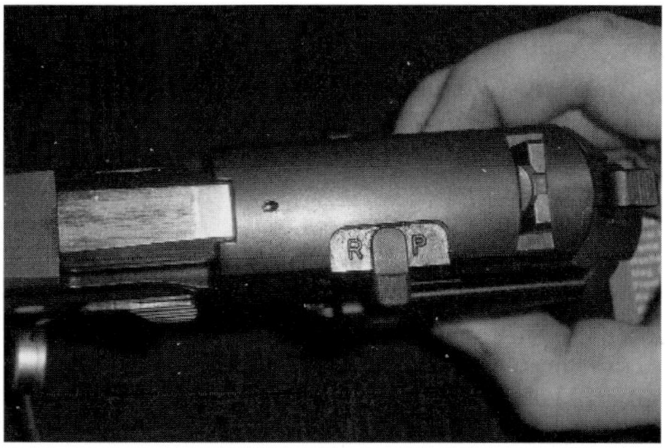

Top view of the CZ-999 slide; the trigger mode switch is in its intermediate (non-working) position, to show both 'P' (Pistol) and 'R' (Revolver) mode markings.

be made either in single or double action and all subsequent shots are made in single action until the decocking lever is used to bring the hammer down. In Revolver mode, the pistol fires all shots in DAO mode, that is, the shooter cocks the hammer for each shot by the trigger pull – unless he manually cocks the hammer after each shot. The hammer will not stay cocked after each shot is fired and the gun cycled. This is exactly the same as in any double-action revolver and the value of this arrangement is questionable to say the least, as it complicates both the design and the manual of arms for this weapon. Otherwise, the CZ-999 is similar to the CZ-99 and for details of its action and disassembly, please refer to the entry above.

Slovak Republic

The Slovak Republic emerged from the dissolved Czechoslovakia in 1993. Unlike many other such events that took place after the fall of the Iron Curtain, there was no conflict between the now-separated Czech and Slovak republics, which, surely, was all for the best. It must be noted that the Czech Republic retained most of the industrial capacity of the former Czechoslovakia and for now the Slovak Army is mostly equipped with older, Czechoslovakia-period arms. However, liberal laws have allowed smaller, private business organizations to develop, manufacture and sell small arms and at least one such company, the Grand Power SRO, makes a service-type pistol, which is certainly worthy of note. At the present time, the K100 pistol, made by Grand Power, is used in limited numbers by Slovak police and special units; it is also exported from the country and sold on the domestic civilian arms market.

K100 Grand Power

Trigger type: Double/single action
Calibre/ammunition used: 9 × 19mm Para (.40 S&W and .45 ACP proposed for 2006)
Weight, empty: 740g (25.9oz)
Length: 203mm (7.9in)
Barrel length: 108mm (4.2in)
Magazine capacity: 17 rounds

The K100 pistol has been in development since 1996, when the first patent was received for its improved rotary-barrel action. The first prototype, designated K1, was built in 1998 and series production

9mm K100 pistol of recent manufacture.

Same gun, another side view.

K100, partially disassembled.

commenced in 2003, under the new designation of K100.

The K100 is a short-recoil-operated, locked-breech pistol with a polymer frame. It uses the relatively rare rotating barrel locking system, with a somewhat unique implementation. Instead of the typical 'lug and cam slot' arrangement, which is usually responsible for barrel rotation upon recoil, the K100 has a cross-pin set into the frame and a helix-shaped flat cut made at the bottom of the chamber area of the barrel. Upon recoil, this helix cut slides over the cross-pin, forcing the barrel to rotate, which unlocks it from the slide. Another unusual feature (at least for a locked-breech pistol) is the method of disassembly, which includes a pivoting trigger guard that acts as a disassembly release. The trigger is double action, with an exposed hammer and an ambidextrous, frame-mounted manual safety. Since 2005, new models have also been available with alternative triggers: DAO (no manual safety) and QA (quick-action, with pre-cocked main spring). An automated firing-pin safety block is also provided. The slide stop release levers are ambidextrous. Magazines are double stack, with the magazine release button located at the base of the trigger guard. Sights are either of fixed or adjustable type; the front sight is pinned to the slide, the rear sight is dovetailed.

For special purposes, the K100 can be fitted with an extended, threaded barrel and a silencer (sound moderator) of indigenous design. To ensure reliable functioning with a heavy silencer attached to the barrel, it has a built-in 'recoil booster' module, which helps to cycle the heavier barrel/silencer unit. Other specialized versions are the K102R and K105R, which have fire mode selectors and can fire in single shots and in two-round (K102), or variable length (K105) bursts.

Manual safety Ambidextrous levers at either side of the frame. Push the lever up to engage the safety; pull the lever down (to the horizontal position) to disengage the safety. The safety system allows the pistol to be carried cocked and locked (hammer cocked, safety on).

South African Republic

During the early post-war period, South Africa was still a British dominion, therefore its armed forces and police mainly used the same weapons as the UK. The South African Republic received its independence in 1961 and in 1962 it was subject to an international trade embargo because of its policy of apartheid. Because of this embargo, enforced by the UN, the South African Army and police forces had either to purchase weapons on international black markets, or build weapons domestically. For example, during the 1970s South Africa obtained significant stocks of Spanish-made Star pistols in 9mm. Probably the first combat-type handgun to be made in South Africa was the 9mm Mamba pistol. This huge, fifteen-round weapon was made of stainless steel. Its design originated from Rhodesia and small numbers were produced in South Africa in around 1977–8; circa 1980 the US-based company Navy Arms bought a

The 9mm Mamba pistol was made in very limited quantities. It is not known if it was ever used as a service pistol in South Africa, although it was certainly suitable for such a role.

manufacturing licence for the Mamba pistol and advertised it extensively in the shooting press for a short time, but it seems that American production never happened, at least on any significant scale.

To fill the niche of a military service pistol, the state-owned Lyttelton Engineering Works (LEW) factory, a part of huge arms-producing DENEL Corporation, began the search for a new 9mm pistol. It appeared that the Italian Beretta 92F suited South Africans the best, but the UN embargo precluded the Italians from selling weapons or manufacturing licences to South Africa. In the end, LEW engineers obtained several copies of the Beretta 92F and simply reverse-engineered them, without any licence. In 1988, the LEW-produced unlicensed copy of the Beretta 92 was adopted in South Africa as the Z-88 pistol. With the fall of the apartheid regime and the subsequent lifting of the embargo, Vektor, the current small arms division of DENEL, obtained an official licence from Beretta and was thus allowed to export Z-88 pistols legally.

In 1992, Vektor produced an improved 9mm version of the Z-88, known as the Vektor SP1. A little later, Vektor also introduced a .40 S&W calibre version of the SP1, known as the SP2. Both pistols were and still are exported to Europe and South America, in several versions. The SP1 pistol is currently in service with the South African armed forces.

9mm Vektor Z-88 pistol, a very close copy of the Italian Beretta 92F. (Courtesy of Bradley W. Taylor)

The 9mm Vektor CP1 pistol was intended as a concealed carry weapon for police and the civilian public. It has an unusual, streamlined design with a polymer frame and an even more unusual safety, located at the front of the trigger guard. The CP1 uses a delayed blowback action, probably borrowed from the German HK P7 pistol. Introduced during the late 1990s, it is still manufactured in limited numbers, mostly for export.

There were also few private companies that made small arms. During the late 1990s, the smaller company Tressitu Ltd produced the 9mm TZ-99, a copy of the Yugoslav CZ-99 pistol. It was a good pistol, but production didn't last for long.

In the late 1990s, the South African police issued a requirement for a compact, concealable pistol chambered for 9 × 19 ammunition. In 1997, the contract for a new police pistol went to the Republic Arms Company, which had developed the RAP-401 pistol, a

9mm Tressitu TZ-99, a Yugoslav design that travelled to South Africa via Israel.

relatively compact sidearm suitable for concealed carry. Later on, Republic Arms also produced a .40 calibre version of the same pistol, intended for export. This was designated RAP-440. It seems that RAP pistols were influenced by the Spanish Astra A-75. RAP pistols are favoured for their reliability and mild recoil; on the other hand, these pistols are somewhat heavy for their size.

It seems that the major competitor for the RAP pistols was the 9mm ADP pistol, produced by the Truvelo Armoury. Designed by Alex Du Plessis, the ADP in some respects leaned towards German designs, such as the Heckler-Koch P7.

9mm ADP Mark II (ADP standing for Advanced Defensive Pistol or Alex Du Plessis, depending on the source) was briefly marketed in the USA as the Heritage Stealth C-1000. Like the Vektor CP1, the ADP also has a polymer frame and gas-delayed blowback action, although in a slightly more angular and traditional-looking package.

Vektor SP1 and SP2

Trigger type: Double/single action
Calibre/ammunition used: 9 × 19mm Para, also .40 S&W in SP2
Weight, empty: 995g (34.8oz)
Length: 210mm (8.2in)
Barrel length: 118mm (4.6in)
Magazine capacity: 15 (9mm) or 11 (.40) rounds

The Vektor SP1 and SP2 pistols are almost the same, except for the calibre and magazines, and are broadly based on the Italian Beretta 92 design. SP stands for 'Service Pistol' and both weapons are full-sized, holster-type pistols of solid construction. Vektor also makes compact versions of both models, marketed as 'General models'; these have shorter barrels, slides and grips.

The Vektor SP1 is a short-recoil-operated, locked-breech pistol. It uses a Walther-type tilting locking piece, located below the straight-recoiling barrel, to lock it to the slide. The frame is made from aluminium alloy; other parts are steel. The trigger is double action, with an exposed hammer and a frame-mounted, ambidextrous safety. Sights are fixed; the front sight is pinned to the slide, the rear is dovetailed. Magazines are double stack; the magazine release button is located at the base of trigger guard.

Vektor SP1 pistol. (Courtesy of Oleg Volk)

Vektor SP1 pistol. (Courtesy of Oleg Volk)

Manual safety Levers on both sides of the frame. Push the lever up to engage the safety; it will lock the hammer, sear and slide; pull it down to disengage the safety; a red dot will appear on the slide.

Field stripping procedure 1) remove the magazine by pressing the magazine release button; 2) check that the chamber is empty; 3) pull the slide all the way back and lock it with the slide-stop lever; 4) rotate the takedown lever (located at the left side of the frame) downwards about 90 degrees; 5) pull the slide slightly back to disengage the slide stop, then carefully push it all the way forwards and out of the frame; 6) remove the return spring by pushing its head slightly forwards and then pulling it out of the detent in the barrel; 7) push the locking lever pin (protrudes backwards from below the barrel breech area), then lift up the barrel from the slide.

Reassemble in reverse order.

RAP-401

Trigger type: Double/single action
Calibre/ammunition used: 9 × 19mm Para, also .40 S&W in RAP-440
Weight, empty: 915g (32oz)
Length: 165mm (6.4in)
Barrel length: 89mm (3.5in)
Magazine capacity: 8 (9mm) or 7 (.40) rounds

The RAP-401 is a short-recoil-operated, locked-breech pistol of all-steel construction. It uses a modified Browning-type locking system, as the barrel engages the slide with a single lug that enters the ejection window. Unlocking is controlled by a cam-shaped lug below the barrel. The trigger is double action, with an exposed hammer and frame-mounted safety, which also doubles as a decocker. Sights are fixed. Magazines are single stack; the magazine release button is located at the base of the trigger guard.

9mm RAP-401 pistol. (Courtesy of Oleg Volk)

Manual safety Levers on both sides of the frame. Pull the lever up to engage the safety; it will lock the hammer either in cocked or lowered position; pull it down to the fixed position to disengage the safety; a red dot will appear on the slide. To decock the hammer, push the safety lever down from the 'fire' position, then release it.

Spain

For several centuries, Spain was home to some of the leading arms makers in Europe. During the early years of the twentieth century, numerous copies and clones of FN/Browning, Mauser and Colt pistols, as well as Smith & Wesson revolvers, were made by a great many companies in Spain, ranging from small workshops up to large, industrial-type factories. Some of these guns were of very good manufacture, some were cheap 'Saturday night specials', while some were average. Many of these guns were used by allied nations during World War I. Of course, there were also several original designs worthy of note, such as the Astra Model 400, known in Spanish service as the Modelo 1921. This was a large blowback pistol, based on the Campo-Giro system of 1916, which fired the relatively powerful 9 × 23mm Bergman-Bayard cartridge more commonly known in Spain as the 9mm Largo.

Between the wars, Spain continued to make clones of foreign designs, such as modified copies of the US M1911 pistol made by Llama and Star, as well as Mauser C96 clones made by Astra and Azul. During World War II, Spain supplied Germany with 9mm military pistols like the Astra 600 (a slightly smaller version of the Astra 400) and the Star Model B, the clone of the Colt M1911 in 9mm Luger calibre. After World War II, Spain continued to export pistols of three brands – Astra, Llama and Star. These pistols were used by West German police (Astra 600/43 and Star B), the South African military (Star), as well as by government and civilian shooters in many other countries, including both Americas.

The Spanish armed forces and police continued to use the 9 × 23 Largo/Bergman ammunition as standard until the mid-1980s. In 1945, the Spanish Army decided to replace their older Modelo 1921 pistols with

The Astra A-60 was a compact, 7.65 or 9mm (short) pistol with a double-action trigger and double-stack magazine. It was popular as a police and civilian weapon.

a new, more modern design. Two pistols were tested in competition, the Star Modelo Super, and the Astra Model 700. Both weapons were based on pre-war designs from the same companies, the Star A and Astra Model 400/M1921 respectively and both were chambered for 9mm Largo. The Star Modelo Super won the trials and in 1946 became the standard Spanish military pistol. During the 1979–82 period, the Spanish Government conducted new trials to replace the aging 9 × 23 Largo pistols with more modern sidearms in the NATO-standard 9 × 19 calibre. As a result of these trials, the Spanish military adopted the Llama M82 pistol, while and the police and *Guarda Civil* adopted the Star 30M, both being double-action, high-capacity pistols chambered for 9mm Luger/NATO ammunition.

Out of dozens of Spanish gunmakers, only three companies are covered by the remit of this book. These are: Unceta y Cia, which operated under the 'Astra' trademark; Llama Gabilondo, which operated under the 'Llama' trademark, and the Bonifacio Echeverria SA, which manufactured guns under the 'Star' trademark. It must be noted that after the installation of Franco's regime in 1939, only these three companies were allowed to produce handguns, under strict Government control; all other companies were prohibited from handgun manufacture.

In the post-war period, Astra first made several pocket-type guns, then started to make modern combat-type pistols in the early 1980s with the introduction of the Astra A-80 pistol, which was heavily based on the SIG-Sauer P220. These pistols were also accompanied by smaller, police-type weapons such as the Astra Constable, Astra A-70 and A-75. The A-80 pistol soon evolved into the A-90 and A-100, but in the late 1990s Astra ceased production and merged with Star under the common name of *Agrupacion Social Trabajadores Armeros, SAL*, or ASTAR in short. However, the ASTAR brand did not survive for long, and the company went out of business completely at the turn of the twenty-first century.

For most of the second part of the twentieth century, Star manufactured a wide variety of pistols, more or less based on the famous Colt M1911 pistol, in a wide variety of calibres, sizes and models. The most famous of these were models A (9mm Largo), B (9mm Luger) and P (.45 ACP), also available in the modified Super versions from 1946 to the mid-1980s. These pistols were used in Spain and also widely exported to Latin America and Africa. Other popular pistols from Star were compacts, also based on the M1911 – the 9mm models BKM (aluminium frame) and BM (steel frame) and the .45 calibre PD. Back in the 1980s, the Star PD (introduced in 1975) was one of the smallest .45 calibre pistols on the market and it created a strong following among American shooters, where this calibre was most favoured. The smaller-calibre, but still quite powerful, Model BM was extensively used by the Spanish *Guarda Civil* and police officers who required concealable sidearms.

From the late 1970s, Star began to make a variety of more modern double-action pistols, starting with the military-type Star 28DA. Of those, the most notable were the Star 30M and 31M, used by Spanish law-

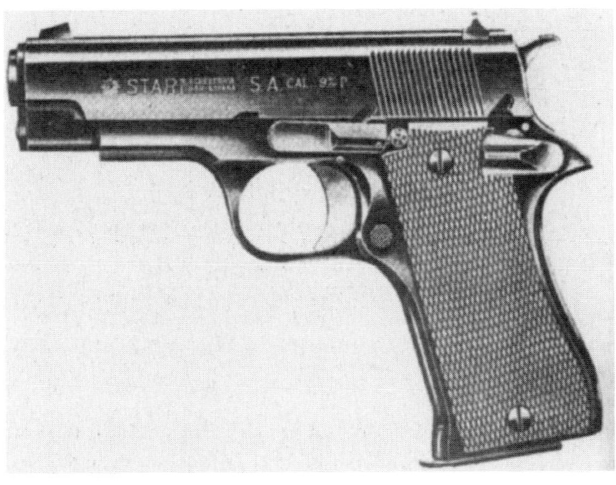

9mm Star BM pistol.

The .45 calibre Star P was a popular and less expensive alternative to the venerable Colt M1911; it has been exported to both Americas.

The 9mm Star 28DA pistol served as a development stage for the more successful Star 30M.

Three last-generation Star brand pistols, from top to bottom: the formidable Megastar in .45 ACP, the service-type 30M in 9 × 19 and the more compact, pocket-type Firestar in the same 9 × 19 chambering.

The 9mm Llama XI-B of post-war manufacture bears a close resemblance to the American M1911 pistol.
(Courtesy of Cherry's Fine Guns)

The Llama Omni pistol had a Browning-type locked-breech system and a double-action trigger; this gun was less than popular because of problematic functioning and high cost and so was only made for a short time. It was available in several major calibres, including 9mm, .38 Super and .45 ACP.

Astra A-80 pistol, with its 'European-type' heel magazine release.

enforcement organizations, as well as a few other service-type handguns, such as the large, high-capacity Star Megastar, chambered for such potent rounds as 10mm automatic and .45 ACP, or the smaller, pocket-type Firestar, that was available in 9mm, .40 S&W and .45 ACP. However, for various reasons, the export of Star guns declined and, after the aforementioned merger with Astra, Star finally went out of business.

In the post-war period, the Llama Company continued to make a variety of M1911 clones in various calibres. In the late 1970s and early 1980s it also joined the double-action club, first with the 9mm Llama Omni, an innovative but unsuccessful design, then with the more successful M82 pistol, which was adopted by the Spanish Army. Until very recently, Llama continued to produce modified M1911 clones under the 'Llama Max' trademark, but by the middle of 2005, the sad news reached everyone concerned: Llama, the last bastion of Spanish pistol-making, finally fell on hard times and closed its doors. It is not known if Spain will ever again make pistols as it used to – sometimes inexpensive, but most often strong and elegant.

Astra A-80, A-90, A-100

The Astra A-80 pistol appeared in 1981. It was heavily based on the contemporary SIG-Sauer P220 pistol, both in appearance and in design. However, only three years later the design was improved with the introduction of the A-90 model, with the addition of an American-type magazine release and a manual ambidextrous safety mounted on the slide in addition to the frame-mounted decocker. Both A-80 and A-90 pistols were made concurrently until 1989. The last of this breed was the A-100, which appeared in 1990. This pistol discarded the manual safety and was a close kin to the famous SIG-Sauer P226 pistol.

The Astra A-80 is a short-recoil-operated, locked-breech pistol of all-steel construction. It uses a modified Browning-type lock; the barrel engages the slide with a single lug which enters the ejection window. Unlike SIG-Sauer pistols, the Astra A-80 and

Astra A-80, A-90, A-100

	A-80	A-90	A-100
Trigger type		Double/single action	
Calibre/ammunition used	9 × 19mm Para, 9 × 23mm Largo, .38 Super, .45 ACP	9 × 19mm Para, .45 ACP	9 × 19mm Para, .40 S&W, .45 ACP
Weight, empty	985g (34.5oz) empty	985g (34.5oz) empty	985g (34.5oz) empty
Length	180mm (7in)	180mm (7in)	180mm (7in)
Barrel length	95mm (3.7in)	95mm (3.7in)	95mm (3.7in)
Magazine capacity	15 (9mm) or 9 (.45) rounds	17 (9mm) or 9 (.45) rounds	17 (9mm), 13 (.40) or 9 (.45) rounds

its later modifications had machined, single-piece slides from the start. The trigger is double-action, with an exposed hammer and an automatic firing-pin block safety. On A-80 and A-100 pistols there are no manual safeties and the pistols are fitted with a decocking mechanism, controlled by the lever at the left side of the frame. Astra A-90 pistols also have a manual safety on the slide, which blocks the firing pin and moves its rear part out of the reach of the hammer, allowing the gun to be carried 'cocked and locked'. The A-90 also retains an automated firing-pin safety and a manual decocking lever. Magazines are double stack; the magazine release is located at the heel of the grip on the A-80 and on the base of the trigger guard on later models. Sights are dovetailed to the slide; the A-80 has fixed rear sights, the A-90 and A-100 have rear sights adjustable for windage.

Manual safety None on the A-80 and A-100; decocking lever at the left side of the frame, in the cut of the grip panel. On the A-90 – ambidextrous lever on the slide; pull it down to engage the safety; push it up to turn the safety off; a red dot will appear on the slide.

Field stripping procedure 1) remove the magazine by pressing the magazine release button; 2) check that the

Astra A-90; note the added manual safety lever on the slide and the magazine release at the base of the trigger guard.

Astra A-100; the manual safety of the previous model was discarded as redundant.

chamber is empty; 3) pull the slide all the way back and lock it there with the slide stop; 4) rotate down the disassembly lever, which is located at the right side of the frame, above the trigger; 5) release the slide and carefully push it forwards and out of the frame; 6) remove the return spring assembly from below the barrel; 7) remove the barrel from the slide.

Reassemble in reverse order.

Llama M82

Trigger type: Double/single action
Calibre/ammunition used: 9 × 19mm Para
Weight, empty: 1,100g (38.5oz) empty
Length: 209mm (8.2in)
Barrel length: 114mm (4.5in)
Magazine capacity: 15 rounds

The Llama M82 pistol is the standard sidearm of the Spanish Army. Its design is heavily based on the Italian Beretta 92, the major differences being the construction of the slide (the M82 has an enclosed slide with a small ejection window, as opposed to the traditional Beretta open-top slide) and the material of the frame (the M82 is of all-steel design, while the Beretta 92 has an aluminium frame). Not surprisingly, the Llama M82 is rather heavy for a modern, military-type pistol. Llama also produced a civilian-only version of the M82, designated M87. It is similar to the M82 except for a tuned, sports-type trigger, an additional frame-mounted safety and adjustable sights. Some variants of the M87 were also fitted with extended barrels and muzzle compensators.

The Llama M82 is a short-recoil-operated, locked-breech pistol which uses a Walther P38-type tilting wedge lock. The trigger is double action, with an exposed hammer. Ambidextrous safety levers are located on the slide. Sights are fixed, with the rear being dovetailed to the slide. Magazines are double stack; the magazine release button is located at the base of the trigger guard.

Manual safety Ambidextrous lever on the slide; pull it down to engage the safety; push it up to turn the safety off; a red dot will appear on the slide.

Star Super

Trigger type: Single action
Calibre/ammunition used: 9 × 23mm Largo (mod. A), 9 × 19mm Para (mod. B), .45 ACP (mod. P)
Weight, empty: 1,080g (37.8oz) empty
Length: 215mm (8.4in)
Barrel length: 122mm (4.8in)
Magazine capacity: 9 (9mm) or 7 (.45) rounds

The Star Modelo Super pistols were based on earlier Star pistols, but with certain improvements. Star Super series pistols were manufactured between 1945 and the mid-1970s (in 9mm Largo and .45) and up until 1989 (in 9mm), concurrently with the similar pistols of pre-war design. The most important improvement was the

Llama 82 pistol.

9 × 23 Largo calibre Star A Super pistol. *(Courtesy of Cherry's Fine Guns)*

9 × 19 Star B Super pistol; note the disassembly lever, mounted co-axially with the slide stop on the right side of the frame, above the trigger.

modified locking system: the Colt/ Browning M1911-type swinging link system was replaced by a modified one, with a solid lug that has a cam-shaped cut in it. The field-stripping procedure has also changed from previous models. The simplest way to distinguish the Modelo Super from earlier designs is to look at the right side of the gun – Super pistols have a disassembly lever, located at the right side of the frame, above the trigger, while the previous type of Star pistols have no such control. Modelo Super pistols were made in several calibres, including 9mm Largo, 9mm Luger and .45 ACP. It must be noted that some of the post-war pistols, especially compact ones, did not follow the Super pattern and retained the M1911-type system with a swinging link, which controlled barrel unlocking and locking on recoil.

Star Super pistols are short-recoil-operated, locked-breech weapons of all-steel construction. They use Browning-type locking; upon recoil, the rear part of the barrel is tipped down out of the engagement with the slide by the cam-shaped cut made in the lug below the breech, which interacts with the frame. The barrel locks to the slide with two lugs that fit into cuts made in the underside of the slide. The trigger is single-action, with an exposed hammer and a manual frame-mounted safety. Magazines are single stack, with the magazine release located at the base of the trigger guard. Sights are fixed.

Manual safety Lever on the left side of the frame, above the grip panel; push it up to engage the safety; pull it down to turn the safety off. The safety can be engaged only when the hammer is cocked. When engaged, the safety locks the hammer and slide.

Field stripping procedure 1) remove the magazine by pressing the magazine release button; 2) check that the chamber is empty; 3) push on the recoil spring plug to unlock the barrel bushing; 4) rotate the barrel bushing, then remove it from the front of the slide; 5) rotate the disassembly lever, which is located at the right side of the frame above the trigger, down and forwards; 6) push the slide forwards and out of the frame; 7) remove the return spring from the slide by pulling its rear (along with guide rod) away from the barrel, then rearwards; 8) pull the barrel from the slide.

Reassemble in reverse order.

Star 30M

	30M	30PK
Trigger type	Double/single action	
Calibre/ ammunition used	9 × 19mm Para	
Weight, empty	1,140g (39.9oz)	860g (30.1oz)
Length	205mm (8in)	193mm (7.5in)
Barrel length	119mm (4.6in)	98mm (3.8in)
Magazine capacity	15 rounds	15 rounds

Star 30M pistols were developed as improved versions of the Star 28DA handguns, which were (unsuccessfully) tested in the USA. Star 30M pistols were made between 1983–94, when they were replaced in production by slightly improved 31M pistols of the same basic design. The standard 30M pistols were mostly used by the Spanish *Guarda Civil*, while the shorter and lighter 30PK pistols (with aluminium frames) were mostly used by police forces.

The Star 30M is a short-recoil-operated, locked-breech pistol. It uses Browning-type locking; the barrel engages the slide with two lugs which fit into recesses on the underside of the slide. Unlocking is controlled by the cam-shaped cut made in the lug below the chamber; this lug interacts with the slide-stop axis pin. The trigger is double-action, with an exposed hammer and a two-part firing pin. The hammer, with its main-spring and sear, are incorporated into one en-block unit, which forms the removable backstrap of the pistol grip. The ambidextrous safety, when engaged, retracts the firing pin from the reach of the hammer and then locks it, while slide, trigger and hammer operations remain unaffected. This allows the gun to be cycled and dry-fired with the safety on. Other safeties include the magazine safety and the firing-pin block.

Manual safety Levers on either side of the slide; pull down to engage the safety; push up to turn the safety off; a red dot will appear on the left side of the slide, just behind the safety lever. Note that when the safety is applied, only the firing pin will be locked.

Field stripping procedure 1) remove the magazine by pressing the magazine release button; 2) check that the chamber is empty; 3) pull the slide back slightly; 4) while holding the slide, push the slide-stop axis pin inwards from the right, then pull it out of the frame towards the left; 5) carefully ease the slide, then push it forwards and out of the frame; 6) remove the return spring from below the barrel; 7) remove the barrel from the slide.

Reassemble in reverse order.

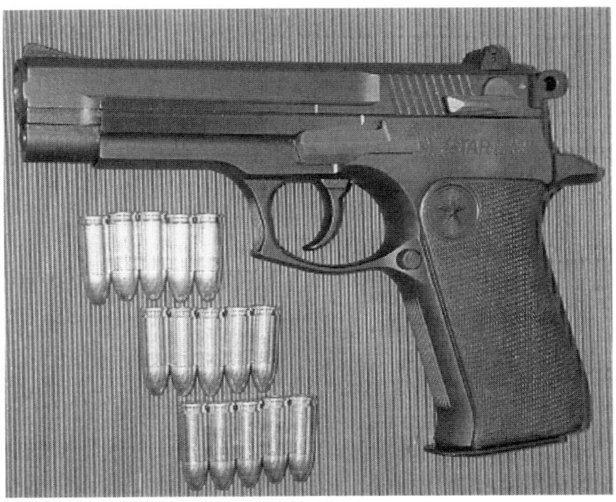

Star 30M.

Star 30M, partially disassembled.

SWITZERLAND

Switzerland has retained its neutrality for the last 200 years, with its well-trained and well-equipped armed forces being one of the basic foundations of its independence. For many decades, Swiss requirements for individual marksmanship were among the most demanding; similarly demanding requirements also applied to small arms. The Swiss Army was among the first to adopt a semi-automatic pistol – in 1900 it adopted the Luger Parabellum in 7.65mm calibre. However, by the early 1940s the Parabellum pistol, even in its modified M1906/29 form as made by *Waffenfabrik Bern*, was more or less obsolete. Just when half of the world entered the bloodbath of World War II, the Swiss Army started its quest for a new, more potent semi-automatic pistol, chambered for 9 × 19 Luger/Parabellum ammunition. Many foreign and some domestic designs were tried and rejected; of foreign designs, only the FN Browning High Power pistol displayed sufficient reliability, but its accuracy was below Swiss standards. The final stages of the trials, which took place between 1943–7, were conducted between weapons made by two major Swiss arms manufacturers – the state-owned *Waffenfabrik*

Bern (W+F) and the privately owned *Schweizerische Industrie-Gesellschaft* (SIG). Early W+F designs were based on modified FN Browning High Power pistols, but in 1947 W+F discarded the Browning locking-breech, tilting-barrel action because of its insufficient accuracy and went to a fixed-barrel system with a gas-delayed blowback action. SIG based its pistols on the patents of Charles Petter, bought from France during the late 1930s. Early SIG pistols displayed the clear influence of Petter designs, being copied from his French Mle.1935 pistol.

In the search for an ideal military pistol, both W+F and SIG also tried both single-stack and double-stack magazine designs. Interestingly enough, the Swiss Army found thick-gripped pistols with double-stack magazines to be less accurate, due to their less comfortable grip, so only single-stack designs competed in the last stage. In the final chase between the W+F 47 and SIG S.P. 47/8, the latter won the trials and in 1949 it was officially adopted as *Selbstladepistole SP/49* (or Pistole 49 for short). Also known as the 'Neuhausen' (after the city where the pistols were made), the S.P. 47/8 (before adoption of the P210 trademark in 1957), or under its current commercial designation of SIG P210, quickly earned the distinction of being the most accurate and most expensive military pistol ever made. Also adopted by the Danish Army, as well as the West German Border Police (*Bundesgrentzschutz*) and some Swiss police units, this pistol was later converted with a series of sport modifications and a great many shooting matches were (and still are) won with this gem of workmanship.

Despite (or, rather, because of) the excellence of the Pistole 49/P210 in accuracy and reliability, this pistol was far too expensive, even for the Swiss Army. SIG began to experiment with less expensive manufacturing methods in around 1966. Early works centred on development of the new slides for P210 pistols, manufactured from stamped-steel halves by welding. Later on, SIG began development of a more modern weapon that employed a stamped slide and a lighter aluminium frame. Also, by the mid-1970s, double-action pistols had become all the rage in the military and police, and in 1975 the Swiss Army decided to adopt the most modern design that emerged from the same SIG Company. The newest SIG P220 pistol had all of the desirable features of a contemporary military pistol – a double-action trigger with decocker, combined with a lightweight aluminium frame. The only potential issue with the Pistole 75 was its single-stack magazine that held 'only' nine rounds, compared with the fifteen rounds of the contemporary Beretta 92 or CZ-75. However, the Swiss Army always emphasized accuracy, so for them this was not an issue at all.

Envisioning a great export potential for the new P220 handgun, but being seriously limited by strict Swiss regulations limiting the export of 'war materiel', SIG bought out the German company J.P.Sauer & Sohn, which then began to produce P220 pistols and its derivatives for worldwide distribution. Today, the Pistole 75 (9mm SIG P220) is a general issue sidearm in the Swiss Army; however, there are still some P210 pistols in the stocks of the Swiss Army, as some officers who were issued these pistols back in the 1960s

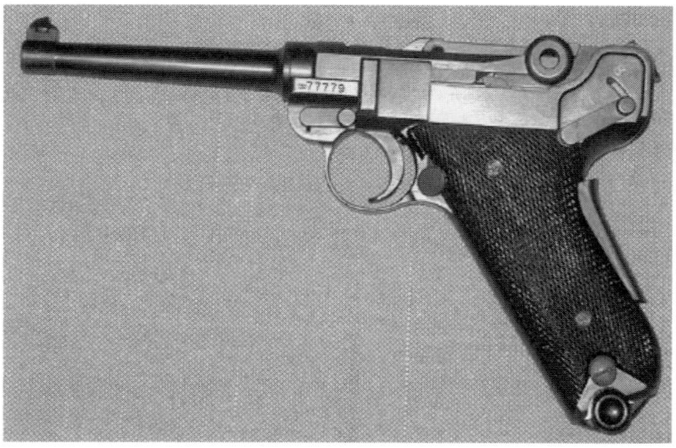

The 7.65mm (7.65 × 21 Luger) Pistole 06/29 was the standard Swiss service pistol until 1949. Manufactured by the Waffenfabrik Bern, this was an improved German Luger Parabellum design.

The 9mm experimental W+F 43 pistol made circa 1943 by Waffenfabrik Bern was broadly based on the FN Browning High Power pistol.

The 9mm experimental W+F 47, on the other hand, was based on the pre WWII-era patent of the American, J. Destree, and used a gas-delayed blowback action with the gas cylinder and piston located below the barrel (à la later HK P7).

9mm experimental pistol made at the SIG factory to the Petter design. Its French ancestry is clearly visible in the overall shape and placement of the safety.

and 1970s are still in active service. It also worth mentioning that the last P06/29 Luger pistols were retired from the Swiss Army only during the 1980s. The Swiss police uses various types of handguns, including several SIG-Sauer pistols such as the P225 and P228, P230/232, as well as some others.

The success of the P220 spawned a whole family of service pistols, starting from more compact P225, developed for the German police trials of the

The 9mm experimental SIG S.P. 44/16 was much closer to the final design – it already had an internal slide rail and a frame-mounted safety. However, the sixteen-round magazine made its grip too wide for the typical Swiss soldier of the time.

Prototype SIG S.P. 44/8 (top) and early production SIG S.P. 47/8 (bottom).

The very early production 9mm SIG-Sauer P226 pistol; it has a stamped-steel slide of a different shape, compared to modern pistols of the same model, as well as a bottom magazine release. Note that it has been manufactured in Germany.

mid-1970s, through the very successful P226 that almost won the US XM9 military pistol trials of the mid-1980s, and up to the relatively new P229 which, currently ranked, is among the most popular police pistols worldwide.

In 1984 the Swiss company ITM obtained a licence from Czechoslovakia for production of the CZ-75 pistol. The first ITM pistols, designated AT-84, were close copies of the CZ-75, made to the highest standards of fit and finish. Later on, this design evolved into the AT-88, which featured an automatic firing-pin safety, designed by the Swiss designer, Martin Thuma. After the merger of Sphinx Systems and ITM in 2000 under the name of Sphinx, these pistols, with further improvements incorporated into the design, continued production as the Sphinx AT2000. Sphinx pistols are made in a variety of calibres and versions. Used by many European sport shooters in disciplines such as IDPA and IPSC, service-type Sphinx pistols are also issued by several police organizations in Belgium, Norway, Switzerland and Venezuela.

The AT-88 pistol, made during late the 1980s and 1990s by ITM of Solothurn, was a high-end clone of the famous Czechoslovak CZ-75 pistol. Later it evolved into Sphinx AT2000.

During the late 1990s, following the introduction of the Belgian small-bore, high-velocity FN Five-seveN pistol, the small private company, Thuma Engineering, led by designer Martin Thuma, developed a high-velocity, 5.6mm calibre, service-type pistol, known as the MTE-224. Available in two basic versions, standard (semi-automatic) and machine (selective fire), the MTE-224 fires proprietary .224V ammunition, developed in the Czech Republic.

SIG P210

Trigger type: Single action
Calibre/ammunition used: 9 × 19mm Para; also 7.65 × 21mm Para and .22LR
Weight, empty: 900g (31.5oz)
Length: 215mm (8.4in)
Barrel length: 120mm (4.7in)
Magazine capacity: 8 rounds

The SIG P210 is a short-recoil-operated, locked-breech pistol of all-steel construction. It is somewhat unusual in that it has inner slide rails; that is, full-length slide rails are machined on the inside of the frame. Most parts are machined from solid steel forgings and then hand fitted. This results in increased durability and accuracy, but also greatly increases the cost of manufacture. The locking system is of the Browning type; the barrel locks into the slide with two lugs which fit into cuts on the underside of the slide. Unlocking is controlled by a cam-shaped cut made in the lug below the chamber area. The trigger is single action, with an external hammer and a frame-mounted safety. In line with older Petter patents, the hammer, its spring and sear are made into a single removable unit. On service-type pistols sights are fixed; both front and rear sight blades are dovetailed to the frame. Magazines are single stack; the magazine release is located at the base of the grip (except for recent production P210-8 pistols made for US market, which have a button-type magazine release at the base of the trigger guard).

The SIG P210-2 was the same gun as the military issue SP/49. It was available either with wooden or plastic grips.

9mm commercial SIG P210-5 pistol of current manufacture, with adjustable target-type sights.

Over time, there have been several modifications of the basic P210 design, distinguished by digits after the dash, for example P210-2, P210-4, P210-5 and so on. While service guns were made in 9mm calibre only, commercial pistols were also available in 7.65 × 21 Luger/Parabellum calibre, as well as in 5.6mm rimfire/.22LR. Calibre conversion units were available, as well as a complete kit that included one frame, two slides (7.65/9mm and .22LR) and three barrels and return springs.

The basic versions of P210 included the following:

- P210-1 – commercial version with polished finish
- P210-2 – same as the service-type SP/49; matte finish
- P210-3 – version for Swiss police, polished finish, fixed sights
- P210-4 – version made specifically for the German border police (about 5,000 made, most in 9mm and a few in .22LR for training); has no lanyard ring on the grip
- P210-5 – sport-type commercial pistol with adjustable sights and standard or extended 150mm barrel

9mm commercial SIG P210-6 pistol of current manufacture, with fixed sights.

SIG P210-5, partially disassembled.

- P210-6 – sport-type commercial pistol with fixed sights; differs from P210-2 with better finish and more expensive grips
- P210-7 – dedicated .22LR version
- P210-8 – 'American' version with side-mounted magazine release.

Manual safety lever on the left side of the frame. Push the lever up into the horizontal position (marked with a white 'S' letter) to engage the safety; pull it down to disengage the safety; the position is marked with a red 'F'.

Field stripping procedure 1) remove the magazine by pressing the magazine release button; 2) check that the chamber is empty; 3) pull the slide back slightly, to align with the disassembly marks on the slide and barrel; 4) holding the slide partially retracted, push the slide-stop axis pin inwards from the right, then pull the slide-stop lever out of the frame towards the left; 5) carefully push the slide forwards and out of the frame; 6) pull the hammer unit up out of the frame; 7) remove the return spring with its guide from below the barrel; 8) pull the barrel out of the slide.

Reassemble in reverse order.

SIG-Sauer P220

Trigger type: Double/single action
Calibre/ammunition used: 9 × 19mm Para and .45 ACP; also .38 Super and 7.65 × 21 Luger
Weight, empty: 800g (28oz) (aluminium frame)
1,100g (38.5oz) (stainless-steel frame)
Length: 198mm (7.7in)
Barrel length: 112mm (4.4in)
Magazine capacity: 9 (9mm) or 7 and 8 (.45) rounds

The SIG-Sauer P220, often considered as the best 'out of the box' .45 calibre double-action pistol, started its life as the 9mm Pistole 75, the new double-action sidearm for the Swiss Army. Developed during the early 1970s by the famous Swiss company, *Sweizerische Industrie-Gesellschaft*, or SIG in short, this pistol was intended to replace the extremely accurate, but overly expensive, SIG P210. The pistol was indeed a very promising design, but strict Swiss laws severely limited the export of 'war materiel', so SIG decided to move the manufacture of its newest pistol abroad. For this reason, SIG Holdings AG in 1976 acquired significant stocks of the J.P. Sauer & Sohn Company, which was located in Germany, and during the late 1970s started the manufacture of the SIG-Sauer P220 pistol in Germany.

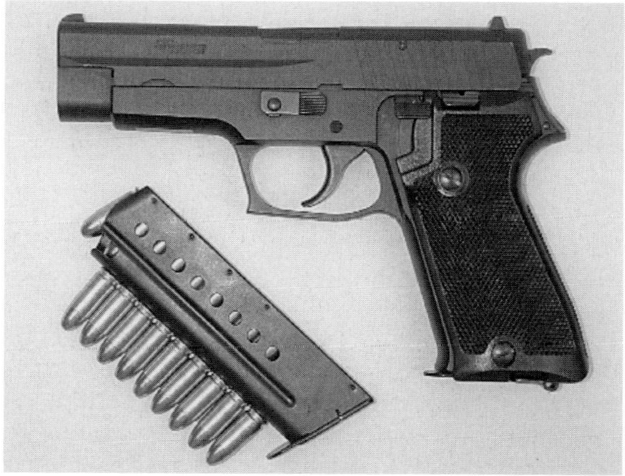

9mm P220 pistol with original 'European' magazine release, located at the heel of the grip.

The P220, as with most of its later siblings, has been an instant and continuous success. Adopted by the Swiss and Japanese armies and by a large number of European police organizations, this pistol also rapidly found its way into the USA, first through the Browning Arms Company of Utah and, since the mid-1980s, through the SIGARMS Company of New Hampshire. Initially made in four calibres, today the P220 is manufactured only in .45 ACP, as its 9mm version has been largely replaced by the SIG-Sauer P226, a pistol of similar size but with a larger magazine capacity. Today, the P220 is made in a number of versions which include both 'service' and 'sporting' modifications, the latter often fitted with extended barrels, compensators and adjustable sights. Service versions may have the Picatinny rail for accessories, as well as double-action or double-action-only triggers. The P220 also could be made with an aluminium frame and steel slide, or with both the frame and slide made from stainless steel. It is worth nothing that a high-capacity version of the .45 calibre P220 was developed during the early 1990s for the US market. Apparently designated P221, this gun was effectively killed by the so-called 'Assault weapons ban', the infamous legislation that limited the magazine capacity of civilian arms to ten rounds, which was enforced in the USA between 1994 and 2004.

The SIG-Sauer P220 is a short-recoil-operated, locked-breech pistol. The locking is of the modified Browning type, where the barrel engages the slide with a single large lug entering the ejection window. The unlocking is controlled by a cam-shaped lug under the

Same 9mm P220, right side.

9mm P220, partially disassembled.

barrel, which interacts with the steel insert set into the aluminium frame. The slide is made from sheet steel by stamping and forming processes, with the separate breechblock pinned into the rear of the slide (machined slides are available on some current production .45 calibre models). The standard trigger is the double-action type, with an automatic firing-pin block safety and a frame-mounted decocking lever. DAO modifications do not have a decocker. Magazines are single stack, with the magazine release located on the heel of the grip on early versions, or at the base of the trigger guard on the .45 calibre versions made for the US market. The standard sights are fixed, with high-contrast inserts, and the rear sight is dovetailed to the slide.

.45 calibre P220R (Rail) pistol of recent manufacture, as made for the US market. This pistol is equipped with a Picatinny rail on the frame for attachments like tactical lights or laser aiming modules and has a magazine release button in the base of the trigger guard.

Manual safety None; the decocking lever is on the left side of the frame, in the cut of the grip panel.

Field stripping procedure 1) remove the magazine by pressing the magazine release button; 2) check that the chamber is empty; 3) pull the slide all the way back and lock it there with the slide stop; 4) rotate down the disassembly lever, which is located at the left side of the frame, above the trigger; 5) release the slide and carefully push it forwards and out of the frame; 6) remove the return spring assembly from below the barrel; 7) remove the barrel from the slide.

Reassemble in reverse order.

SIG-Sauer P225

Trigger type: Double/single action
Calibre/ammunition used: 9 × 19mm Para
Weight, empty: 740g (25.9oz) empty
Length: 180mm (7in)
Barrel length: 98mm (3.8in)
Magazine capacity: 8 rounds

The SIG-Sauer P225 is a compact version of the P220 pistol, described above. It is entirely similar in design and function and is widely used by numerous European

SIG-Sauer P225.

Same pistol; the slide is locked open.

P225, partially disassembled.

police organizations, including certain Swiss and German police departments, as well as the Canadian Army and even a few US police departments. The earliest versions of the P225 were equipped with a magazine release located at the bottom of the grip, but current production pistols have the magazine release button at the base of the trigger guard.

Manual safety None; the decocking lever is on the left side of the frame, in the cut of the grip panel.

Field stripping procedure 1) remove the magazine by pressing the magazine release button; 2) check that the chamber is empty; 3) pull the slide all the way back and lock it there with the slide stop; 4) rotate down the disassembly lever, which is located at the left side of the frame, above the trigger; 5) release the slide and carefully push it forwards and out of the frame; 6) remove the return spring assembly from below the barrel; 7) remove the barrel from the slide.

Reassemble in reverse order.

SIG-Sauer P226

Trigger type: Double/single action
Calibre/ammunition used: 9 × 19mm Para, also .357 SIG and .40 S&W
Weight, empty: With aluminium frame: 870g (30.45oz) (9mm), 920g (32.2oz) (.357, .40)
with stainless steel frame: 1,180g (41.3oz)
Length: 196mm (7.6in)
Barrel length: 112mm (4.4in)
Magazine capacity: 15 (9mm) or 12 (.40 and .357) rounds

9mm P226 pistol.

The SIG-Sauer P226 pistol was developed as an improved, high-capacity version of the P220, especially for the US XM9 trials of 1982–4. The P226 was one of just two pistols that nominally passed all of the technical requirements for these trials, but at the last moment it lost to the other design, the Beretta 92SB-F, on the basis of the cost. Despite this, the P226 quickly became one of the most popular service pistols, being used by police, military and security forces, as well as by many civilian shooters around the world. Initially available in 9mm, it was subsequently adapted to .40 S&W and .357 SIG calibres, which are extremely popular in the USA. In fact, the P226 was the first pistol to be chambered for the .357 SIG, a high-velocity, high-penetration 9mm cartridge, preferred by many American highway patrol services as well as by the US Security Service.

The SIG-Sauer P226 is basically similar to the P220 described above, except for the slightly different shape and double-stack magazines of greater capacity. It is available in a number of versions, with standard aluminium or stainless-steel frames, with or without a Picatinny rail. Triggers are either a typical SIG-Sauer double-action arrangement with a frame-mounted decocker and an automated firing-pin block, or a double-action-only set-up without a decocker. 9mm and .40/.357 versions of this pistol also may differ in the design of the slide – while 9mm pistols may have a stamped-steel slide, larger calibre pistols have heavier slides, machined from a solid bar of steel.

Manual safety None; the decocking lever is at the left side of the frame, in the cut of the grip panel.

9mm P226 Stainless pistol of recent manufacture, equipped with integral Picatinny accessory rail on the frame.

Field stripping procedure 1) remove the magazine by pressing the magazine release button; 2) check that the chamber is empty; 3) pull the slide all the way back and lock it there with the slide stop; 4) rotate down the disassembly lever, which is located at the left side of the frame, above the trigger; 5) release the slide and carefully push it forwards and out of the frame; 6) remove the return spring assembly from below the barrel; 7) remove the barrel from the slide.

Reassemble in reverse order.

SIG-Sauer P228/P229

	P228	P229
Trigger type	Double/single action	
Calibre/ ammunition used	9 × 19mm Para	.40 S&W, .357 SIG, 9 × 19mm Para
Weight, empty	825g (28.9oz)	905g (31.7oz)
Length	180mm (7in)	180mm (7in)
Barrel length	98mm (3.8in)	98mm (3.8in)
Magazine capacity	13 rounds	13 (9mm) or 12 (.40 or .357) rounds

.40 calibre P229 DAO Stainless, one of the recent models made specifically for the US law-enforcement market.

The SIG-Sauer P228 pistol was developed from the larger P226 in 1989 in the same way that the earlier P225 pistol had been developed from the P220 – which means that the designers basically chopped some length from the barrel and slide, as well as from the grip. The overall shape of the P228 was also made slightly less angular and more suited for concealed carry. In around 1991, the P228 was adopted by the US armed forces as the M11 pistol and is issued to the military police and some other units. The P228 is also a highly popular police and civilian sidearm and is widely issued to law-enforcement forces in Europe and the USA. The P229, which was developed especially for the US market and is assembled in the USA using both German and US-made parts, is technically similar to the P228 except for the heavier slide, which is made by machining a bar of steel, as opposed to the P228 slide, which is made by stamping and forming from sheet steel. The heavier slide on the P229 is necessary because of the more powerful ammunition used in this pistol. P229 pistols also can be adapted to any of three calibres (9mm, .357 and .40) by switching the barrel and return spring; P228 pistols are 9mm only and cannot be converted to other calibres.

Like the larger P226, the P228 and P229 pistols are available in aluminium frame/steel slide or stainless-steel frame and slide versions, with standard DA or DAO triggers.

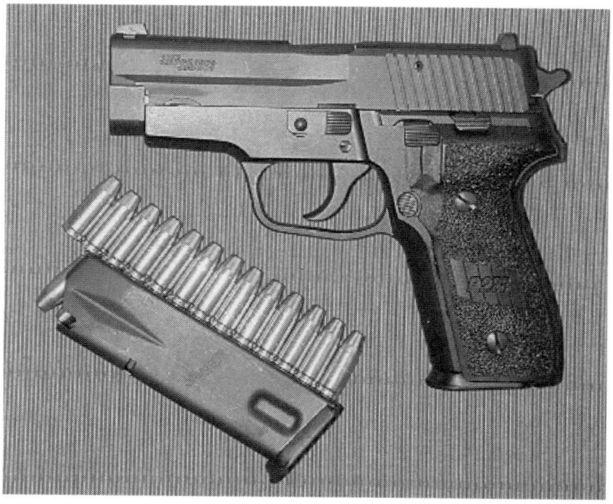

9mm SIG-Sauer P228.

9mm SIG-Sauer P228, partially disassembled.

Manual safety None; decocking lever at the left side of the frame, in the cut of the grip panel.

Field stripping procedure 1) remove the magazine by pressing the magazine release button; 2) check that the chamber is empty; 3) pull the slide all the way back and lock it there with the slide stop; 4) rotate down the disassembly lever, which is located at the left side of the frame, above the trigger; 5) release the slide and carefully push it forwards and out of the frame; 6) remove the return spring assembly from below the barrel; 7) remove the barrel from the slide.

Reassemble in reverse order.

.357 SIG P229 Stainless pistol, fitted with custom grips. *(Courtesy of Ken Lunde)*

SIG-Sauer P230/P232

	P230	P232
Trigger type	Double/single action	
Calibre/ ammunition used	7,65 × 17SR 9 × 17 9 × 18 Police	7,65 × 17SR 9 × 17
Weight, empty	460g (116.1oz) (580g [20.3in] in SL version)	465g (600g [21oz] in SL version)
Length	168mm (6.6in)	168mm (6.6in)
Barrel length	92mm (3.6in)	92mm (3.6in)
Magazine capacity	7 (9mm) or 8 (7.65mm) rounds	7 (9mm) or 8 (7.65mm) rounds

The SIG-Sauer P230 is a compact, self-defence pistol, which was quite popular in European police circles during the 1970s and 1980s. Well-suited to concealed carrying, this pistol is quite accurate and reliable and made to a high standard of fit and finish. However, today its role is more as a concealed-carry defensive weapon for civilian users, or as a back-up gun for police officers. The SIG-Sauer P232 is an improved version of the P230, with a very slightly altered shape and some changes to the manufacturing technologies. The P232 replaced the P230 in production in 1996.

9mm (9 × 17 Short) P230SL pistol, made from stainless steel.

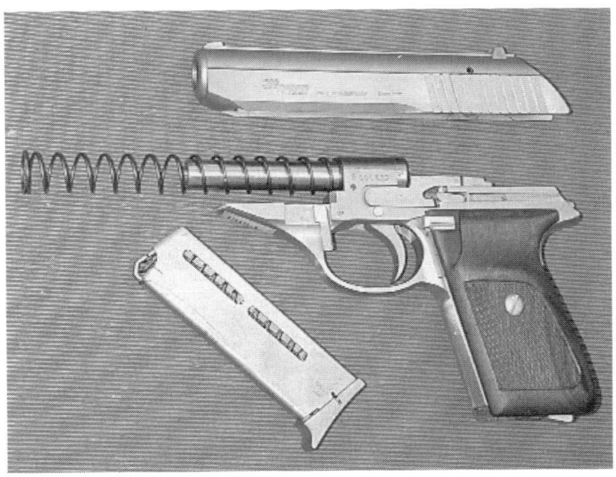

P230 pistol, partially disassembled.

Rare version of the P230 with a manual safety (above the grip panel) and a lanyard ring, possibly made for the Japanese police. (Courtesy of Cherry's Fine Guns)

P232SL stainless-steel pistol.

The SIG-Sauer P230 is a blowback-operated pistol with a fixed barrel. In the standard version, the frame is made from aluminium alloy and the slide is of stamped steel with a separate breechblock pinned in place. The P230SL version has a stainless-steel frame and slide and thus is somewhat heavier. The same applies to the P232 pistols. The trigger is double-action, with an external hammer, an automatic firing-pin block safety and a frame-mounted decocker. The Double Action Only version without the decocker is also available. A rare version with an added manual safety also exists; it was apparently developed for the Japanese police but a few specimens ended up on the civilian market. Magazines are single stack, with the magazine release catch located at the base of the grip. The slide stop has no external release and once the slide is locked back after the last shot, the only way to close it is to pull it back and then release it. Sights are fixed, with the rear sight being dovetailed to the slide.

Manual safety None; decocking lever at the left side of the frame, in the cut of the grip panel.

Field stripping procedure 1) remove the magazine by pressing the magazine release button; 2) check that the chamber is empty; 3) rotate down the disassembly lever, which is located at the left side of the frame, above the front of the trigger guard; 4) pull the slide all the way back, raise its rear side off the rails and slowly release it forwards and off the frame; 5) remove the return spring from the barrel.

Reassemble in reverse order.

SIG-Sauer SIGPro

Trigger type: Double/single action
Calibre/ammunition used: 9 × 19mm Para, .357 SIG, .40 S&W
Weight, empty: 760g (26.6oz) empty
Length: 187mm (7.3in)
Barrel length: 99mm (3.9in)
Magazine capacity: 15 (9mm) or 12 (.357 and .40) rounds

The SIGPro pistol is a true result of modern globalization. Its frames are made in Switzerland, while the slides are machined and the pistols assembled either in the USA at the SIGARMS factory in Exeter, New Hampshire (for the US market), or in Germany at the Sauer factory for European markets. The fact that this pistol has been designed as a dedicated .40 calibre platform for the hot-selling American .40 S&W cartridge, and later adapted to the 'smaller' 9mm ammunition, leads the author to conclude that it was an 'America-special' project. However, the biggest sales so far achieved by this pistol have been made in Europe, where France ordered more than 200,000 9mm SIGPro 2022 pistols for the *Gendarmerie* and other law-enforcement forces. In the USA, this pistol so far remains in the shadow of the metal-framed 'classic' SIG-Sauer pistols.

The basic version of the SIGPro, known as the Model SP2340, entered production in 1998; the 9mm model SP2009 entered production in the following year and the improved SP2022 has been manufactured since 2003. Compact versions of the

2340 and 2009 pistols are also offered for police and civilian use.

SIGPro pistols are locked-breech, short-recoil-operated weapons that use modified Browning-type locking. The barrel locks to the slide by means of a single large lug that engages the ejection window; unlocking is controlled by a cam-shaped cut made in the lug under the barrel, which interacts with the slide-stop axis pin. The frame is made from impact-resistant polymer. The trigger is of double-action or double-action-only type, with a modular hammer unit which can be replaced by a qualified gunsmith or at the factory. The DA trigger unit is equipped with a decocking lever; the DAO unit has no decocker. Each pistol is equipped with an automatic firing-pin block safety.

There was also a rare modification of the SIGPro, fitted with a manual safety lever mounted on the slide instead of the frame-mounted decocker. Apparently, this version was developed for a large potential contract with a South American country, but the contract fell through and only a few guns of this type were actually made and sold through commercial channels. The grip of the SIGPro is also modular, with a detachable grip panels made from polymer. Each gun is shipped with two panels of slightly different shape and size, so the gun can be tailored for the particular grip preferences of the specific user. Other optional

ABOVE: *SIGPro SP2340 pistol chambered for .40 S&W cartridge.* *(Courtesy of Cherry's Fine Guns)*

ABOVE RIGHT: *.40 calibre SIGPro SP2340 pistol with DAO trigger (no decocking lever).* *(Courtesy of Cherry's Fine Guns)*

RIGHT: *9mm SIGPro SP2022, made for French Gendarmerie and police.*

grip panels are available from the manufacturer if required. Sights are fixed, with white or luminous inserts; both front and rear sights are dovetailed into the slide. Magazines are double-stack.

Manual safety None; decocking lever at the left side of the frame, in the cut of the grip panel.

Field stripping procedure 1) remove the magazine by pressing the magazine release button; 2) check that the chamber is empty; 3) pull the slide almost all the way back, to align the square notch on the slide with the slide-stop lever axis; 4) push the slide-stop lever axis inwards from the right, then pull the slide-stop lever completely from the frame to the left; 5) release the slide and carefully push it forwards and out of the frame; 6) remove the return spring assembly from below the barrel; 7) remove the barrel from the slide.

Reassemble in reverse order.

9mm P250DCc showing its DAO working; note that when the trigger is partially pulled, the spurless hammer is being cocked.

SIG-Sauer P250

Trigger type: Double-action only
Calibre/ammunition used: 9 × 19mm Para
Weight, empty: 720g (25.2oz)
Length: 180mm (7in)
Barrel length: 102mm (4in)
Magazine capacity: 15 rounds

The SIG-Sauer P250 is the most recent development of the J.P. Sauer company. First presented during the IWA-2004 exhibition (Germany, March 2004), at the time of writing this pistol was available only in Europe. Development was initiated in around 2001, in accordance with the latest German police guidelines for a service pistol, which were issued in 2000. Currently, only one version is available, the model 250 DCc – Defence Concept, compact. In the future, other versions are promised, such as full-size pistols and examples with conventional SA/DA or DAO trigger systems, as well as versions in other calibres.

The SIG-Sauer P250 is a short-recoil-operated, locked-breech pistol with a polymer frame. It uses a modified Browning-type locking mechanism, with a barrel that locks to the slide by a single large lug that engages the ejection window. The pistol features an external hammer and so far the only available

9mm P250DCc (compact) pistol.

Same gun, slide locked open.

Defence Concept trigger is a modified double-action-only system with a short trigger reset for the second and all subsequent shots. There are no manual safeties nor decocking levers, and slide stop release levers are provided on both sides of the frame. Both the front and rear sights have white inserts and are dovetailed to the slide. The front part of the frame, under the barrel, is shaped to provide an integral Picatinny-type accessory rail.

Manual safety None.

Field stripping procedure 1) remove the magazine by pressing the magazine release button; 2) check that the chamber is empty; 3) pull the slide all the way back and lock it there with the slide stop; 4) rotate down the disassembly lever, which is located at the left side of the frame, above the trigger; 5) release the slide and carefully push it forwards and out of the frame; 6) remove the return spring assembly from below the barrel; 7) remove the barrel from the slide.

Reassemble in reverse order.

Sphinx AT2000

	AT2000	AT2000P
Trigger type	Double/single action	
Calibre/ ammunition used	9 × 19mm Para, 9 × 21mm IMI, .40 S&W	
Weight, empty	1,000g (35oz)	940g (32.9oz)
Length	204mm (7.96in)	184mm (7.2in)
Barrel length	115mm (4.5in)	93mm (3.6in)
Magazine capacity	15 (9mm) or 11 (.40) rounds	13 (9mm) or 10 (.40) rounds

Sphinx AT2000 pistols are short-recoil-operated, locked-breech weapons of all-steel design. The barrel locks into the slide with a single large lug that enters the ejection port; unlocking is controlled by a cam-shaped cut made in the lug under the breech of the barrel. The slide runs on the full-length rails machined into the inside of the frame. The trigger is double-action, with an exposed hammer and frame-mounted safety, as well as an automatic firing-pin safety block. Pistols are available with left-side or ambidextrous safety levers as factory options; other options include decock-only safety arrangements, as well as sport-type, tuned single-action triggers on competition models. Magazines are double-stack; the magazine

9mm Sphinx AT2000.

Sphinx AT2000, partially disassembled.

9mm Sphinx AT2000PS with decocking-only safety arrangements.

release button is located at the base of the trigger guard. Sights are fixed on combat models.

Sphinx pistols are available in a wide variety of finishes and sizes; most suitable for service use are the full-size AT2000 or semi-compact AT2000P, or their decocking-only versions (marked with an added suffix 'S', for example AT2000PS).

Manual safety Lever on the left side of the frame (ambidextrous optional). Push the lever up into the horizontal position to engage the safety; pull it down to disengage the safety; a red dot will appear on the frame.

Field stripping procedure 1) remove the magazine by pressing the magazine release button; 2) check that the chamber is empty; 3) pull the slide back slightly, to align the disassembly marks on the slide and barrel; 4) holding the slide partially retracted, push the slide-stop axis pin inwards from the right, then pull the slide-stop lever out of the frame towards the left; 5) carefully push the slide forwards and out of the frame; 6) remove the return spring with its guide from below the barrel; 7) pull the barrel out of the slide.

Reassemble in reverse order.

Sphinx 3000 Tactical. *(Courtesy of Sphinx)*

Sphinx 3000 Standard.

SPHINX 3000

	3000	3000 Tactical
Trigger type	Double/single action	
Calibre/ ammunition used	9 × 19mm Para, 9 × 21mm IMI, .40 S&W	
Weight, empty	1,040g (36.4oz)	950g (33.3oz)
Length	215mm (8.4in)	195mm (7.6in)
Barrel length	115mm (4.5in)	95mm (3.7in)
Magazine capacity	16 (9mm) or 12 (.40) rounds	

Sphinx 3000 pistols have been manufactured since 2003. In some respects, these pistols are an evolution of the previous Sphinx AT2000, but some features are

Sphinx 3000, partially disassembled. *(Courtesy of Sphinx)*

5.56mm MTE224VA machine pistol.

different. Basic, full-size or Standard pistols are mostly intended for sport shooters in disciplines such as IPSC or IDPA; the more compact Tactical model is better suited for duty or self-defence carry, as it is somewhat shorter and lighter. However, there is so far no information about the actual service use of Sphinx 3000 pistols; this could be attributed to the relatively heavy weight of the all-steel pistol, as well as to its significant price, dictated by the extensive precision machining and close fitting of each gun.

Sphinx 3000 pistols are short-recoil-operated, locked-breech weapons of all-steel design. One interesting aspect is the two-part frame, which is assembled from a grip section and a top section using extra strong glue. The barrel locks into the slide with a single large lug that enters the ejection port; unlocking is controlled by a cam-shaped cut made in the lug under the breech of the barrel. The slide runs on the full-length rails machined on the inside of the frame. The trigger is double-action, with an exposed hammer and either an ambidextrous frame-mounted safety or decocking levers, as well as an automatic firing-pin safety block. Magazines are double stack; the magazine release button is located at the base of the trigger guard. Sights are fixed on combat models; all Sphinx 3000 pistols are fitted with an integral Picatinny rail on the frame below the barrel.

Manual safety Levers on both sides of the frame. Push the lever up into the horizontal position to engage the safety; pull it down to disengage the safety; a red dot will appear on the frame.

Field stripping procedure 1) remove the magazine by pressing the magazine release button; 2) check that the chamber is empty; 3) pull the slide back slightly in order to align the disassembly marks on the slide and the barrel; 4) holding the slide partially retracted, push the slide-stop axis pin inwards from the right, then pull the slide-stop lever out of the frame towards the left; 5) carefully push the slide forwards and out of the frame; 6) remove the return spring with its guide from below the barrel; 7) pull the barrel out of the slide.

Reassemble in reverse order.

MTE 224V/224VA

Trigger type: Double/single action
Calibre/ammunition used: .224V (5.56 × 24mm)
Weight, empty: 1,045g (36.6oz)
Length: 260mm (10.1in)
Barrel length: 165mm (6.4in)
Magazine capacity: 16 or 26 rounds

The MTE 224V pistol is broadly based on the design of the ITM AT-88/Sphinx AT2000 pistols, developed during the 1980s by Martin Thuma. However, the MTE 224V uses unique small-calibre, high-velocity ammunition designated .224V and developed in the Czech Republic by the Calibre Prague Organization. This ammunition is based on the 7.62 × 25mm case necked down to accept a pointed, jacketed bullet of .224 (5.56mm) calibre. The obvious intent of this ammunition and weapon is to provide an alternative to the Belgian FN Five-seveN pistol, with its low recoil, high penetration against body armour and flat bullet trajectory. The MTE 224VA is the selective-fire 'machine pistol' version of the semi-automatic MTE 224V. Both weapons are offered by the Martin Thuma Engineering company of Switzerland, but so far there have been no recorded sales of these weapons to any law-enforcement or military organizations.

The MTE 224V is a simple blowback-operated weapon with a fixed barrel. The pistol is almost entirely made from steel. The trigger is double-action, with an exposed hammer and a frame-mounted safety. Barrels are extended and fluted for better cooling; some versions are also equipped with muzzle brake/compensators. Sights are fixed and the bottom of the frame below the barrel is shaped to provide an integral accessory rail. Magazines are double stack; sixteen-round magazines are a flush fit into the grip, while twenty-six-round magazines extend downwards. MTE 224VA pistols can be provided with an accessory magazine holder that clamps to the rail under the barrel; when inserted into this holder, the spare magazine also serves as a front grip to provide better control of the weapon during fully automatic fire.

Ukraine

After the dissolution of the Soviet Union, the independent Ukraine inherited from the USSR a significant potential in military industry, although there were no small-arms production plants. On the other hand, the Ukraine also retained significant stocks of older, Soviet-era weapons, including large reserve stores full of obsolete firearms dating back to both world wars. Regardless of that, the police and armed forces of the new country required something more modern and potent than the available (and well-worn) Makarov pistols. As national pride and historical issues precluded Ukraine buying arms from Russia, and recognizing the need for a domestically produced, service-type pistol, the state-owned scientific and production organization NPO 'FORT' started development of a new weapon, chambered for widely available 9 × 18 Makarov ammunition. As there were no previous facilities for small arms manufacture, all necessary machinery was purchased from the Czech Republic; barrels for the new pistols were also purchased from Czech manufacturers until around 2002.

First introduced in 1998, the new Fort-12 pistol was an amalgam of previously known ideas. Early pistols were plagued with many teething problems, including poor reliability and short service life, but these pistols are now well established and issued to many Ukrainian law-enforcement organizations. Later on, NPO 'FORT' continued the development of service pistols and tried to produce a more powerful weapon, chambered for 9 × 19 ammunition. Designated Fort-14, this weapon was later converted to 9 × 18 PM ammunition, for unpublished reasons (the most probable causes being either the lack of domestically made ammunition in this calibre or the impossibility of producing a gun strong enough to handle the 9 × 19, or both).

9mm (9 × 18) Fort-17 pistol.

Since 2003, the 9 × 18 Fort-14 has been issued to the Ukrainian police. The most recent product of NPO 'FORT' is the polymer-framed Fort-17 pistol, which is very close to the original Fort-12 except for the frame material. Surprisingly, so far nobody seems to have told the 'FORT' designers about the virtues of a decocking function in a double-action pistol, even if the gun is intended to be carried 'cocked & locked' at times. Other than that, the Fort-17 (as well as its predecessors) looks like a useful police weapon (public ownership of handguns is not permitted in the Ukraine).

While designs from NPO 'FORT' are sound and practical, if a little outdated, there have been more exotic Ukrainian attempts in the same field. The most intriguing small arms 'producer' in Ukraine was the KB-ST (Design Bureau for Special Techniques). This highly controversial organization was established in 1993 and during the late 1990s announced a series of a 'revolutionary' small arms designs, such as sub-machine guns and sniper rifles of outstanding, if not unbelievable, characteristics. However, no foreign

PSh – Shevchenko pistol prototype, 9 × 18 version.

experts ever were able to test these 'wonder weapons', and by now it seems that KB-ST has folded and gone out of business.

One such wonder weapon widely advertised in the Ukrainian literature as 'unique' and 'superior' was the *Pistolet Shevchenko*, or PSh in short. This overly complicated and strange-looking weapon was also an amalgam of previously known ideas, some of which had been tried and rejected fifty or even 100 years ago. To start with, the PSh used a two-stage feed system, borrowed from the 1900-period Gabbet-Fairfax Mars pistol of British origin. This system, in which the cartridge is first pulled back from the magazine and placed on a lifter, then raised in-line with the barrel and fed forwards and into the chamber, was necessary because of the elongated barrel, which protruded back into the action. To further complicate the construction, its designer, Shevchenko, used a very rare gas-operated locking system first (and probably last) used in a WWII-era prototype Horn assault rifle from Germany. This system employs two small gas pistons that are forced sideways from the frame and into locking recesses in the slide. Another strange feature was a hand-cocking device, probably taken from an HK P7 pistol, but, in a twist, instead of a full-hand cocking lever, the PSh employed some form of a second trigger, located below the 'primary' trigger and used to cock the striker when pressed by one middle finger. If you still have not had enough of this design, it also had a ribbed, aluminium jacket on the barrel and a very special sort of slide stop that locked the slide open once there was *one* cartridge left in the gun. The claimed purpose of this arrangement was to allow the shooter

9mm Khortitsa-125 pistol.

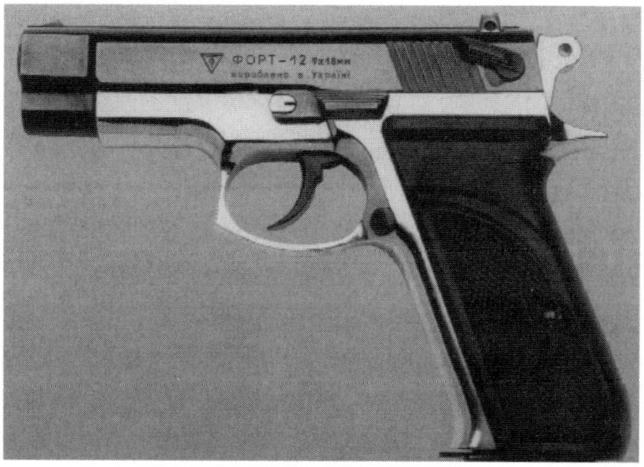

Fort-12.(Courtesy of NPO 'FORT')

to make his 'deliberate last shot' – in his own head, probably. Not surprisingly, this weapon, 'available' in 9 × 18 PM, 9 × 19 and .45 ACP, never went past magazine pages and, probably, a handful of prototypes. Nevertheless, the author decided to include it, if just for the amusement of the readers.

Another design that also never went past the prototype stage was more conventional, but also departed from mainstream design. The *Khortitsa-125* modular pistol was built around a gas-delayed blowback action, borrowed from the German HK P7 pistol, combined with a modular trigger system available in SA, DA/SA and DAO configurations. Developed circa 1998–9 by another state-owned organization, the *Radiopribor* plant, this pistol was designed in calibres such as 9 × 18 PM and 9 × 19. According to available information, the *Khortitsa-125* never went past the prototype or pre-production stage and is not manufactured.

Fort-12

Trigger type: Double/single action
Calibre/ammunition used: 9 × 18 PM
Weight, empty: 830g (29.1oz)
Length: 180mm (7in)
Barrel length: 95mm (3.7in)
Magazine capacity: 12 rounds

The Fort-12 is a blowback-operated pistol of all-steel construction. The barrel is fixed and pinned to the frame and the recoil spring is located below the barrel, around the full-length guide rod. The trigger is double action, with an exposed hammer and a slide-mounted safety. Magazines are double stack; the magazine release button is located at the base of the trigger guard.

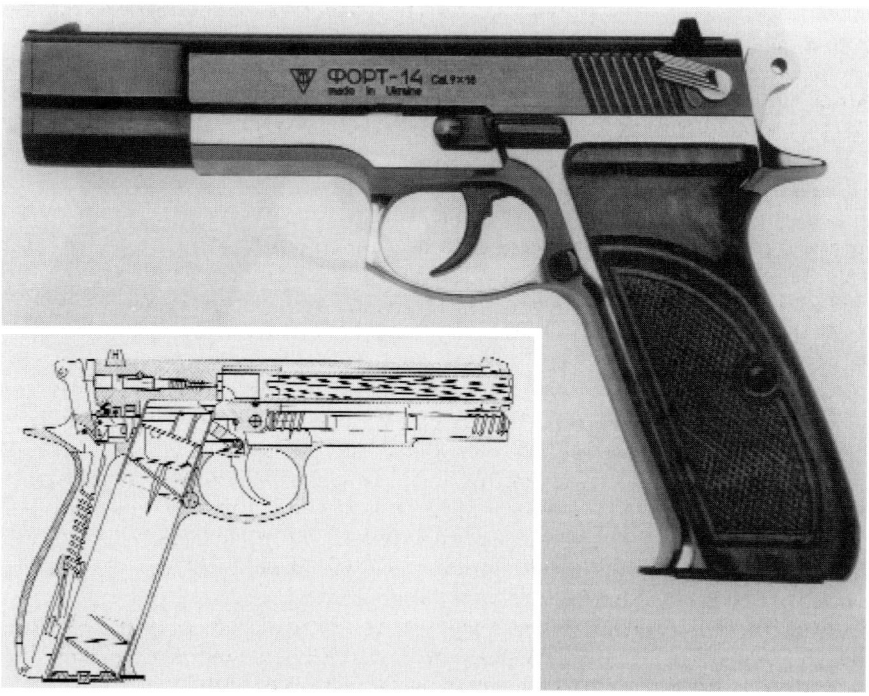

Fort-14 photo and diagram. *(Courtesy of NPO 'FORT')*

Sights are fixed; the front sight is integral to the slide, the rear sight is dovetailed.

Fort-12 pistols are available in a wide variety of finishes, including gold-plated and engraved models that are officially used as presentation or Government award pieces.

Manual safety Lever on the left side of the slide. Push the lever up into the horizontal position to engage the safety; pull it down to disengage the safety; a red dot will appear on the slide. Note: the safety locks the hammer either in the cocked or lowered position.

Field stripping procedure 1) remove the magazine by pressing the magazine release button; 2) check that the chamber is empty; 3) push the slide-stop lever pin inwards from the right, then pull the slide-stop lever out of the frame towards the left; 4) push the slide forwards and out of the frame; 5) remove the return spring with its guide from the slide.

Fort-14

Trigger type: Double/single action
Calibre/ammunition used: 9 × 18 PM
Weight, empty: 920g (32.2oz)
Length: 210mm (8.2in)
Barrel length: 123mm (4.8in)
Magazine capacity: 16 rounds

The Fort-14 is a blowback-operated pistol of all-steel construction. Its barrel is fixed when fired, but can be easily removed during basic disassembly – a feature inherited from the early 9 × 19 locked-breech prototypes. The return spring is located below the barrel. The trigger is double action, with an exposed hammer and a slide-mounted safety. An additional safety feature is provided in the form of an automatic firing-pin block. Magazines are double stack, the magazine release button being located at the base of the trigger guard. Sights are fixed; the front sight is integral to the slide, the rear sight is dovetailed. The Fort-14TP version is provided with an integral accessory rail on the frame, below the barrel. The pistol can also be supplied with an extended, threaded barrel that will accept a proprietary silencer from the same manufacturer.

Manual safety Lever on the left side of the slide. Push the lever up into the horizontal position to engage the safety; pull it down to disengage the safety; a red dot will appear on the slide. Note: the safety locks the hammer either in the cocked or lowered position.

USA

In contrast with the interwar period, the last sixty years have been rather turbulent for the handgun world in the USA. Before looking at this history, sometimes ironic, sometimes rather sorry, we must notice that there are three large segments of the small arms market in the USA. These are the civilian market, the police market and the military market, which, in turn, can also be split into several different sub-segments with their own specific requirements. These market segments are linked in a rather complicated manner and definitely have a significant influence on each other. For example, the military pistol trials of 1985 included weapons which were, for the most part, already commercially available to the civilian shooting public in the USA; the winner of these trials was not only awarded a huge military contract, but also gained significant increases in sales in both civilian and police markets, because of the general opinion that 'the US Army buys the best guns available' (in fact, however, in recent years it has usually bought the least expensive ones).

But let's start at the beginning. The end of World War II left the US military with huge stocks of .45 calibre M1911A1 pistols and a variety of .38 calibre revolvers, most of which were in serviceable condition. The subsequent formation of NATO resulted in a requirement for a new military pistol, compliant with the newly adopted NATO standards and the most modern tactical requirements. In 1947 the US Army Ordnance Technical Committee published a set of requirements for a new military pistol. These requirements included the following basic specifications:

- Ammunition: 9 × 19mm NATO
- Trigger type: Double action
- Weight, empty: No more than 710g (24.9oz)
- Length: No longer than 178mm (6.9in)
- Magazine capacity: 9 rounds minimum

Three companies were asked to submit their prototypes – Colt's Patent Firearms Manufacturing, High Standard Manufacturing and Smith & Wesson Firearms Inc. Of these three, S&W initially refused to participate, considering itself to be more of a revolver specialist; however, a change in the company's top management in 1948 resulted in a change of mind and S&W joined the trials.

During the early 1950s, the US Army began extensive tests in a search for a new 9mm pistol. The standard, against which all new pistols were tested, was

.45 calibre M1911A1 is the world's oldest service pistol still in use. (Courtesy of Cherry's Fine Guns)

the venerable .45 calibre M1911A1 pistol. The following designs were tried:

- modified M1911A1 pistol with shortened barrel and slide, chambered for .45 ACP and 9 × 19.
- experimental 9mm lightweight Browning High Power made in Canada by John Inglis Company.
- 9mm High Standard T3 pistol with a simple blowback action and a radially grooved chamber. This gun was of all-steel construction and thus rather heavy, weighing about 910g (32oz), plus it had a magazine capacity of only seven rounds.
- 9mm Colt T4 pistol, with a short-recoil, locked-breech action and a lightweight aluminium frame. This pistol weighed about 790g (28oz) and had a magazine capacity of thirteen rounds.
- 9mm Smith & Wesson X-series pistol, with an aluminium frame, a single stack eight-round magazine and a locked-breech action.

These trials continued until 1955, when it was decided to terminate them on the grounds of economy, as there were 'enough M1911A1 pistols available in serviceable condition'. The original intention of NATO standardization was ignored and it was decided to keep the existing sidearms for an indefinite period of time. However, these trials had a significant impact on other market segments, as two of the three US companies continued their developments and later brought new pistols to the civilian and police markets. The first was Colt, which introduced its Commander version of the M1911A1 in around 1951. The first Commander pistols had aluminium frames and shortened barrels and were chambered for 9mm ammunition. Later on, Colt introduced Commander-type pistols chambered for other calibres, such as .38 Super, .45 ACP and a rarely encountered 7.65 × 21 Luger (made for the European police market). Surprisingly, Colt did not pursue its promising T4 design any further, being

9mm High Standard T3 prototype.

9mm Colt T4 prototype.

9mm X-series Smith & Wesson double-action prototype.

entirely satisfied with its famous Browning design of 1911 fame.

Smith & Wesson came second, bringing its S&W Double Action Automatic pistol to the market in around 1955 (renamed Model 39 in 1957). Unlike Colt, S&W showed a serious interest in innovation and also manufactured versions of this pistol with an all-steel frame (under the same Model 39 designation) and with a single-action trigger (as the Model 44). However, it took more than ten years before the S&W Model 39 pistol was adopted by any major police force – in around 1967, after extensive trials, Illinois State Police replaced its .357 Magnum S&W Model 13 revolvers with about 2,000 9mm S&W Model 39 semi-automatic

American police officer tries a VP70M machine pistol made in Germany by Heckler & Koch (photograph circa 1970). Note that the duty weapon worn by this officer is a revolver.

pistols. Still, in spite of this, by the mid-1970s most US police departments still carried .38 or .357 calibre revolvers.

It was the war in Vietnam which stimulated several new developments in the field of semi-automatic pistols. First, the Navy SEALs requested a silenced 9mm pistol. Developed on the basis of the S&W Model 39, the new weapon, designated WOX-13 or Mk22 Model 0, had an extended barrel and raised sights so that it could be fired with a silencer attached. These pistols also had their automatic decocking feature disabled, so the guns were able to be carried with the hammer cocked and the safety engaged. Successful field tests of this weapon resulted in a requirement for a more potent pistol, with a higher magazine capacity and rust-resistant construction, which came out circa 1968. To meet this request S&W took off its shelf the design for a high-capacity double-action pistol, based on its M39, which had lain idle since around 1964, and put it into production in 1971 as the Model 59. This pistol emerged too late to see any significant use in South-East Asia, but it served as a platform for many subsequent weapons from the same company.

During the early 1960s a small company, MB Associates, founded by Robert Maynard and Arthur Bill, brought out a new weapon which was intended to revolutionize the world of small arms. Known as the Gyrojet, it was sort of a hand-held, multi-shot rocket launcher. It was advertised as silent and capable of being fired underwater, but in reality Gyrojet weapons turned out to be extremely unreliable and dramatically inaccurate; also, these weapons were rather ineffective at point-blank distances, gaining their full power only at ranges of about 20 (65ft). A few of the Gyrojet 'pistols' made their way to the jungles of Vietnam in the hands of US military officers, who bought them from their own funds, but no successful reports ever appeared and by the late 1960s the Gyrojet project was over.

The year 1972 saw one of the less publicized adoptions – that year, the US Army finally replaced the long-obsolete Colt Model M (M1903 or M1908, in .32 ACP/7.65 × 17 and .380 ACP/9 × 17 respectively) General Officer pistols with the only slightly more modern, but certainly more potent, M15. The M15 pistol was developed at the US Army's Rock Island Armory; each gun was handmade and fitted from the same M1911A1 pistol components that were used to build National Match pistols for the Army. These pistols had cut-down slides, barrels and frames, and packed six rounds of .45 ACP ammunition. The left grip panel of each M15 pistol had a brass plate insert, which was used to engrave the name of the general to whom this gun was issued.

RIA M15 General Officer pistol; note the brass plate ready to accept the engraved name of the owner.

Starting in the early 1960s, the US Air Force (USAF) decided to revert from semi-automatic M1911A1 pistols to revolvers; there were several reasons for that, including the lighter weight of the more recent revolvers with aluminium frames, as well as a simplified manual of arms. Another, although minor, reason for this step back was the fact that existing pistols were worn out. During the Vietnam War, pilots and other personnel of the USAF and US Navy were armed with a wide variety of revolvers, made mostly by Colt and Smith & Wesson. These included 2in and 4in barrel versions, as well as six- and five-shot models, with steel and aluminium frames, and some even with aluminium cylinders. As the available commercial .38 Special ammunition was of rather marginal effectiveness, the USAF developed its own .38 Special round, loaded with a jacketed ball bullet to conform to international conventions. Designated PGU-12/B, this round was still relatively weak in terminal effects and in 1977 the USAF requested additional funds to develop an even more effective revolver load. However, instead of authorizing new expenditure, Congress requested the USAF to carry out a survey of the entire situation with regard to handguns in the US armed forces. The resulting analysis showed that out of about 590,000 assorted handguns listed in US military inventories, about 120,000 were revolvers of various types; also, only 430,000 (about 73 per cent) of those sidearms were in serviceable condition. More than ninety types of handgun ammunition were also listed. As a result of these findings being reported to Congress, it was decided to authorize adoption of an entirely new, modern 9mm semi-automatic pistol for all of the armed forces.

In 1978 the USAF ordnance research laboratory at Eglin AFB began a test and evaluation process in the search for a new military sidearm. There were no formal requirements published for these trials; the USAF just tried to find a better gun from all available pistols, which were tested against .45 calibre M1911A1 semi-automatics and .38 calibre S&W M15 revolvers. The pistols tested included the following 9mm models: the Beretta 92S-1 from Italy; the Colt SSP (Stainless Steel Pistol) from the USA; the FN GP (Browning High Power) and the FN GP Fast Action from Belgium; the Heckler-Koch VP70 and P7 from Germany; the S&W Model 459 from the USA; and the Star 28DA from Spain. From these pistols the USAF selected the Beretta 92S-1 as the winner, but the US Army declared that pistols were an Army prerogative, not Air Force, and also that the USAF tests were 'not scientific'. The results of the initial trials were rejected. Instead, after some bureaucratic work, the US armed forces initiated the Joint Services Small Arms Program (JSSAP), led by the US Army. The objective of JSSAP was to get a new 9mm pistol for all branches of the US Military.

The first obvious step for the JSSAP was to create a set of selection criteria; these included as many as seventy mandatory requirements, plus a number of additional desirable requests. There was nothing wrong in principle with such detailed requirements – at least at first sight. The key problem was that many specific requirements had no practical basis or tactical scenario behind them, appearing to have been conjured out of thin air. Some requirements also were either too specific or too ambiguous. Also, there were no margins in many 'numerical' requirements – for example, if the mandatory durability requirement was 5,000 rounds, then a 5,001-round durability was an acceptable result, but 4,999 rounds was considered as a failure in the test and therefore in the entire trials.

Basic Requirements for the Proposed XM9 Pistol, 1981

Calibre/ammunition	9 × 19mm NATO
Trigger type	Double action
Safety	Ambidextrous manual safety mandatory
Length	220mm (8.6in) maximum
Barrel length	102mm (4in) minimum
Weight	1,300g (45.8oz) loaded maximum
Magazine capacity	10 rounds minimum

In February 1981 the US Army announced that no contender met *all* of the mandatory requirements; the overall favourite was the S&W M459 pistol, with the Beretta M92 running a close second. As a result, the US Army requested more funds for spare parts and ammunition for existing .45 calibre pistols. At the same time, some officials proposed a 'cheap' 9mm alternative in the form of retrofitting the existing M1911A1 pistols with 9mm barrels, slides, springs and magazines.

However, Congress rejected the Army's request for '.45 calibre' funds and requested that the Army should continue trials until a winner could be chosen. Following that decision, the JSSAP slightly relaxed its requirements and in November 1983 new trials were organized. It must be noted that another major flaw of the test procedure was that only a few pistols of each make and model were tested – some tests included

The 9mm Colt SSP (Stainless Steel Pistol) was designed during the late 1960s and then shelved until the XM9 trials. It was then voluntarily withdrawn from the trials by Colt and, like the T4 pistol of 1951–5, was never put into production. This was just another serious mistake on the part of the Colt management. (Courtesy of Cherry's Fine Guns)

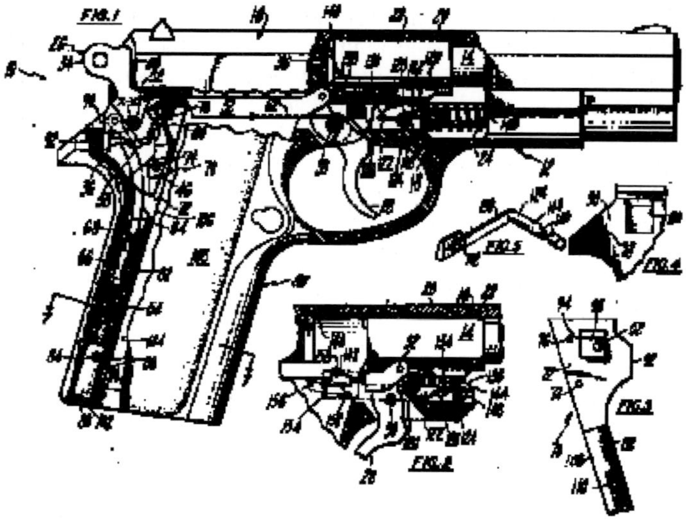

Patent diagram of the Colt SSP.

about three to five guns from every contender and that was certainly not enough to draw statistically viable conclusions about overall reliability, durability and other properties of each system. For comparison, the final stages of the previous US Army pistol trials, which had taken place in 1910–11, included field tests of 200 of each of the competing models and these tests were mostly carried out by actual troops in the field rather than in laboratories.

This stage of trials included pistols from Beretta (Model 92SB-F), Colt (SSP), Fabrique Nationale (HP-DA), Heckler & Koch (P7A13), SIG-Sauer (P226), S&W (M459), Steyr (GB) and Walther (P88). After completion of all the tests, only two designs were selected as 'acceptable' – the Italian Beretta 92SB-F, presented by Beretta USA, and the German-made SIG-Sauer P226, presented by SACO Defence. The third contender, which had failed two critical tests, was the Smith & Wesson 459M; it lost the firing-pin energy test by a margin so small that it was probably the result of an arithmetical rounding error when converting from the NATO-standard metric to US-standard Imperial systems, and the average service life test (the US Army regarded an 'expected' service life of 5,000 rounds as a required minimum for every tested gun, not an average for all tested guns). Because of this, S&W was not allowed to participate in the last stage of trials – the bidding. In January 1985 both SACO and Beretta USA submitted their last and best offers, with Beretta under-bidding SACO by just US $12.85 per pistol, magazines and spare parts kit, or about US $4 million for the entire contract of 315,930 pistols. Subsequently, it was named as the winner of the XM9 trials and Beretta USA received a contract for production of '9mm US Pistol, M9' on April 1985.

For some time, there were many complaints and much speculation about this decision. Smith & Wesson objected in court to its removal from the final stage of the trials, but lost the case on a formal basis. Also, there were serious suspicions that someone either leaked SACO/SIG-Sauer's best and final price offer to Beretta, or, alternatively, that there was a secret agreement between the USA and Italy, so that the Italian company got the US pistol contract in exchange for Italy allowing the US to place its cruise missiles in Sicily.

In short, the investigation conducted by the US General Accounting Office (US GAO) between September 1985 and June 1986 resulted in the following conclusions:

- The pistol trials of 1977, 1981 and 1984 were all flawed in their method and execution; also, the first trials, conducted by the USAF, were possibly biased towards Beretta.
- There was no evidence supporting either the leakage of pricing information to Beretta, or of a secret Italy–USA agreement.
- The elimination of the Smith & Wesson pistol from the final stage of the trials was unfair.
- It was not clear whether the Army should continue to buy the newly adopted Beretta pistols or re-evaluate the trials results.

For those interested in more detailed information on this subject, the author recommends reading the US GAO report GAO/NSIAD-86-122 (dated June 1986), available at no charge from the US GAO website.

The conclusion of the XM9 trials, however, did not end the troubles. During 1987 and 1988, several new M9 pistols experienced terminal slide failures. Some of these failures resulted in injuries to operators. The first two pistols to have such failures were guns from Navy stocks. Several safety warnings were subsequently issued by the Navy and Army regarding the service life of M9 slides, which was initially set at 7,000 rounds, then lowered to 3,000 and finally to 1,000 rounds until replacement. Extensive testing showed that these failures appeared on pistols made both in Italy and in the USA, when anywhere between 4,000 and 25,000 rounds had been fired; also, some pistols which failed were fed only with 9mm NATO ammunition, while some others were used with commercially available 9mm ammunition.

As a result of these failures, which gained much publicity, the Army decided to start the next trials, the XM10, which were initiated in 1989. These trials

With the adoption of the 9mm M9 pistol as the standard sidearm of the US Army, the .45 calibre M15 General Officer pistol was replaced in service with the 9mm M9 General Officer pistol, which is the same as the full-size M9 (US-made Beretta 92FS), but with a brass plate set into the left grip panel to accept the name of the owner and with a special serial number with a 'GO' prefix. (Courtesy of the US Army)

US marines practising with M9 pistols. (Courtesy of the USMC)

included modified Smith & Wesson M459 pistols, new Ruger P85 pistols and current production Beretta 92F/M9 pistols (as Beretta refused to resubmit new samples on the basis that its pistols had already passed the tests). After some testing, the Army confirmed the M9 as the winner and signed another contract with Beretta for more pistols. In the meantime, Beretta had designed a quick fix in the form of an enlarged hammer-pin head, which was used to stop a failed slide from flying into a shooter's face. Also, during the same period, the Phrobis Company, at the request of the US Navy, designed a new, replacement slide with a reinforced 'closed top' design. This design was never put into production, but, considering the money spent on this project, as well as on safety tests and the upgrading of existing M9 pistols to 92FS standard, it seems that the 'lower bidder wins' rule sometimes (really, quite often, if not always) results in higher overall costs in the long term.

In the meantime, some US Navy SEAL units decided to adopt non-standard pistols. During the 1970s, SEAL units supplemented their issue M1911A1 pistols with HK P9S pistols in .45 calibre. Since the late 1980s, Navy SEALs had also bought SIG-Sauer P226 pistols, which were apparently more tolerant of the extra-powerful non-standard ammunition often used by SEAL teams. There were some issues with aluminium frames cracking due to excessive recoil, but at least these problems caused no direct harm to the shooter, unlike some of the early Beretta cracks. Another issue with the new pistols, which the SEALs were quick to note, was insufficient stopping power of

Early version of the HK OHWS .45 pistol, with the original HK-made silencer and LAM/tactical light unit attached. *(Courtesy of Heckler & Koch)*

Colt .45 calibre SOCOM OHWS prototype, also with silencer and LAM. This pistol failed the trials and was never put into production, like several other earlier Colt military-type pistols.

This side-by-side comparison between Mark 23 and USP-C 45 pistols, both made in .45 calibre by Heckler & Koch of Germany for the US market, clearly shows why many US special forces operators prefer the latter pistol over the former. In many cases, the advantages of the longer barrel and the 50 per cent greater magazine capacity (twelve rounds vs eight) are simply not worth the additional pound of weight that the user would have to carry on long missions behind enemy lines. Both pistols are shown with silencers made by Advanced Armaments Co. (USA). *(Courtesy of Oleg Volk)*

the 'new' 9mm ammunition, when compared with the old and venerable .45 ACP. Recognizing the particular and specific needs of various special forces operators, US Special Operations Command (US SOCOM) in around 1991 issued a new requirement for an Offensive Handgun Weapon System, or SOCOM OHWS in short. The new pistol was to combine a modern double-action system with the stopping power of any factory .45 calibre ammunition, as well as extreme reliability, durability and other special capabilities such as ease of silencing.

After initial solicitations, two companies were selected for development of the OHWS – the US-based Colt and the German-based Heckler & Koch. After the first phase of trials, a contract for the second phase of development was awarded to Heckler & Koch and in 1995 US SOCOM officially adopted the HK OHWS pistol as the 'Mark 23 Model 0 pistol'. This large handgun possessed outstanding reliability and durability; it packed serious firepower with its twelve-round magazine and long barrel; and it was capable of firing both standard .45 calibre ball ammunition and JHP +P ammunition without any problems. It was also capable of mounting a tactical light, laser designator and a silencer made in the USA by Knight's Armament Co. However, these features came at the price of excessive bulk and weight. Although looking extremely attractive on magazine pages or at the range, this huge and heavy pistol quickly became cumbersome when carried on the hip during long missions, along with many pounds of other equipment. Not surprisingly, many operators gradually switched to the smaller and lighter HK USP Tactical pistol, which has almost the same firepower and mounts the same sound suppressors, but at a significantly smaller weight. Further field experience showed that an even more compact and lightweight pistol was desirable, with the result that HK USA quickly responded with the recent HK USP Compact Tactical, a comparatively small and light handgun that still holds eight rounds of .45 ACP and has an extended, threaded barrel to accept the standard-issue silencer.

Interestingly enough, this urge for the most modern pistols did not kill the venerable M1911A1. Highly regarded due to its accuracy potential and extreme reliability, the M1911A1 is still going strong in the US Marine Corps, which issues the M1911A1 MEU (SOC) pistols to its forward operations and recon units – MEU (SOC) stands for Marine Expeditionary Unit (Special Operations Capable). These pistols are built on the basis of a standard M1911A1, with several added commercial off-the-shelf components like ambidextrous manual safeties, match triggers and barrels. M1911A1 MEU (SOC) pistols are built at the USMC base at Quantico.

The most recent large-scale operations of the US armed forces in Afghanistan and Iraq also brought major complaints about the standard issue M9 pistols. Many problems were traced to magazines which caused malfunctions in the field, but the major gripe was the insufficient stopping power and lethality of jacketed 9mm bullets. For several years, various experts discussed the possibility of adopting a new military pistol in a more effective calibre, such as .40 or .45, until in the summer of 2005 the US Army officially published a Request for Proposal (RFP) for a new military sidearm, designated the Joint Combat Pistol (JCP). This RFP called for a commercially available, off-the-shelf weapon chambered for none other than the .45 ACP ammunition; it was to be available with a double-action trigger, in versions with and without manual safety, and with standard (eight rounds

Beretta's modified compact version of the military 92FS/M9, the 92FS compact Type M with military-type finish and polymer grips, was one of the contestants for the US XM11 compact pistol trials.
(Courtesy of Bradley W. Taylor)

minimum) and high-capacity (ten rounds minimum) magazines. Other requirements were the possibility of fitting a silencer and a modular grip of adjustable size. It seemed that the Army wished to replace all pistols now in military inventories with a single joint combat pistol, suitable for both defensive (general issue) and offensive (special operations) use. However, the requirement was 'postponed indefinitely' in September 2006.

The last military sidearm worthy of note is the M11 9mm pistol, adopted in 1991. This pistol, which is a US-made SIG-Sauer P228 of German origin, has been adopted as a more compact alternative to the rather bulky M9/Beretta 92FS pistol. It is mostly issued to personnel in the military police and Intelligence, to those who are required to carry their sidearms concealed. It is probably worth nothing that these trials received much less attention from the general public, compared with the previous XM9 trials. The pistols tested against the SIG-Sauer entrant were Beretta's compact 92 Type M with a single-stack magazine and Glock's model 19 (which was dropped in the early stages of the trials).

The police market began its movement towards semi-automatics during the late 1960s and early 1970s, when several large police departments replaced their trustworthy 'sixguns' with more modern double-action semi-automatics in 9mm calibre, such as the S&W Model 39 or Heckler & Koch P7. The adoption of a 9mm pistol by the US military spurred this rearmament; many police departments adopted Beretta Model 92 pistols, or Smith & Wesson second-generation 9mm pistols like the full-sized Models 459/559 or compact Model 469. The mid-1980s also saw the rocketing success of Glock pistols. Thanks to their outstanding reliability, durability and simplicity, as well as very aggressive marketing and clever usage of the 'lower bidder wins' rule, the Austrian company quickly captured a significant portion of the American law-enforcement market.

However, the 'revolution of Wonder Nines' did not last for long. The infamous Miami massacre of 1986, when FBI agents armed with 9mm pistols failed to stop two heavily armed bank robbers, resulted in a search for a more effective round. After serious examination, the FBI selected the 10mm Auto (10 × 25) as its next service calibre. This ammunition had been on the market since about 1983, initially made by Norma of Sweden, and was first used in a double-action Bren Ten pistol, made by Dornhaus & Dixon Co. Production and management problems plagued Bren Ten manufacture during its short production life and for several years the only factory that made a 10mm pistol was the Colt with its Delta Elite pistol, which was based on the tried and proven M1911 design. However, with the FBI decision to adopt the 10mm as the next standard issue calibre, Smith & Wesson quickly adopted this round and offered several pistols in its 'third generation' line of semi-automatics. Of these, the most famous were the full-sized Model 1006 and the compact Model 1076, which was selected as a duty gun for FBI agents. Both were excellent weapons, although somewhat heavy. Also, with standard Norma-made loads the recoil was quite stout and so, in order to achieve faster and more accurate shooting, by the late 1980s the FBI had adopted a seriously 'downloaded' version of the 10mm cartridge.

Since the 25mm case of the standard 10mm Auto cartridge was overly big for such a load, S&W, along with Winchester/Olin, simply cut down the 10mm case by about 3mm (0.12in), to create a shorter round, which still fired a 10mm/.40 calibre bullet with an energy and stopping power superior to most 9mm ammunition on the market. This cartridge, which was dubbed the .40 Smith & Wesson, filled the niche between 'big' .45 and 'small' 9mm; it allowed for a relatively high magazine capacity and serious stopping power, creating about an ideal (or, it would be better to say, about the best compromise) police and self-defence round. Smith & Wesson quickly brought out pistols in this calibre; Beretta and Glock also followed suit (actually Glock offered its first .40 calibre pistol several months before the S&W), as well as most other makers. It is worth nothing that the .40 S&W, as well as the .357 SIG round derived from it several years later, are very similar in the size and ballistic properties to the almost forgotten .41 AE and 9mm AE rounds, developed in the mid-1980s by Evan Wilding of Action Arms Inc.

As a result, by the late 1990s most American law-enforcement officers (LE officers) were armed with semi-automatic pistols. Interestingly, most prevailing brands were of non-American origin – Glock, Beretta and SIG-Sauer (although the two latter companies manufactured guns in-country) – although some police departments still issue (or authorize for individual officers' purchase) Smith & Wesson pistols and, in a few cases, Rugers. Only a few police departments still authorize the issue of single-action pistols for general duty use. On the other hand, the specially trained and equipped SWAT (Special Weapons and Tactics), HRT (Hostage Rescue Teams) and other such units still in many cases prefer single-action pistols for their extreme first shot accuracy and, sometimes, sheer 'cool' factor. Most of these single-action semi-automatics are modern recreations of the eternal John Browning design of 1911, usually with all the 'bells and whistles' of a modern high-end target gun. The never-ending cult of the 1911 can be attributed both to the outstanding properties of the original design and to the typical American love for self-created 'American legends'. Most surprisingly, the Colt Company, which brought out this gun in 1911 and still makes it, over time lost much of its market to more successful manufacturers. Today, dozens of companies make and import versions of the 1911, from the inexpensive copies made in the Philippines to the very expensive and fine-tuned offerings from both large and small companies like Les Baer, Wilson Combat, Kimber and so on. But, even worse than that, the Colt Company lost most of its innovations; many promising combat pistol designs were shelved and never brought to market after their failure with the military; and those that were manufactured turned out to be very poor contenders, like the Double Eagle or the infamous All American 2000.

The late 1990s and first years of the twenty-first century saw some changes in American law enforcement with regard to pistols. First, the popularity of Austrian Glock pistols gradually declined, under pressure from two other European-based makers, Heckler & Koch and SIG-Sauer. The key reason for this decline was, in the opinion of the author, the overly simplistic manual of arms that typically accompanied Glock pistols, combined with the inefficient firearm training provided to many American LE officers. This resulted in a considerable number of accidental or negligent discharges, often with fatal results. Introduction of the so-called 'New York trigger package', which increased the weight of the trigger pull in Glock pistols, did not

The 10mm Bren Ten pistol by Dornhaus & Dixon was the first production pistol chambered for the 10mm automatic cartridge (10 × 25). Its design was based on the famous Czechoslovak CZ 75 pistol.

The 10mm Colt Delta Elite pistol was just another variation of the proven US M1911A1 design, chambered for the then-new cartridge, which never really caught on with service users. On the other hand, the 10mm Auto has a strong following among some civilian US shooters. (Courtesy of Cherry's Fine Guns)

help their case either, as both HK and SIG-Sauer had refined their DAO pistols to the point where they provided accuracy at least no worse than that of the Glock, but with the increased safety of the longer trigger pull typical of DAO pistols. There were other reasons as well, such as a lack of innovation on Glock's part during the last decade, as well as some problems with their .40 calibre pistols (traced to the fact that .40 calibre Glock pistols were based on their 9mm designs, while many competitors designed their guns in .40 and then rebarrelled them down to 9mm, producing potentially stronger and more durable guns).

The most recent (2005) huge orders for pistols, awarded by the US Department of Homeland Security to Heckler & Koch and SIGARMS, worth in total more than 50 million dollars, could produce the impression that America has all but lost its pistol manufacturing industry. Of course, this impression would be false. While the Colt Company still makes only 1911-style pistols, Smith & Wesson continues to produce excellent service-type firearms in a wide variety of calibres. In mid-2005, this company announced its newest combat-type pistol, the Military and Police, which is to carry on this legendary name, inherited from the most successful double-action revolver ever made (also from S&W). This service-type pistol possesses all the modern features, including a polymer frame, a DAO-type trigger with a very smooth pull, a modular design and so on; it remains to be seen whether this company will regain the share of the American police market it once held. The other player that must not be forgotten is the Sturm, Ruger & Company, which also makes very strong, sturdy, and relatively inexpensive guns. Ruger centrefire pistols are serious tools of trade, although they may look a little less 'modern' or 'sexy' than many rivals. These pistols are issued to several police departments and other law-enforcement organizations in the USA, as well as in some other countries.

So far, we have mentioned mostly large companies, both domestic and foreign, but a characteristic of the US firearms market is that there are also hundreds of smaller players, from mid-size companies down to one-man gunsmithing workshops. Some of these companies make pistols of more or less indigenous design, with the intention of selling them to civilian shooters as well as to law-enforcement officers who are permitted to carry their own weapons on duty, or for off-duty carry. Some companies are just engaged in the repair, servicing and tuning of weapons made by larger companies. For example, there is a huge industry at work in the '1911' market, which makes parts for, assembles and customizes pistols of M1911A1 pattern. While most of this industry is targeted at the civilian sector of the market, some service users can also benefit from this, as in the case with the USMC M1911

The .45 calibre Springfield TRP Operator is the standard sidearm of the FBI Hostage Rescue Team. Based on the old and faithful M1911A1 design, this pistol is known for great accuracy and reliability.
(Courtesy of Springfield Armory)

Imitation is the sincerest form of flattery. Two former rivals of Colt now make 1911-pattern pistols. Top is the SW1911 from Smith & Wesson; bottom is the GSR from SIGARMS.

The Smith & Wesson SW99 pistol is an example of modern globalization. Made partly in Germany and partly in the USA, this is a slightly modified version of the Walther P99. Unlike the original German gun (which is chambered in 9mm and .40 S&W), the SW99 is also available in .45 calibre. (Courtesy of Cherry's Fine Guns)

The .45 calibre Ruger P345 pistol is the most recent product of Sturm, Ruger & Co. This weapon can serve either as a duty weapon, or as a self-defence sidearm for anyone who needs a weapon that is potent, yet relatively light and comfortable to carry and use. (Courtesy of Sturm, Ruger & Co.)

MEU (SOC) pistols that use many commercially available parts.

During recent decades there have been some attempts by smaller companies to produce service-type pistols. One such attempt worthy of note was the ASP pistol, developed by the New York gunsmith Paris Theodore at the request of an undisclosed 'US Government agency', which required a compact, yet powerful, sidearm for use by undercover agents worldwide. Paris Theodore started with the commercially available S&W Model 39 pistol, then cut it down in almost every dimension. The pistol was made as compact and snag-proof as possible, with patented Guttersnipe sights and translucent polymer grip panels. It is not known how many (if any) ASP pistols were actually shipped to the US Government, but in around 1983 Armament Systems & Procedures Co. produced several hundreds of ASP pistols for the open market.

Another interesting and equally rare conversion was the product of Louis Seecamp, made for full- or Commander-sized Colt M1911 pistols. In 1973 Seecamp patented his conversion that turned single-action Colts into the more 'trendy' double action. This conversion included a second trigger, which was placed ahead of the shortened standard single-action trigger. Through the linkage, set into a cut-out on the right side of the frame, the front trigger was linked to the hammer and, once pulled, it cocked the hammer

9mm ASP pistol was a slim and 'aerodynamically shaped' weapon, based on the proven Smith & Wesson M39 semi-automatic pistol cut down to size. The patented one-piece Guttersnipe sight is shown next to the gun.

.45 calibre Colt Commander pistol converted to double action by Louis Seecamp. Note that it actually has two triggers, one in front of another.

until it was engaged by the standard sear. Once the hammer was cocked, a further pull on the front trigger pushed the 'rear' trigger to release the hammer in a typical single-action manner. Such conversions were made by the Omega Defense Industries Company, and sold under the Viking Combat trademark, as well as under Seecamp's own name.

Browning BDM

Trigger type: Double/single action or double-action only, selectable
Calibre/ammunition used: 9 × 19mm Para
Weight, empty: 870g (30.5oz)
Length: 200mm (7.8in)
Barrel length: 120mm (4.7in)
Magazine capacity: 15 rounds

The Browning BDM bears the distinction of being one of slimmest high-capacity 9mm handguns. It is also unusual in that it is the product of neither of John Browning's heritage strongholds, Colt of the USA or FN of Belgium. The BDM is the creation of the US-based Browning Arms Company, which, by the late 1980s, was an importing agent for FN Herstal. Unfortunately, at that time FN dropped its BDA9 double-action pistol, despite there being a strong market for such weapons. Recognizing the niche, Browning Arms decided to bring its own design to the US market, handing the task to the Czechoslovak-born designer Peter Sodoma, who designed the new pistol between 1988 and 1991. Initially developed and manufactured by Arms Technology Inc. of Utah under contract from Browning Arms, it was later made at FN Manufacturing, Inc., the US-based subsidiary of FN Herstal, which at that time was making M16 rifles and M240 and M249 machine guns for the US Government. With this pistol, Browning Arms tried to cover all bases within the service pistol market, as it permits either traditional DA/SA or revolver-type DAO fire, without any modification. However, this complicated feature was also one of the weak points of the design; to cure this, during the late 1990s Browning Arms tried to market BRM and BPM pistols, which were variants of the BDM (which stood for Browning Double Model), limited to only one mode each – DAO (BRM) or DA/SA (BPM). Neither design succeeded, however, and by the turn of century these guns had been discontinued. There were rumours that the US Secret Service looked seriously into adopting this pistol during the early 1990s, but finally rejected it for undisclosed reasons.

The Browning BDM is a short-recoil-operated, locked-breech pistol of all-steel construction. It uses a modified Browning-type locking system; the barrel engages the slide with a single large lug that enters the ejection window to lock. Unlocking is controlled by a cam-shaped projection below the breech part of the barrel. The trigger and safety arrangements were the most unusual features of this pistol. The BDM has a double-action trigger with an exposed hammer and frame-mounted ambidextrous safety/decocker; switching between DA/SA and DAO modes is achieved by rotating a switch, made in the form of a screw head, built into the left side of the slide. The switch has two

Browning BDM.

positions, marked with the letters 'P' (Pistol, DA/SA) and 'R' (Revolver, DAO). To change the mode, the shooter uses a flat screwdriver or another object with a similar flat tip. The sights are fixed, with the rear dovetailed and the front pinned to the slide. Magazines are double stack, with the magazine release button located at the base of the trigger guard.

Manual safety Ambidextrous levers at both sides of the frame. Pull the lever down to engage the safety; push the lever up to disengage the safety; a red dot will appear on the frame. When applied, the safety also drops the hammer down to half cock.

Field stripping procedure 1) remove the magazine by pressing the magazine release button; 2) check that the chamber is empty; 3) pull the slide all the way back and lock it there with the slide stop; 4) rotate downwards the disassembly lever, which is located at the left side of the frame, above the trigger; 5) release the slide and carefully push it forwards and out of the frame; 6) remove the return spring assembly from below the barrel; 7) remove the barrel from the slide.

Reassemble in reverse order.

.45 ACP US issue M1911 pistol. (Courtesy of Cherry's Fine Guns)

Colt Government Model/M1911A1

Generally known as the 1911, this pistol was developed by John Moses Browning and Colt's Patent Firearms Manufacturing Company between 1905 and 1911, for the US armed forces. Adopted in 1911 as the US Pistol, .45 calibre, Model of 1911, it served as a standard sidearm to the US military until 1985, with only minor modifications applied in 1926 (adopted as the M1911A1). Before 1941, these pistols were made mostly by Colt and the Government-owned Springfield Armory; Colt made pistols for both military and commercial consumption, while Springfield worked only for the military. These pistols were also widely exported for foreign military sales, as well as manufactured under licence in Argentina and Norway. During World War II, the US Government made contracts with several other manufacturers, such as Remington-Rand, Singer Sewing Machines, Ithaca, and so on. In total, more than one million M1911A1 pistols were made between 1941–5, with the result that, after the war, the US military never bought any more of these guns. It must be noted that the US police was always a minor user of M1911A1-type pistols, known to non-military markets as the 'Government model'; most police forces switched from their revolvers directly to double-action semi-automatics.

During this time, Colt produced a wide variety of pistols based on the basic M1911A1 design, in a number of calibres, finishes and sizes. First to come was the Commander model that resulted from the US Army trials of 1951–5 to select a 9mm pistol. Early Commanders resembled the standard M1911A1, but with a lightweight, aluminium alloy frame and with shortened barrels and slides. Initially available in 9 × 19, these were later also manufactured in other calibres

Colt Government Model/M1911A1

	M1911A1/Government model	Commander	Officers model
Trigger type	Single action		
Calibre/ammunition used	.45 ACP; also 10mm Auto, .38 Super and 9 × 19mm		
Weight, empty	1,075g (37.6oz)	809g (28.3oz)	700g (24.5oz)
Length	216mm (8.4in)	197mm (7.7in)	171mm (6.7in)
Barrel length	127mm (4.95in)	108mm (4.2in)	76mm (2.96in)
Magazine capacity	7 rounds	7 rounds	6 rounds

.45 ACP M1911A1.
(Courtesy of Cherry's Fine Guns)

.45 ACP Colt Government model Series 80 Stainless.

Colt Government model Series 80, partially disassembled.

such as .45, .38 Super and even the relatively rare 7.65 × 21 Luger. Later on, following the introduction of the M15 pistol, designed by the Rock Island Armory in the early 1970s, Colt brought out the even shorter Officers model pistols with a shortened grip that held a six-round magazine, as well as the Defender model, which has an Officers-style short barrel with a standard-size grip that holds a seven-round magazine. For many decades, Colt was the major maker of 1911-style pistols in the USA; however, various problems plagued this company in the 1980s, which resulted in a number of other companies bringing their own 1911-style pistols to the market. As of today, a great many companies both in the USA and abroad make pistols to the M1911 pattern; by any measure, the M1911 could be considered as one of the most popular, successful and influential pistol designs of the twentieth century.

As said above, a great many versions of the M1911 were and still are manufactured in the world and it is almost impossible to describe all modifications and deviations from the 'true M1911' pattern. So only a few of the most important versions of the Colt will be mentioned below.

The first modification, the M1911A1, included mostly ergonomic changes, based on initial experiences with the troops. These included a shorter trigger, along with relief cuts on the frame for the trigger finger; an improved front sight; an elongated hammer spur; and an arched mainspring housing. For many years, this remained the standard, but in the early 1970s Colt introduced the collet-type barrel bushing in its Series 70 pistol. This was designed to improve the accuracy of the pistol through the more consistent fit between the barrel and the bushing. In 1983, Colt introduced its Series 80 pistols, which were fitted with an automatic firing-pin safety, and in around 1991 Colt brought out the M1991 series of pistols, which were similar in design to the Series 80 pistols but relied on more modern (and less expensive) production techniques.

To mention just a few of the companies that make M1911-type pistols now, this is the list in alphabetical order: Armscor (Philippines), Colt, Dan Wesson, IMBEL (Brazil), Kimber, Les Baer, Para-Ordnance (Canada), SIGARMS, Smith & Wesson, Tanfoglio (Italy), Taurus (Brazil), Valtro (Italy), and Wilson Combat.

Countries which have issued M1911-type pistols as standard military sidearms include Argentina, Brazil, Norway and the USA. Also, many countries used M1911-type pistols as non-standard-issue weapons; for example Britain received many such pistols during both world wars; Tsarist Russia bought several thousands of Colt-made M1911 pistols for police use before 1917 and during World War II the USSR received many M1911A1 pistols through the Lend-Lease programme (although these pistols apparently were never issued to Soviet troops and may still lie somewhere in the stocks of the Russian or Ukrainian armies).

The M1911A1 pistol is a short-recoil-operated, locked-breech weapon of all-steel construction. It uses the classic Browning system of locking, in which the barrel engages the slide with two lugs, machined on the

.45 ACP Colt Commander pistol.

(Courtesy of Cherry's Fine Guns)

top of the barrel, that enter appropriate recesses on the underside of the slide. Barrel locking and unlocking upon recoil is controlled by the swinging link, which connects the barrel to the frame (via the axis pin of the slide stop release lever). The muzzle of the barrel is supported in the slide by the removable barrel bushing. The trigger is of the single-action type, with an exposed hammer and a frame-mounted manual safety. Additional safety is provided by an automatic grip safety and an out-of-battery disconnector safety. Colt Series 80 and M1991 pistols are also fitted with an automatic firing-pin safety. Magazines are single-stack, the magazine release button being located at the base of the trigger guard. The sights on most service models are fixed; on some modern guns the rear or both sights can be dovetailed to the slide.

Manual safety Lever at the left side of the frame (ambidextrous on some late production guns). Push the lever up to engage the safety; pull the lever down to disengage the safety. When applied, the safety also locks the slide.

Field stripping procedure 1) remove the magazine by pressing the magazine release button; 2) check that the chamber is empty; 3) push the recoil spring (located just below the muzzle) inwards, until the barrel bushing is free to rotate; 4) rotate the barrel bushing clockwise until it will clear the exit for the recoil spring plug; take care to stop the plug from flying forwards out of the slide; 5) remove the return spring and its plug by pulling them forwards out of the slide; 6) push the slide-stop axis pin inwards from the right side of the frame, then pull the slide-stop lever out of the frame to the left; 7) push the slide forwards and out of the frame; 8) turn the barrel bushing counter-clockwise, then pull it out; 9) while holding the slide upside-down, fold the barrel link forwards (towards the muzzle), then pull the barrel forwards and out of the slide by the muzzle end.

Reassemble in reverse order.

Colt Double Eagle

Trigger type: Double action
Calibre/ammunition used: .45 ACP, also .40 S&W and 10mm Auto
Weight, empty: 1,205g (42.2oz)
Length: 216mm (8.4in)
Barrel length: 127mm (4.95in)
Magazine capacity: 8 rounds (.45)

The Colt Double Eagle pistol was manufactured between 1990 and 1997. Unfortunately for the company, the Double Eagle came to the market far too late and was regarded as an outdated design, with its heavy, all-steel construction and single-stack magazine. It was available in standard full-size, as well as in the more compact Commander and Officers versions, in several calibres, but none of the versions succeeded. One ironic aspect of this otherwise unremarkable design is that it became quite popular as a pattern for making tear-gas and blank-firing copies in Germany and Italy, as well as airsoft copies (toy air guns that fired 6mm plastic ball bullets) in South-East Asia.

The Colt Double Eagle is short-recoil-operated, locked-breech pistol. It uses the Browning-type locking system, copied from the Colt M1911 pistol,

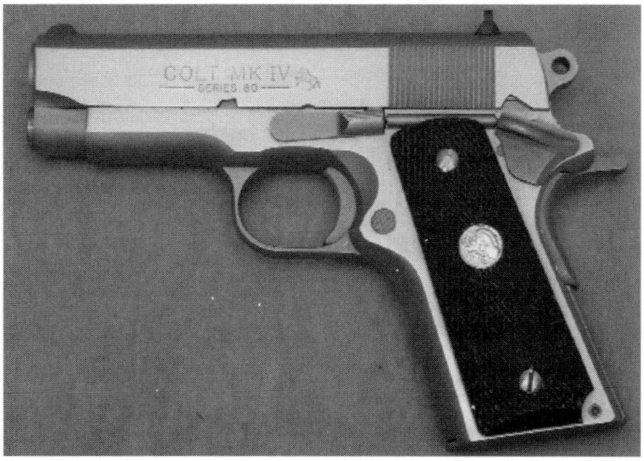

.45 ACP Colt Officers model pistol. (Courtesy of Cherry's Fine Guns)

Colt Double Eagle pistol in .45 ACP. (Courtesy of Cherry's Fine Guns)

All American Model 2000 pistol. (Courtesy of Cherry's Fine Guns)

with its swinging link. The trigger is double action, with an exposed hammer and a frame-mounted decocking lever. Additional safety is provided by the automatic firing-pin block. Magazines are single-stack, the magazine release button being located at the base of the trigger guard. Sights are either fixed or adjustable, dovetailed to the slide.

Manual safety None; decocking lever at the left side of the frame, in the cut of the grip panel.

Field stripping procedure 1) remove the magazine by pressing the magazine release button; 2) check that the chamber is empty; 3) push the recoil spring plug (located just below the muzzle) inwards, until the barrel bushing is free to rotate; 4) rotate the barrel bushing clockwise until it will clear the exit for the recoil spring plug; take care to stop the plug from flying forwards out of the slide; 5) remove the return spring and its plug by pulling them forwards out of the slide; 6) push the slide-stop axis pin inwards from the right side of the frame, then pull the slide-stop lever out of the frame to the left; 7) push the slide forwards and out of the frame; 8) turn the barrel bushing counter-clockwise, then pull it out; 9) while holding the slide upside down, fold the barrel link forwards (towards the muzzle), then pull the barrel forwards and out of the slide by the muzzle end.

Reassemble in reverse order.

Colt All American 2000
Trigger type: Double-action only
Calibre/ammunition used: 9 × 19mm Para
Weight, empty: 800g (28oz)
Length: 197mm (7.7in)
Barrel length: 114mm (4.5in)
Magazine capacity: 15 rounds

The All American 2000 pistol was designed by Reed Knight, Jr and Eugene Stoner (of M16 fame), at the time both working at Knight's Armament Co. The design of the Knight's Model 2000 pistol was purchased by Colt in an attempt to bring to the market a more or less up-to-date 9mm design, suitable for police use. However, the pistol, as introduced in 1991, was plagued by serious reliability and safety issues, combined with a very heavy double-action-only trigger. Two types of Model 2000 pistol were made, with a polymer or aluminium frame, but neither sold well. As a result of poor sales, and in the view of its financial problems, Colt discontinued the All American 2000 in 1993, so the initially advertised .40 and .45 calibre versions never emerged.

The All American 2000 is a short-recoil-operated, locked-breech design using rotating barrel locking. The barrel has five radial locking lugs that securely engage the slide; upon recoil, a separate lug on the underside of the barrel follows a cam-shaped cut in the frame insert, thus rotating the barrel to unlock it from

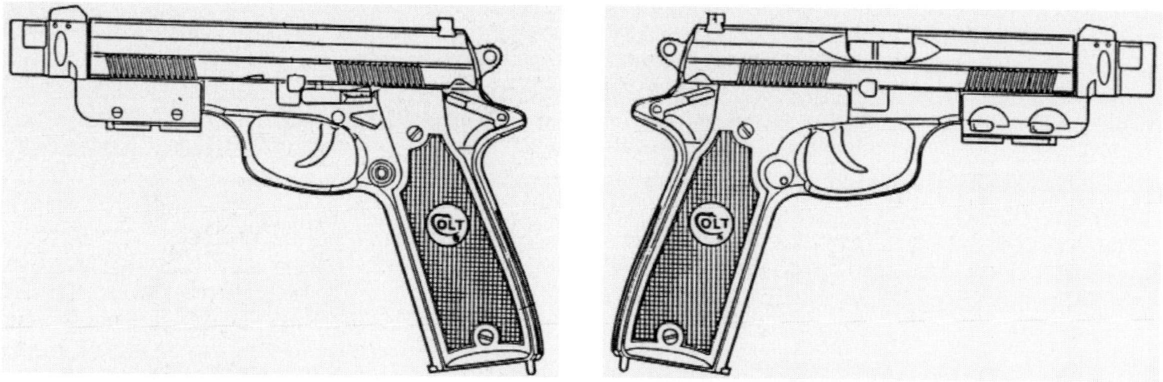

.45 calibre Colt OHWS pistol with muzzle brake/compensator installed.
(Courtesy of Colt, drawing from the original manual)

the slide. The trigger is of the double-action-only type, striker-fired, with an automatic trigger safety. Despite obvious attempts to improve the trigger pull by introducing roller bearings to the design, this pistol has a very poor (heavy, long and uneven) trigger pull, resulting in poor accuracy. Magazines are of double-stack design, the magazine release button being located at the base of the trigger guard. Sights are fixed, the rear sight blade being dovetailed to the slide.

Manual safety None

Colt SOCOM OHWS

Trigger type: Double/single action
Calibre/ammunition used: .45 ACP
Weight, empty: 1,510g (52.9oz) empty
Length: 249mm (9.7in)
Barrel length: 121mm (4.7in)
Magazine capacity: 10 rounds

This prototype pistol was developed during the early 1990s under the US SOCOM OHWS – Offensive Handgun Weapon System – programme. It combined features from several other Colt designs, such as the M1911 (manual safety), Double Eagle (double-action trigger with decocking) and All American 2000 (rotating barrel locking). There were also a few new features, but overall the weapon was heavy and bulky and it eventually lost to the other OHWS candidate, from the German company Heckler & Koch.

One interesting feature of the Colt OHWS was that various muzzle attachments (including the silencer, as required by OHWS documents, and muzzle compensator) were actually fixed to the frame rather than the barrel, because the rotating barrel action is quite sensitive to the weight of the moving parts. For this reason, the frame under the barrel has been fitted with special rails, which could accept a special mount for a muzzle compensator or silencer; in turn, the bottom of this mount has been provided with its own mounting points to attach tactical lights or laser aiming modules.

The Colt OHWS is a short-recoil-operated, locked-breech pistol made almost entirely from stainless steel. It uses rotating barrel locking, similar to that found in the All American 2000 pistol described above. The trigger is double action, with an exposed hammer and separate manual safety and decocking levers. Sights are fixed, with the rear sight dovetailed into the slide and the front sight mounted on the muzzle device, which was attached to the frame. Magazines are single stack and, as the required minimum magazine capacity is ten rounds, the grip length is significant to say the least. The magazine release button is located at the base of the trigger guard.

Manual safety Ambidextrous levers located at the frame, above the grip panels. Push the lever up to engage the safety; pull the lever down to disengage the safety. The safety works only with the hammer in the cocked position. Also, the pistol is fitted with a separate decocking lever, located in a cut made in the left grip panel, behind the trigger.

Smith & Wesson Models 39, 59 and Later

The Smith & Wesson Company, previously known for its excellent line of revolvers, joined the world of semi-automatic pistols in the mid-1950s, when the company

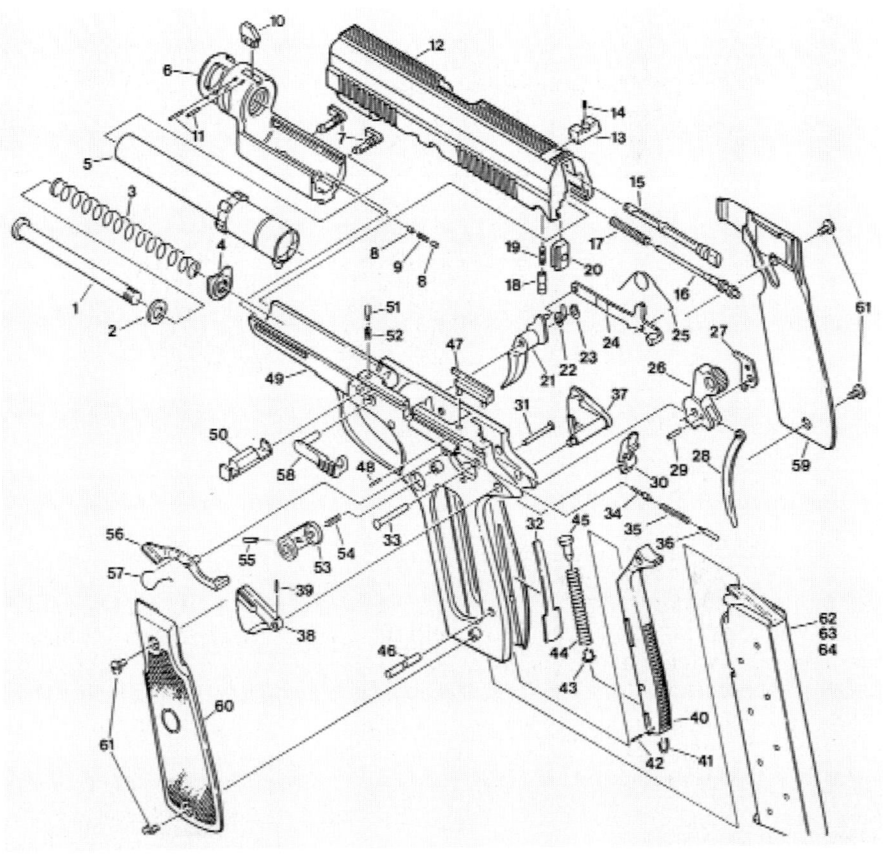

Colt OHWS diagram. *(Courtesy of Colt, drawing from the original manual)*

The 9mm S&W Model 39 pistol represented the first generation in the long and successful line of post-war semi-automatic pistols from the Smith and Wesson company. *(Courtesy of Cherry's Fine Guns)*

9mm S&W Model 59, a first-generation pistol with a double-stack magazine. *(Courtesy of Cherry's Fine Guns)*

Technical Data for Some S&W Pistols

Model	Generation	Calibre	Length	Barrel length	Weight, empty	Magazine capacity, rounds	Comments
39	First	9 × 19mm	192mm (7.5in)	102mm (4in)	780g (27.3oz)	8	
59	First	9 × 19mm	192mm (7.5in)	102mm (4in)	840g (29.4oz)	14	
459	Second	9 × 19mm	192mm (7.5in)	102mm (4in)	n/a	14	
559	Second	9 × 19mm	192mm (7.5in)	102mm (4in)	n/a	14	
659	Second	9 × 19mm	192mm (7.5in)	102mm (4in)	n/a	14	
469	Second	9 × 19mm	175mm (6.8in)	89mm (3.5in)	730g (25.6oz)	12	compact
669	Second	9 × 19mm	175mm (6.8in)	89mm (3.5in)	745g (26.1oz)	12	compact
457	Second	.45 ACP	184mm (7.2in)	95mm (3.7in)	812g (28.6oz)	7	
645	Second	.45 ACP	225mm (8.8in)	127mm (5in)	1,064g (37.6oz)	8	
908	Third	9 × 19mm	173mm (6.8in)	89mm (3.5in)	670g (23.5oz)	8	compact
915 910	Third	9 × 19mm	187mm (7.3in)	102mm (4in)	790g (27.7oz)	15 (10)	
3913	Third	9 × 19mm	171mm (6.7in)	89mm (3.7in)	700g (24.5oz)	8	compact
5906	Third	9 × 19mm	190mm (7.4in)	102mm (4in)	1,086g (38oz)	15	
5903	Third	9 × 19mm	190mm (7.4in)	102mm (4in)	819g (28.7oz)	15	
4013	Third	.40 SW	171mm (6.7in)	89mm (3.5in)	760g (26.6oz)	9	compact
4006	Third	.40 SW	190mm (7.4in)	102mm (4in)	1,072g (37.5oz)	11	
4003	Third	.40 SW	190mm (7.4in)	102mm (4in)	808g (28.3oz)	11	
1006	Third	10mm auto	225mm (8.8in)	127mm (5in)	n/a	9	
1076	Third	10mm auto	200mm (7.8in)	102mm (4in)	1,136g (39.8oz)	9	
4513	Third	.45 ACP	196mm (7.6in)	95mm (3.7in)	811g (28.4oz)	7	compact
4566	Third	.45 ACP	200mm (7.8in)	108mm (4.2in)	1,108g (38.8oz)	8	
4563	Third	.45 ACP	200mm (7.8in)	108mm (4.2in)	856g (30oz)	8	

decided to bring to the open market the nine-millimetre double-action weapon, which it had initially developed for the US Army trials. This pistol, soon designated Model 39, was marketed in several minor versions until about 1980. Other than the basic double-action Model 39 with an aluminium frame, S&W also briefly marketed the same model but with a heavier steel frame (fewer than 1,000 of these pistols were made between 1955–67), as well as the even more rare Model 44 with a single-action trigger. In 1971, S&W brought out a high-capacity version of the Model 39, designated Model 59. This was almost identical to the Model 39 except for a thicker grip with a straight backstrap, which accommodated a double-stack fourteen-round magazine. The same gun with a steel frame has been made in very limited numbers (slightly more than 100 pieces total) as Models 47 or 147.

By 1980, Smith & Wesson had replaced these first-generation guns with the second generation of pistols. Recognizable by the three-digit model numbers, the second generation introduced various materials for the frames, slides and other parts, including the then-new stainless steel. Also, compact (shorter frame and barrel), yet high-capacity weapons were added

Model 39 pistol, partially disassembled.

The 9mm Model 659 pistol was an all-stainless service-type second-generation gun with a double-stack magazine; it was quite popular as a police duty weapon. (Courtesy of Cherry's Fine Guns)

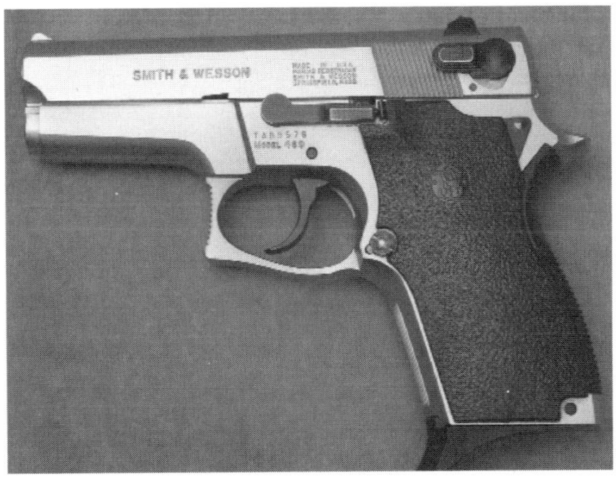

The 9mm Model 649 pistol also represented the second generation of S&W semi-automatic pistols. This compact weapon was fairly popular as a police and civilian sidearm, well suited for concealed carry.
(Courtesy of Cherry's Fine Guns)

The 10mm Model 1076 was developed for FBI use; this was a formidable weapon, but somewhat heavy by modern standards. Note that instead of the S&W-typical slide safety/decocking lever it has a frame-mounted decocker.

Another third-generation S&W semi-automatic pistol, the Model 4506. This large, service-type weapon fires .45 calibre ammunition, very popular in the USA.

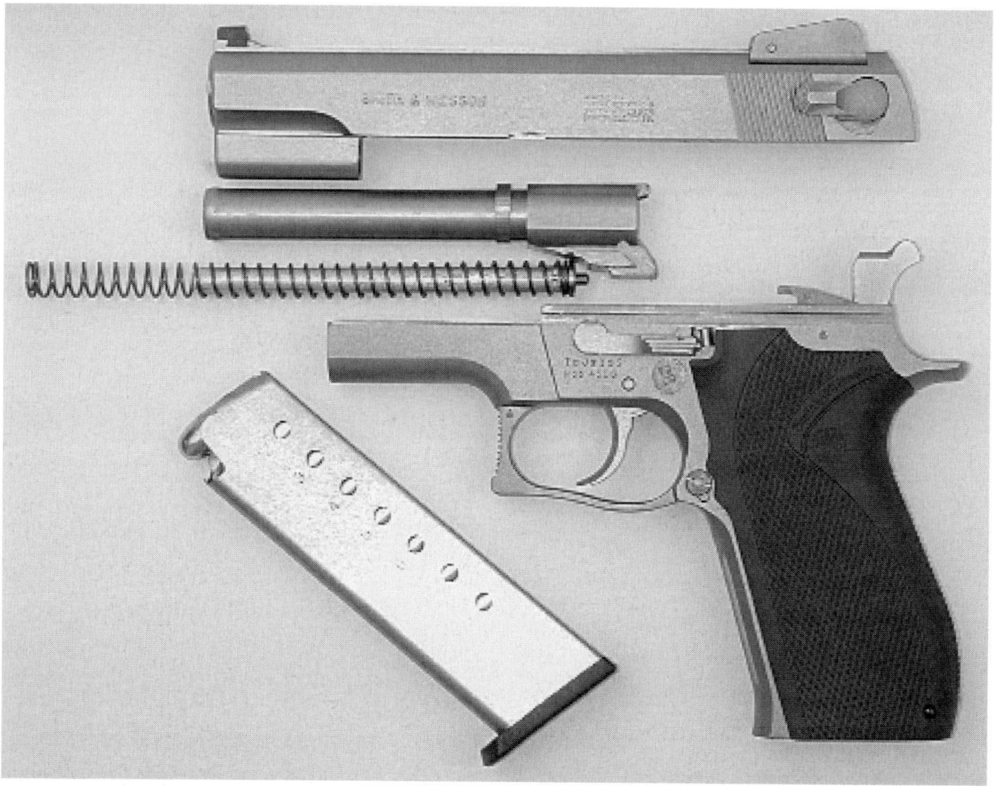

S&W Model 4506, partially disassembled; note that it has no separate barrel bushing, unlike pistols of the first and second generation.

to the line. Other changes to the basic design were rather minor, and included reshaped and enlarged trigger guards, different sight options and finish variations.

The key to second-generation model designations is as follows:

- First digit – frame material:
  - 4xx – aluminium frame
  - 5xx – carbon steel frame
  - 6xx – stainless-steel frame.

- Second and third digits – frame size and magazine capacity:
  - x59 – high-capacity frame, fourteen-round magazine
  - x39 – standard capacity frame, eight-round magazine
  - x69 – high-capacity compact frame, with twelve-round magazine (also will accept fourteen-round magazines).

The third generation of Smith & Wesson pistols was announced in 1988. It brought even more calibres, including the 10mm Auto, .40 S&W and .45 ACP, as well as more options in the trigger versions (traditional DA/SA with safety, DA/SA with decocker, DAO). The basic design also went through some upgrades, including the omission of the earlier barrel bushing. This third generation included more than twenty models in four calibres; of those, about ten are still in production. Third-generation pistols have four-digit model numbers, with the exception of the so-called 'value' series of pistols, which have three-digit numbers like Models 908, 915, 411 or 457. These pistols are made using less expensive finishes and with non-ambidextrous safeties, but in essence are still the same third-generation weapons.

The key to third-generation model designations is as follows:

- First two digits: calibre and frame type (for 9mm models):
  - 39xx: 9mm with single-stack eight-round magazines
  - 59xx: 9mm with double-stack fifteen-round magazines
  - 69xx: 9mm compact, with double-stack twelve-round magazines
  - 40xx: .40 S&W
  - 10xx: 10mm auto
  - 45xx: .45 ACP.

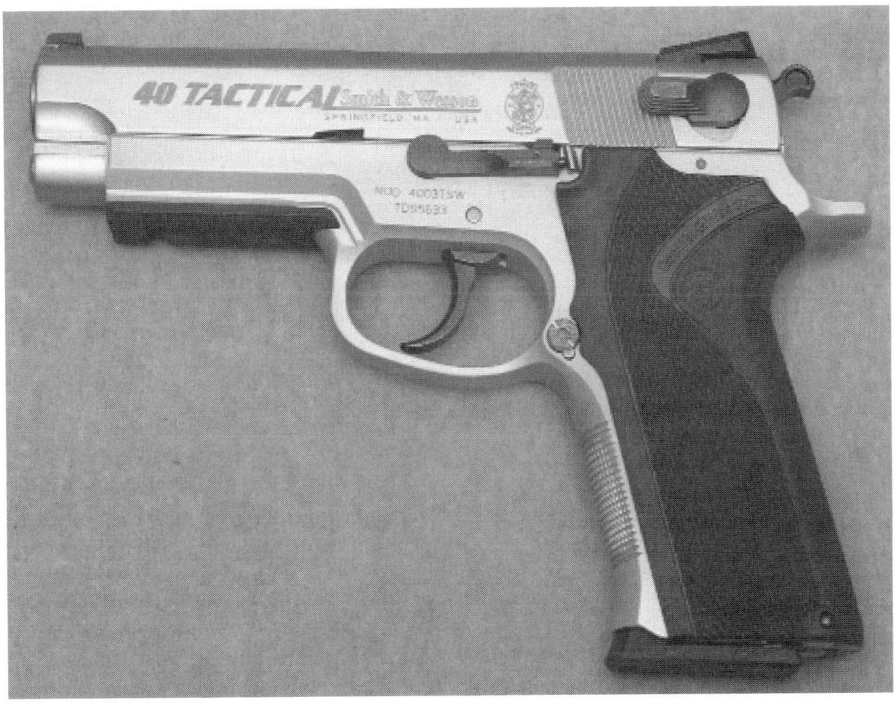

.40 S&W calibre S&W Model 4003 TSW (Tactical Smith & Wesson), with an accessory rail pinned to the frame. (Courtesy of Cherry's Fine Guns)

- Third digit: trigger/safety type and, occasionally, frame type:
  - xx4x: DAO
  - xx5x: DAO, compact
  - xx8x: DAO
  - xx2x: DA with frame-mounted decocker
  - xx3x: DA with frame-mounted decocker
  - xx7x: DA with frame-mounted decocker, compact
  - xx0x: DA with slide-mounted safety/decocker
  - xx1x: DA with slide-mounted safety/decocker, compact
  - xx6x: DA with slide-mounted safety/decocker.
- Fourth digit: frame material (slides and barrels are stainless steel on all third-generation pistols except for three-digit numbered 'value' models that have carbon steel slides and barrels)
  - xxx3: aluminium alloy frame
  - xxx6: stainless-steel frame.

The third-generation pistols are more or less similar in design to earlier pistols, except for a few differences. First, as noted above, third-generation pistols have no barrel bushing and the muzzle of the barrel enters the slide directly. Second, they have discarded the separate grip backstrap and two-piece grip panels in favour of single-piece, wrap-around grips made from polymer. Third, various trigger variations were introduced, along with ambidextrous safety levers. Other changes were of less importance and included variations in sights, shape of the parts and other such details. The following is the technical description of the third-generation pistols.

The 9mm S&W Model 910/915 pistol is a 'value' (less expensive) version of Model 5903.

Smith & Wesson semi-automatic pistols are short-recoil-operated, locked-breech weapons. These pistols use the Browning locking system, where the barrel engages the slide with a single lug, located ahead and above the chamber; this lug enters a recess on the underside of the slide. Unlocking is controlled by a cam-shaped lug below the barrel, which enters the appropriate recess in the frame upon recoil. The trigger is double-action, with an exposed hammer. 'Standard' models have slide-mounted safeties that also double as automatic decockers when the safety is engaged. DAO models have no manual safeties and plugs are installed in place of the safety levers in the slide; hammer spurs are discarded. Models with frame-mounted decocking levers and no manual safeties are no longer made. Automatic firing-pin safety is provided on all third-generation pistols. Most S&W pistols are fitted with automatic magazine safeties, which block firing when the magazine is removed; however, some 10mm pistols made for the FBI contract during the early 1990s have no magazine safety. Magazines are either single or double stack, depending on the model and calibre; the magazine release button is located at the base of the trigger guard. Sights are of various types; most common are either 'low profile combat' or tritium-filled 'night' ones, dovetailed to the slide. Various types of adjustable sights have also been used over time.

Manual safety Ambidextrous levers located on the slide (left side only on first- and second-generation pistols and on third-generation 'value line' models). Pull the lever down to engage the safety; push the lever up to the horizontal position do disengage the safety; a red dot will appear on the frame. When applied, the safety will automatically decock the hammer.

Field stripping procedure 1) remove the magazine by pressing the magazine release button; 2) check that the chamber is empty; 3) retract the slide partially, until the slide stop notch on the left side of the slide is aligned with the front, rounded part of the slide-stop lever; 4) push the slide-stop lever axis pin inwards from the right, then pull the slide-stop lever towards the left and out of the frame; 5) carefully release the slide, then push it forwards and out of the frame; 6) remove the return spring and its guide from below the barrel, by pushing the spring guide head forwards and out of its seat in the barrel lug, then out of the slide; 7) pull the barrel out of the slide.

Reassemble in reverse order.

.40 S&W calibre Sigma 40F pistol, pre-1999 version. (Courtesy of Cherry's Fine Guns)

.40 S&W calibre Sigma SW40P, current production version. Note the shortened slide and barrel and accessory rail on the frame. (Courtesy of S&W)

Smith & Wesson Sigma

Data for Current (Post-1999) Production Pistols
Trigger type: Double-action only, pre-cocked
Calibre/ammunition used: 9 × 19mm and .40 S&W
Weight, empty: 690g (24.2oz)
Length: 184mm (7.2in)
Barrel length: 102mm (4in)
Magazine capacity: 16 (9mm) or 12 (.40) rounds

Smith & Wesson brought out its first polymer-framed pistol, called the Sigma, in 1994. This was a desperate attempt to compete with the aggressively marketed Glock pistols of Austrian manufacture. S&W's designers apparently tried to improve on the Austrian design, but came too close to the original in a number of ways, with the result that Glock promptly filed a patent infringement suit against S&W. Indeed, many parts were actually interchangeable between the Austrian and American guns and S&W ended up paying a significant fine to Glock. Nevertheless, the Sigma pistols, which were received by the market rather coolly, remained in production. In 1999, Smith & Wesson introduced an improved version of the Sigma line, with the barrel and slide shortened by 12.5mm, (0.5in) as well as providing an integral accessory rail and slightly improved grip.

When compared with Glock pistols of the same calibre and size, the S&W Sigma feels much more comfortable in the hand, thanks to its more rounded grip. Also, Sigma pistols were designed for the .40 S&W calibre ammunition and then adapted for the smaller and lighter 9mm ammunition, and not vice versa as in the Glock's case. The only drawback to the current Sigma design, often pointed out by anyone who tries it, is a relatively heavy trigger pull. Nevertheless, this pistol is capable of acceptable accuracy in combat conditions and is quite reliable.

The S&W Sigma is a short-recoil-operated, locked-breech pistol with a polymer frame. The barrel locks to the slide using a single large lug that engages the ejection window; unlocking is controlled by a cam-shaped lug below the barrel, which interacts with a steel insert in the frame. The trigger is a double-action-only type, striker fired. The striker is pre-cocked during each recoil cycle, then brought to full cock and released by the trigger pull. The pistol is fitted with an automatic firing-pin block safety and trigger safety; no manual safeties are provided. Magazines are double stack, the magazine release button is located at the base of the trigger guard. Sights are fixed; both front and rear sights are dovetailed to the slide.

Manual safety None.

Field stripping procedure 1) remove the magazine by pressing the magazine release button; 2) check that the chamber is empty; 3) point the gun in a safe direction and pull the trigger once to release the striker from the sear; 4) retract the slide for about 6mm (0.2in) and, while holding it there, depress the takedown buttons, located at both sides of the frame, above the trigger; 5) while depressing the takedown buttons, release the slide carefully, then push it forwards and out of the frame; 6) remove the recoil spring from below the barrel; 7) pull the barrel out of the slide.

Reassemble in reverse order.

Smith & Wesson Military and Police

Trigger type: Double-action only
Calibre/ammunition used: 9 × 19mm, .357 SIG and .40 S&W
Weight, empty: 700g (24.5oz)
Length: 190mm (7.4in)
Barrel length: 114mm (4.5in)
Magazine capacity: 17 (9mm) or 15 (.357 and .40) rounds

.40 S&W calibre Military and Police semi-automatic pistol. (Courtesy of S&W)

The Military and Police pistol is the latest offering from the Smith & Wesson Company. First announced in the summer of 2005, this pistol is obviously intended for police and civilian markets. The Military part of the name, in the opinion of the author, is more of an advertising ploy, as it seems that the US Army strongly prefers pistols with manual safeties and the M&P has no such devices. However, this pistol seems to have every feature necessary for success among police and civilian shooters. The Military & Police name was a trademark of S&W for some eighty years, being permanently linked to the most successful double-action revolver from this company (and probably, the most successful DA revolver ever built in the world), which is currently known as the S&W Model 10. It has yet to be seen if the new M&P will be as successful as the old one.

The S&W Military and Police pistol is a short-recoil-operated, locked-breech pistol that uses Browning-type locking. The frame is made from high-strength polymer; all other major components (including the barrel and slide) are made from stainless steel. The pistol is striker fired, with a double-action-only trigger. Since most striker-fired pistols require the trigger to be pulled during disassembly (which has been the source of a great many accidental discharges with other striker-fired pistols), the S&W has a neat sear release lever which can be used to release the striker safely without pulling the trigger and with the slide locked back open. The sear release lever is accessible only through the ejection port opening.

The S&W Military and Police is fully ambidextrous, with slide release levers located on either side of frame and a reversible magazine release button. For improved ergonomics, the rear part of the grip is easily replaceable and the gun comes with three grip panels of different size and shape. The front part of the frame, under the barrel, is shaped to form the Picatinny rail for accessories. Both front and rear sights are dovetailed to the slide and can be easily adjusted or replaced by optional tritium night sights.

Manual safety None.

MBA Gyrojet

Trigger type: Single action
Calibre/ammunition used: 13 × 50mm rockets
Weight, empty: About 620g (21.7oz)
Length: About 300mm (11.7in)
Barrel length: No data
Magazine capacity: 6 rounds

The Gyrojet was one of the rare attempts to develop an entirely different weapon with some unique properties. This weapon tried to achieve several goals, including a low firing noise and underwater fire capabilities, along with significant firepower and penetration. The result was very different from any conventional firearm; in fact, it was a hand-held, multi-shot rocket launcher that fired 'armour piercing' projectiles, made from steel and with a pointed nose. Each projectile contained its own rocket engine and means for stabilization – either in the form of retractable fins or inclined jet nozzles which forced the missile to rotate, thus providing a gyroscopic effect for stabilization. Each rocket 'motor' burned for about one-tenth of a second; the maximum velocity of about 380m/s (~1,250fps) was achieved at ranges about 20m from the muzzle; at the muzzle, the velocity was as low as 30m/s (~100fps). Because of that slow acceleration the Gyrojet was almost useless as a weapon at short range, within 5–7m.

Another problem inherent to this weapon was poor reliability (even in ideal conditions this gun misfired about once in every 100 rounds; in actual use it did so much more often than that). Add to this very poor accuracy, because of insufficient stabilization of the projectiles during the early stages of trajectory, and you have more of curiosity and collectible piece than a combat weapon. Probably the only advantage of this weapon was that it was indeed more or less silent – when fired, it produced a short hiss, sort of a pierced-tyre sound, clearly different from any typical gunshot. However, it deserves its place in the history of small arms as one of the very rare unconventional combat weapons that worked, even if only to some extent.

Gyrojet 'pistol'.

It must be noted that there were two 'marks' of Gyrojet weapons, distinguishable by the calibre of projectiles. Original 'Mark I' weapons fired 13mm rockets; 'Mark II' weapons fired slightly smaller 12mm rockets to conform to the US Gun Control Act of 1968.

The Gyrojet launchers were made in the form of a large pistol (or carbine). The frame was made from aluminium alloy, with a perforated 'barrel' which served as a launching guide for the rockets. Up to six rockets were stored in an integral, single-stack magazine located in the grip; the magazine was loaded from the top, through the slide-open cover. After loading, the top rocket remained in the 'barrel', resting against a fixed firing pin set into the rear of the barrel tube. Once the trigger was pressed, it released the hammer, which rotated back and hit the nose of a rocket, pushing it towards the fixed firing pin. The primer in the base of the rocket ignited the rocket charge and sent the projectile accelerating down the 'barrel' tube. This movement cocked the hammer and also freed the space for the next rocket to rise from the magazine into the firing position. To provide the necessary means for cocking the hammer for the first shot, it was fitted with a side extension that projected outside of the frame from the curved cut on the left side of the frame. A manual safety was fitted at the left side of the grip and the magazine follower also had a projection that permitted it to be lowered manually for loading.

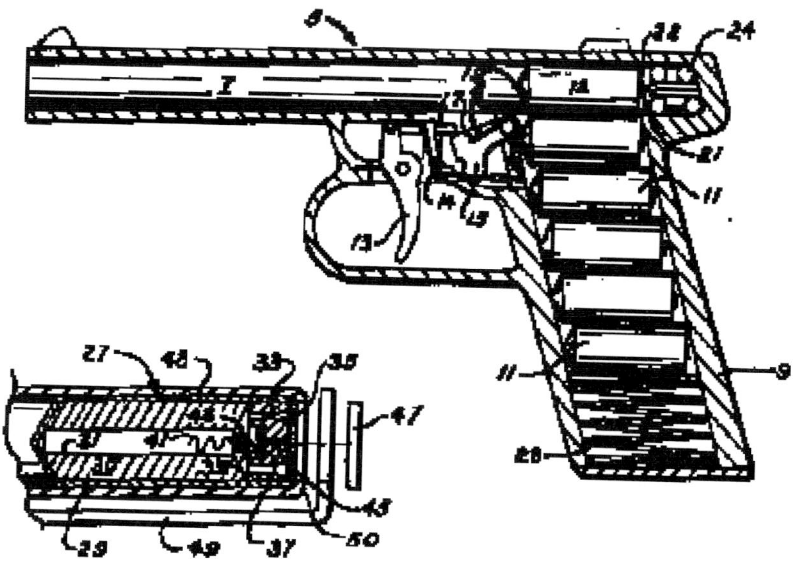

Diagram from one of Gyrojet patents.

9mm Ruger P85. (Courtesy of Sturm, Ruger & Co.)

Ruger P85 – P94

The American company Sturm, Ruger & Co. is among the younger manufacturers of firearms – it was founded after World War II to produce small-bore pistols designed by William Ruger. For many years this company concentrated on small-bore pistols, as well as various revolvers, leaving centrefire pistols for others. Only in the mid-1980s did Bill Ruger decide to use his extensive experience in firearms design and novel manufacturing methods to make a service-type 9mm pistol. The first Ruger centrefire pistol, designated P85, appeared in 1987, too late for the infamous XM9 trials, but it participated in the following XM10 trials. As it experienced some teething problems, the basic design was modified in several ways to become the P89, which replaced the P85 in production in 1990. Two years later Ruger introduced a .45 calibre derivative of the P89 design, known as the P90, and about a year later – a .40 calibre P91 (now discontinued). The P89 and P90 are still manufactured to date and both are fairly popular because of their excellent durability and reliability, as well as moderate price.

These pistols are often referred to as being 'built like tanks' by their users; while being entirely adequate as holster-type weapons, the P89 and P90 are somewhat too big and bulky for concealed carry. Recognizing this issue, Ruger developed a line of more compact and streamlined pistols, which include the short-barrelled 9mm P93 and the average-sized P94, which is available in two calibres – 9mm P94 and .40 S&W P944. These pistols are also quite popular among civilian shooters, as they retain the traditional reliability and strong construction of earlier Ruger designs. Service users of Ruger pistols include several US police departments and other law-enforcement services (such as prison guards), as well as some foreign military and police organizations. The design of all the models from the P89 to the P94 is quite similar; the pistols mostly differing in the shape and size of some parts – most

Ruger P85 – P94

	P85	P89	P90	P91	P93	P94/P944
Trigger type *	DA	DA/DC/DAO	DA/DC	DC/DAO	DC/DAO	DA/DC/ DAO
Calibre/ ammunition used	9 × 19 Para	9 × 19 Para	.45 ACP	.40 S&W	9 × 19 Para	9 × 19 Para/ .40 S&W
Weight, empty	1,077g (37.7oz)	900g (31.5oz)	980g (34.3oz)	900g (31.5oz)	870g (30.5oz)	930g (32.6oz)
Length	200mm (7.8in)	200mm (7.8in)	200mm (7.8in)	200mm (7.8in)	184mm (7.2in)	191mm (7.5in)
Barrel length	114mm (4.5in)	114mm (4.5in)	114mm (4.5in)	110mm (4.3in)	99mm (3.7in)	108mm (4.2in)
Magazine capacity	15 rounds	15 rounds	8 rounds	10 rounds	15 rounds	15 (9mm) or 10 (.40) rounds

* trigger types:
DA = Double/single action with manual safety.
DC = Double/single action with decocker only.
DAO = Double-action only, no manual safeties.

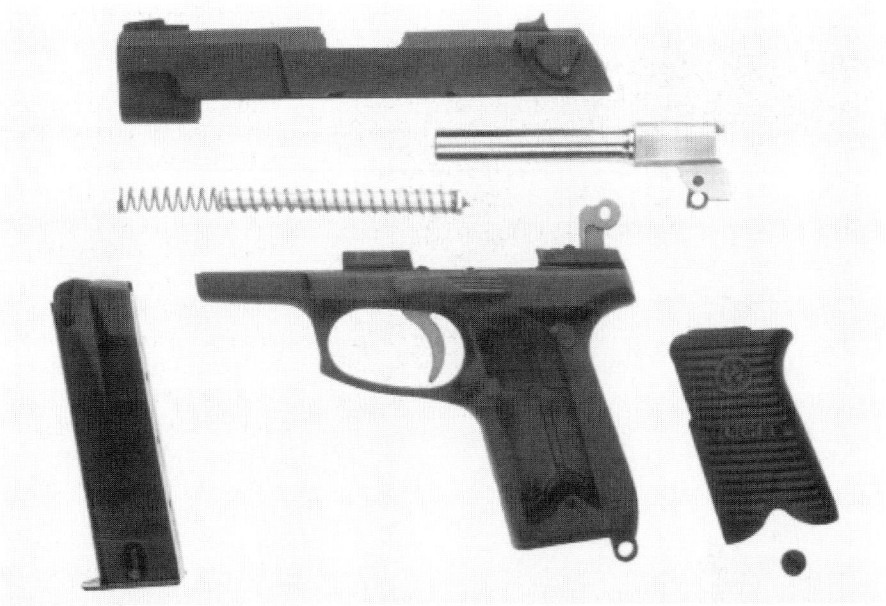

Ruger P85, partially disassembled. (Courtesy of Sturm, Ruger & Co.)

notably slide, barrel and frame. Also, these pistols can be had with carbon or stainless-steel slides and barrels, in various finishes and with three basic trigger types. Most metallic parts are precision cast from either steel or aluminium alloy and the Ruger company is particularly known for bringing this process to a fine art.

Ruger P85-series pistols are short-recoil-operated, locked-breech weapons. Locking is achieved by a single lug on the barrel that engages the ejection port in the slide. Unlocking is controlled by a Browning-type swinging link, which connects the barrel breech with the frame through the slide-stop axis pin. The frames of all Ruger pistols are made from aluminium; other parts (depending on the model) are either carbon or stainless steel. The trigger is double action, with an exposed hammer. The P85 pistol was made in only one version, with ambidextrous slide-mounted safety levers and an automatic firing-pin safety block. Later models introduced two additional versions – one with slide-mounted decocking levers and another with a

9mm Ruger P89DC, with decock-only lever. (Courtesy of Cherry's Fine Guns)

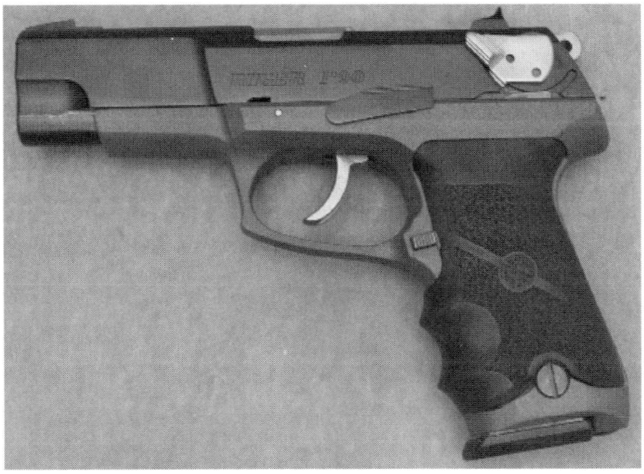

45 ACP Ruger P90, with conventional manual safety and Hogue grip panels. (Courtesy of Cherry's Fine Guns)

9mm P93DC, with decock-only lever. (Courtesy of Cherry's Fine Guns)

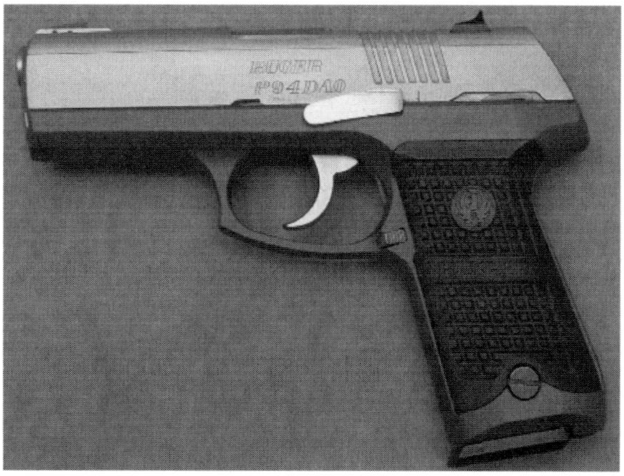

9mm Ruger P94DAO with no manual safeties at all. (Courtesy of Cherry's Fine Guns)

double-action-only trigger and no safety or decocking levers. Magazines are double stack on all models except for the .45 calibre P90, which has single stack. The magazine release button is located at the base of the trigger guard. Sights are fixed – the front sight is pinned to the slide and the rear sight is dovetailed.

Manual safety Standard models: ambidextrous levers on either side of the slide. Pull the lever down to engage the safety; the letter 'S' will appear on the slide above the lever and a white dot will be visible through a small hole in the lever. Push the lever up to the horizontal position to disengage the safety; a red dot will be visible through the hole in the lever. Note – once applied, the safety automatically decocks the hammer.

DC (Decocker) models: ambidextrous decocking levers on either side of the slide. When a lever is pushed, it automatically decocks the hammer, then returns to its 'fire' position once released.

Field stripping procedure 1) remove the magazine by pressing the magazine release button; 2) check that the chamber is empty; 3) retract the slide all the way back and lock it open with the slide-stop lever; 4) while holding the slide-stop lever in its upper (slide lock) position, insert the finger of the other hand through the ejection port and depress the ejector down and forwards until it locks itself; a failure to lower the ejector will make further disassembly impossible; 5) remove the finger from the ejection port; grasp the slide and release the slide-stop lever. Ease the slide forwards until the disassembly marks on the slide and frame are aligned; 6) push the slide-stop axis pin inwards from the right, then pull the slide-stop lever outwards towards the left as far as it will go – do not try to withdraw it from the frame completely! 7) push the slide forwards and out of the frame; 8) turn the slide upside-down and carefully withdraw the return spring and its guide rod from below the barrel; 9) pull the barrel out of the slide.

Reassemble in reverse order.

Ruger P95, P97

	P95	P97
Trigger type *	DA/DC/DAO	DC/DAO
Calibre/ ammunition used	9 × 19 Para	.45 ACP
Weight, empty	780g (27.3oz)	860g (30.1oz)
Length	185mm (7.2in)	185mm (7.2in)
Barrel length	99mm (3.9in)	99mm (3.9in)
Magazine capacity	15 rounds	8 rounds

* trigger types: see previous entry.

Ruger expanded its centrefire pistol line with its first polymer-framed model in around 1995. The new 9mm P95 pistol looks much like the earlier aluminium-framed P93, except that it has grip panels integral to the frame. However, the change of frame material resulted in internal changes, the most important one being in the locking system. Ruger discarded the Browning-type link between barrel and frame in favour of the later Browning-type cam locking. Otherwise, the P95 is much the same as the P93, with a similar double-action trigger, available in three basic versions as previously described. The .45 calibre P97 came about two years later and it is similar to the P95 except for the calibre and the single-stack magazine, which is interchangeable with that of the aluminium-framed P90 pistol of the same calibre. As of mid-2005, Ruger P95 pistols were still in production, offered in two trigger versions (standard DA/SA with safety and DA/SA with decocker) and in a variety of finishes, including the most recent version which is fitted with an integral Picatinny-type accessory rail on the frame, below the barrel. By 2005 the P97 had been replaced in production by the more modern Ruger P345 pistol that is described below.

Since the P95 and P97 are generally similar to the earlier Ruger pistols, only the differences will be mentioned here. Other than the polymer frame, already mentioned above, the P95 and P97 differ in the method of barrel locking and unlocking. The barrel engages the slide with a single lug that enters the ejection window. Unlocking is controlled by a cam-shaped projection below the barrel that enters in a cut made in the base of the recoil spring guide. This system is claimed to decrease the peak recoil transferred to the frame upon firing, as the area of contact between the part that accepts recoil (the return spring guide base/cam block and slide stop pin) is increased. Also, these pistols have no separate slide rail inserts, as the slide rails are moulded into the top of the polymer frame and are integral to it.

9mm Ruger P95DC pistol; the version with the decocker. (Courtesy of Cherry's Fine Guns)

Manual safety Standard models: ambidextrous levers on either side of the slide. Pull the lever down to engage the safety; the letter 'S' will appear on the slide above the lever and a white dot will be visible through a small hole in the lever. Push the lever up to the horizontal position to disengage the safety; a red dot will be visible through the hole in the lever. Note – once applied, the safety automatically decocks the hammer.

DC (Decocker) models: ambidextrous decocking levers on either side of the slide. When the lever is pushed, it automatically decocks the hammer, then returns to its 'fire' position once released.

Field stripping procedure 1) remove the magazine by pressing the magazine release button; 2) check that the chamber is empty; 3) retract the slide all the way back and lock it open with the slide-stop lever; 4) while holding the slide-stop lever in its upper (slide lock) position, insert the finger of the other hand through the ejection port and depress the ejector down and forwards until it locks itself; failure to lower the ejector will prevent further disassembly; 5) remove the finger from the ejection port; grasp the slide and release the slide-stop lever. Ease the slide forwards until the disassembly marks on the slide and frame are aligned; 6) push the slide-stop axis pin inwards from the right, then pull the slide-stop lever outwards towards the left as far as it will go; do not try to withdraw it from the frame completely! 7) push the slide forwards and out of the frame; 8) turn the slide upside-down; pull the cam-block/recoil spring guide assembly up to disengage it from the barrel lug, then withdraw it from the slide to the rear; 9) pull the barrel out of the slide.

Reassemble in reverse order.

Ruger P345

Trigger type: Double/single action
Calibre/ammunition used: .45 ACP
Weight, empty: Approximately 820g (28.7oz)
Length: 197mm (7.7in)
Barrel length: 107mm (4.2in)
Magazine capacity: 8 rounds

The Ruger P345 series of pistols represents a further development of the earlier P97 pistols; available since 2004, the P345 has replaced the P97 in production and is now available in several versions. These include two trigger options (traditional DA/SA with either safety or decocker only), several finish options (including

.45 calibre P345R pistol with an integral Picatinny-type accessory rail under the barrel.

carbon or stainless-steel parts) and a frame either with or without an integral Picatinny-type accessory rail. The basic design is similar to earlier polymer-framed Ruger pistols, except for the changes described below.

First, the P345 has a slimmed and more rounded frame, with a grip that certainly more comfortable to hold and fire. The slide is fitted with a loaded chamber indicator at the top, just behind the ejection port. The trigger and safety arrangements are the same as on previous guns, except for an added magazine disconnect safety, which blocks firing when the magazine is removed. Another added safety feature is an integral safety lock with separate keys; the lock is accessible through the small hole in the right safety lever.

Manual safety Standard models: ambidextrous levers on either side of the slide. Pull the lever down to engage the safety; the letter 'S' will appear on the slide above the lever and a white dot will be visible through a small hole in the lever. Push the lever up to the horizontal position to disengage the safety; a red dot will be visible through the hole in the lever. Note – once applied, the safety automatically decocks the hammer.

DC (Decocker) models: ambidextrous decocking levers on either side of the slide. When the lever is pushed, it automatically decocks the hammer, then returns to its 'fire' position once released.

Field stripping procedure 1) remove the magazine by pressing the magazine release button; 2) check that the chamber is empty; 3) retract the slide all the way back and lock it open with the slide-stop lever; 4) while holding the slide-stop lever in its upper (slide lock) position, insert the finger of the other hand through the ejection port and depress the ejector down and forwards until it will lock itself; failure to lower the ejector will result in impossibility of further disassembly; 5) remove the finger from the ejection port; grasp the slide and release the slide-stop lever. Ease the slide forwards until the disassembly marks on the slide and frame are aligned; 6) push the slide-stop axis pin inwards from the right, then pull the slide-stop lever outwards towards the left as far as it will go; do not try to withdraw it from the frame completely! 7) push the slide forwards and out of the frame; 8) turn the slide upside-down; pull the cam-block/recoils spring guide assembly up to disengage it from the barrel lug, then withdraw it from the slide to the rear 9) pull the barrel out of the slide.

Reassemble in reverse order.

Appendix One

The Design and Functioning of a Semi-Automatic Pistol

Basic Elements of Pistol Design

A modern pistol consists of several major parts and sub-assemblies that are identified on the image (right). Shown at the top is the classic single-action Colt M1911A1; in the middle is the modern double-action SIG-Sauer P220 (European model); and below is another modern model, the Glock Model 20 with a pre-cocked double-action trigger system.

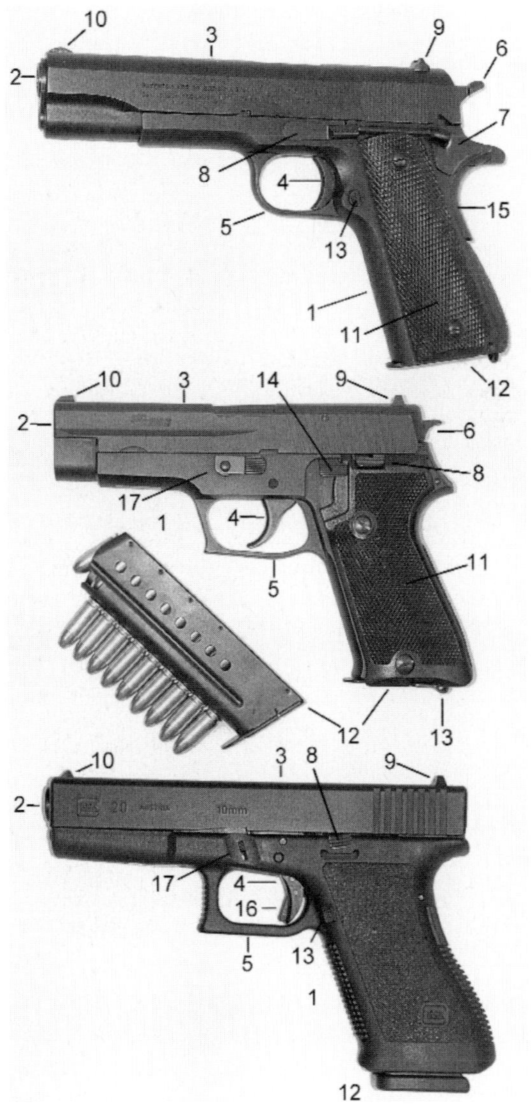

Key

1) frame
2) barrel
3) slide
4) trigger
5) trigger guard
6) hammer
7) manual safety lever
8) slide-stop lever
9) rear sight
10) front sight
11) grip panels
12) magazine
13) magazine latch
14) decocking lever
15) automatic grip safety
16) automatic trigger safety
17 disassembly latch

Major Units and Assemblies

Feed Unit

The feed unit is responsible for feeding fresh cartridges into the barrel. It usually consists of a magazine, a ramp up which the cartridge is guided from the magazine to the chamber and a lug on the slide or breechblock that pushes the cartridge out of the magazine and up the ramp.

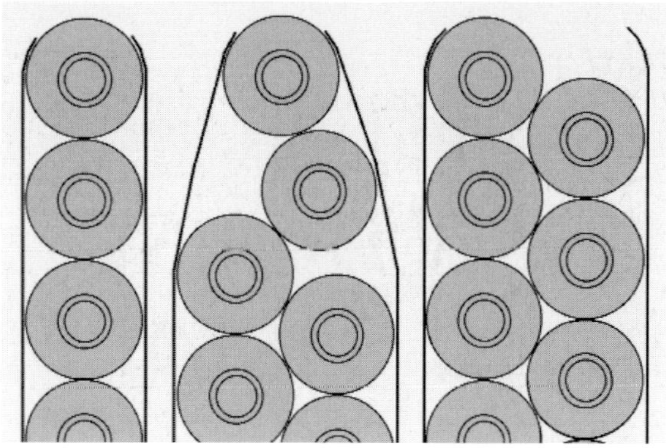

Diagram showing the basics of various types of box magazines, left to right: single-stack magazine; double-stack, single-feed magazine; double-stack, double-feed magazine.

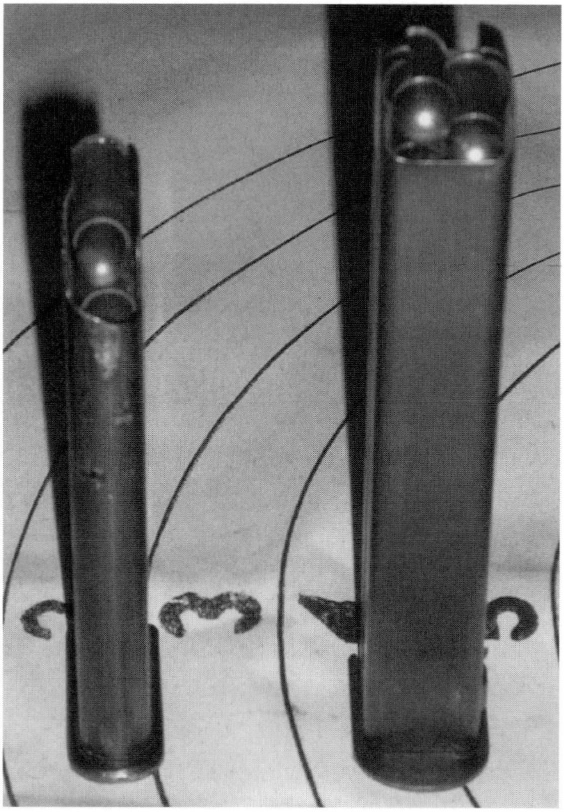

Handgun magazines: left is a single-stack eight-round magazine (for the Russian Tokarev TT pistol); right is a double-stack, double-feed twenty-round magazine (for the Russian Stechkin APS pistol).

In modern pistols, the magazine is a removable box-type unit that is open at the top and holds several cartridges ready for feeding into the chamber, one by one. Mostly, the magazines are inserted into the grip. The cartridges are stored under the pressure of the magazine spring and held in place by feed lips, which are formed by bending in the top of the magazine sides, so that the cartridge at the top is prevented from being pushed out of the magazine; however, the magazines are designed to allow the topmost cartridge to slip forwards and out of the magazine when it is pushed forwards by the breechblock of the firearm or by a finger (when unloading the magazine).

There are two major types of pistol magazines – single stack and double stack. The single-stack magazine holds cartridges in a single column; such magazines are slim but their capacity is limited and most often such magazines hold a maximum of eight or nine rounds. Double-stack magazines hold rounds in a double column, staggered to conserve the length of the magazine. Such magazines usually hold thirteen to fifteen, and sometimes up to twenty, rounds. However, double-stack magazines are thicker and thus require thicker grips to hold them. Furthermore, double-stack magazines can be divided into two sub-types – single feed and double feed. Single-feed magazines have a bottlenecked top, where the cartridges are squeezed from two rows into one, so each round is fed from the top of magazine and into the barrel via the same feed path. Double-feed magazines do not have such bottlenecks and cartridges are fed alternatively from the left and right column. Such magazines have advantages of slightly greater capacity and easier loading, but require a more complicated shape of feed ramp. It is worth mentioning that most pistols that use double-stack, double-feed magazines were designed in the USSR and Russia.

Another useful device usually accommodated into the feed system is the slide stop, or slide hold-open device. This system usually consists of a lever with a hook-shaped projection, which can engage a notch on the slide when it starts its closing movement after recoiling. Once the magazine feeds the last cartridge into the chamber and the magazine follower (which fits on top of the magazine spring) reaches its uppermost position, the follower pushes the slide-stop lever up to engage the slide and hold it open. This system has two major uses: first, it gives the shooter a clear signal that his magazine is empty; and second, it allows for faster reloading, as, once the empty magazine is replaced with a fresh one, it is only necessary to push the slide-stop lever to allow the slide to fly forwards and

The feed system of Heckler & Koch's P8 (USP) pistol, with the slide removed. It shows the open chamber with the feed ramp at the left, the inserted magazine (double stack and single feed) and the fixed ejector behind the magazine. The slide-stop lever is visible at the left side of the frame, along the slide rail.

chamber the first round. In pistols with no such device, the user has to retract the slide manually to chamber the first round from the magazine. A few modern pistols also may have automatic slide-stop disengagement, which unlocks the slide stop once a fresh magazine is inserted into the pistol. However, this cannot be considered a very safe system.

Ejector and Extractor

These two devices are tasked with removing the spent case or unfired round from the chamber (extract) and then throwing it out of the action (eject). The first part of the job is usually done by the extractor, which is a part with a hook-shaped projection that engages the rim or extraction groove of the cartridge once it is chambered. Extractors can be made internal or external, that is, completely enclosed in the channel within the slide or located in a slot cut in the wall of the slide. Most modern pistols use external extractors, which have a separate spring that holds the hook in engagement with the cartridge rim. Some older guns like the Colt M1911 or early FN High Power had internal extractors made from flexible steel, which do not require additional springs but need to be properly curved and heat treated to provide the necessary tension.

Ejectors can be made either fixed or spring-loaded and often they represent a projection put into the path of the cartridge case, which is being extracted from the chamber by the slide. Once the base of the case hits the ejector, it deflects through the opening of the slide and out of the gun. Various pistols may throw cases to the side, or rearwards and over the shoulder of the shooter. Side ejection is preferred as it minimizes the chances

How an extractor works (in this example on a Makarov PM pistol).

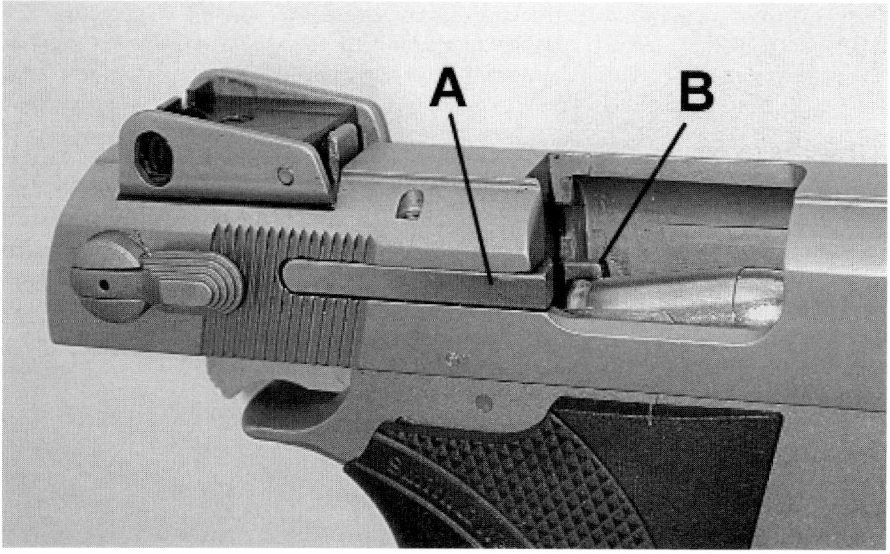

Extractor of the external type (A) and a fixed ejector (B), as made on Smith & Wesson Model 4506 pistol.

of the shooter being hit in the face with the hot cartridge case.

It must be noted that most jams in semi-automatic pistols occur during ejection and extraction; it is necessary to maintain this unit properly to ensure reliable functioning of the pistol. It is advisable to check periodically the tension of the extractor spring, the proper location of the ejector and the proper shape of the extractor claw.

Actions and Barrel Locking Types

The basic reloading and recocking actions of pistols occur during the recoil and return cycle of the pistol's slide and breechblock. In semi-automatic and automatic pistols these actions are driven either by using the energy generated during the previous discharge, or, for the first shot, by the shooter's hands.

The key source of energy used from reloading comes from the pressure of hot burning gases, which acts on the cartridge from inside and, while pushing the bullet forwards up the barrel, also pushed the cartridge case rearwards, against the breechblock. It is necessary for the cartridge case to be supported by the chamber walls for a fraction of a second after firing, as the strength of the cartridge case material (brass or mild steel) cannot withstand the peak pressure generated by the burning powder charge. The opening of the breech must therefore be delayed or retarded somehow to keep the fired case in the chamber until the bullet has left the muzzle and the pressure in the barrel has fallen to a safe level.

There are three basic methods to achieve this goal, known as blowback action, delayed (or retarded) blowback action and locked-breech action.

Blowback Action

The first method, known as blowback action, is to provide a breechblock heavy enough so that its inertia, combined with the strength of the return spring, will resist the initial rearward movement of the case being fired. This method is mostly used with low-powered, low-pressure cartridges; it is technically possible to use a blowback system with more powerful cartridges like the 9 × 19 Luger, .40 S&W or .45 ACP, but in such cases it is necessary to use either a slide which is too heavy to be convenient in a pistol, or a return spring too strong to be compressed manually during loading. Blowback systems are therefore used in conjunction with more powerful cartridges mostly in sub-machine guns, which allow for heavier breechblocks (bolts) to be used without impairing the function of the weapon. Some designs use radially grooved chambers to provide additional friction between the cartridge case and the chamber during the high-pressure period of the discharge. However, this method works mostly with cartridge cases made of brass; less tensile steel cases are poorly suitable for such a system and, furthermore, in time grooves may collect fouling and powder residue and thus lose effectiveness.

Delayed Blowback Action

The second method is usually called delayed or

retarded blowback. In this type of action, the initial opening movement of the breechblock or the slide is retarded by some sort of brake. The three most popular retarded blowback systems are named after their inventors. The first is the Savage-Searle, the second is the Destree/ Barnitzke and the third is the Vorgrimler system. All of these systems have the advantage of a fixed barrel, which is good for accuracy, and also usually have the additional benefit of smoothing the peak recoil of the firearm. Of course, these systems also have their own disadvantages. One of the more common is the lack of primary extraction (that is, the fired case is not pulled out of the chamber by the recoiling breechblock), which may result in cases sticking in the chamber when using high-pressure cartridges. One way to avoid this problem is to use a fluted chamber (with longitudinal grooves in it) that allows some hot powder gases to seep back to surround the case, to resist its expansion under the high internal pressure. Some authorities draw a distinction between the delayed and retarded types: the retarded type is as described above, while the delayed type has a locked-breech which is unlocked after firing, after which blowback is used to extract the case and power the reloading cycle. This distinction, however, mostly is applied to larger weapons like machine guns and automatic cannons.

Savage-Searle Rotating Barrel Delayed Blowback Action

This system was invented in the USA circa 1905 by Elbert Searle, who at that time worked for the Savage Firearms Company. Originally intended for a .45 calibre military-type pistol that participated in the US Army trials of 1907–11, this system was later used in pocket-type Savage pistols made between 1907–28 and, after WWII, in a few military-type pistols, including the Chinese QSZ-92-5.8, the French MAB P15 and the US/Jordan Viper.

In this system the barrel of the firearm has at least two radial lugs. One lug, located below the barrel, engages a perpendicular slot in the frame that allows the barrel to rotate but not to recoil. Another lug engages a curved slot, made on the underside of the slide. The cut in the slide is so shaped that upon its initial opening it will first force the barrel to rotate

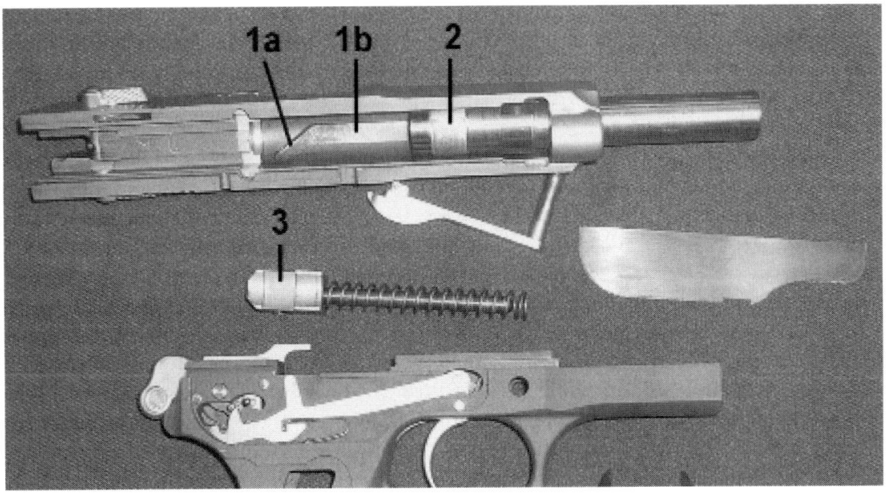

Jordan Viper pistol disassembled to show the key parts of its Savage-Searle delayed blowback action.
1a – inclined part of the slot in the slide that causes the barrel to rotate
1b – straight part of the slot which is followed by the upper lug of the barrel once the bullet has left the barrel
2 – lower lug on the barrel which engages the cut in the frame insert to let the barrel rotate but not recoil
3 – the cut made in the frame insert (base of recoil spring guide); when the gun is assembled, this part is fixed to the frame by the axis pin of the slide stop

through a short arc and then it will ride straight back without further rotation of the barrel. The operating principle is to add the inertia of the rotating barrel and the torque generated by the bullet travelling down the rifled barrel (which will tend to rotate the barrel in the opposite direction), to the inertia of the slide during its initial opening stages. Once the bullet leaves the barrel and the pressure drops, there is no further need to retard the opening of the breech and therefore the top lug on the barrel exits the curved part of the slot (that caused the rotation) and enters the straight part of the slot, so the slide is free to recoil under its inertia and the residual pressure in the barrel without any added braking force.

Destree/Barnitzke Gas-Retarded Blowback Action

The first gas-retarded blowback system was patented in the USA in around 1930 by a Joseph Destree. It is not known if Destree ever made any firearm to his design, but this system gained some popularity during the second half of the twentieth century in pistols such as the prototype Swiss W+F 47 pistol, the German HK P7, as well as the Type 77B from China, and the ADP and Vektor CP1 (both from South Africa). This system employed a separate gas cylinder located below the barrel, with the gas piston being linked to the breechblock (slide). In this system the pressure of the powder gases, tapped from the barrel, is used to resist the initial opening of the mechanically unlocked slide. Once the bullet leaves the barrel, the pressure in the barrel and the associated gas chamber drops rapidly, thus allowing the slide to recoil freely without additional braking.

A similar system was developed by the German engineer Karl Barnitzke working at the Gustloff Werke towards the end of WWII. It was used in at least one semi-automatic 7.92 × 33 *Volksgewehr* (people's rifle) and in a few prototype *Volkspistole* pistol designs. After the war, a slightly modified Barnitzke system was used in the Steyr GB pistol made in Austria.

In the Barnitzke system, hot gases are fed through one or more holes in the barrel into a surrounding cylinder, which is linked to the slide or breechblock to form a variable-volume annular gas chamber around the barrel. Hot gases enter this chamber and act against its front part to push it (along with the slide/breechblock) forwards, against the opening movement of the slide under the pressure inside the cartridge case.

The gas-retarded blowback system has the significant disadvantage of being sensitive to fouling and residue produced by burning propellant, as well as to

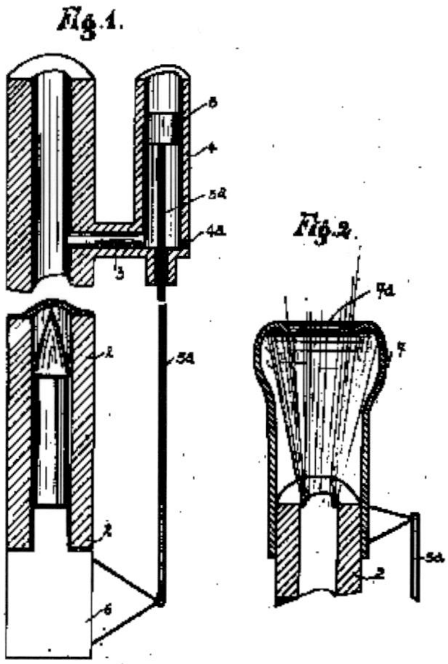

Original patent drawing of Destree gas-retarded blowback action.

Cross-section diagram of the HK P7 pistol, designed to use a slightly modified Destree action. It clearly shows the gas cylinder below the barrel, which is connected to the bore by a vent near the chamber, and the gas piston, attached to the front of the slide.

FIG.1

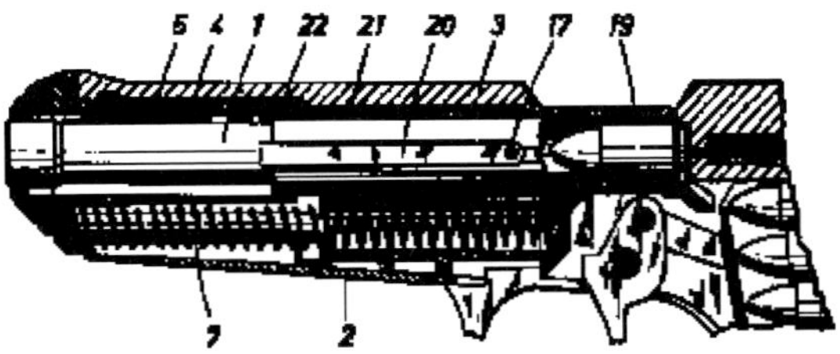

FIG.2

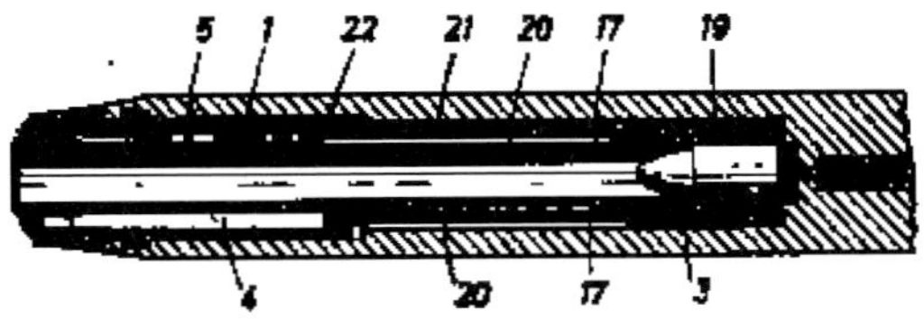

Patent diagram explaining the Barnitzke-type gas-delayed action of the Steyr GB pistol. Some key items on the diagram are: 1 barrel; 3 slide; 4 annular gas cylinder formed by barrel and barrel bushing; 5 barrel bushing; 17 gas ports that allow powder gases to enter the gas cylinder.

lead particles and other fragments that may be separated from the bullet during its path through the rifling in the bore. Clear advantages are relatively smooth recoil and good potential for accuracy, thanks to the fixed barrel. The other plus side of this system is ease of adaptation to the use of heavy silencers, and that there is automatic compensation for ammunition of varying pressures: the higher the pressure generated, the stronger the resistance to breech opening.

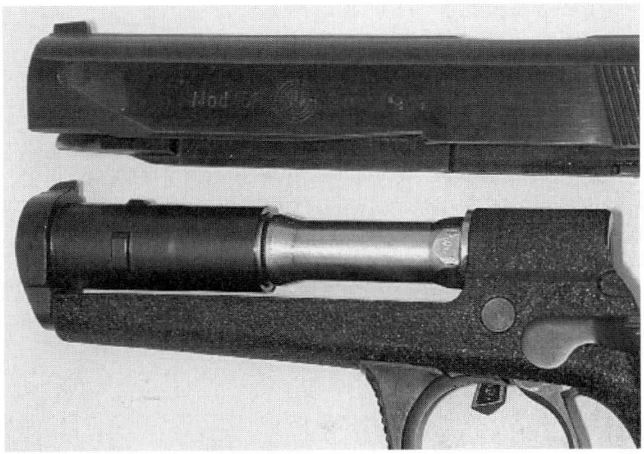

The actual Barnitzke-type gas-delayed action of the Steyr GB pistol; the barrel bushing is mounted on the barrel to form the gas chamber/cylinder. Note the lugs on the bushing that are used to lock it to the slide, which is removed to provide a clear view.

Vorgrimler Roller-Delayed Blowback Action

This system was also developed in Germany towards the end of WWII, by Ludwig Vorgrimler, who was then working on a new assault rifle at the Mauser Werke. After the war, Vorgrimler continued development of this system in Spain, which led to the line of Spanish CETME assault rifles; these in turn were licensed to the German company Heckler & Koch. During the 1960s, Heckler & Koch perfected this system and used it in a full range of small arms, from assault rifles and machine guns down to sub-machine guns (the famous HK MP5) and pistols (HK P9).

This system employs a two-part bolt with two rollers. The front part of the bolt (the bolt head) is relatively light and has a bolt face with an extractor on it. It also has a hollow cavity at the rear, in which the inclined forward end of the rear part of the bolt (the bolt body) is inserted. The system features two rollers in the sides of the bolt head, which are able to move in and out and rest on the inclined forward end of the bolt body. When the gun is fired, the pressure begins to move the cartridge back against the bolt face. The rollers, which are extended into the recesses in the barrel extension, begin to move inwards into the bolt head, due to the inclined shape of the recesses. This movement translates into a faster rearward movement of the heavier bolt body, so, at the initial instant of firing when the pressure in the chamber is still high, the bolt face moves relatively slowly. As the pressure drops to a reasonable level, the rollers gradually leave the recesses in the barrel extension; as soon as the rollers are completely pushed into the bolt, the bolt head and bolt body begin to move backwards at the same speed, extracting and ejecting the spent case and chambering a fresh cartridge on the way back.

It must be noted that this system is the most complicated and requires the breechblock (bolt) to be made as a separate unit mounted within the slide of a

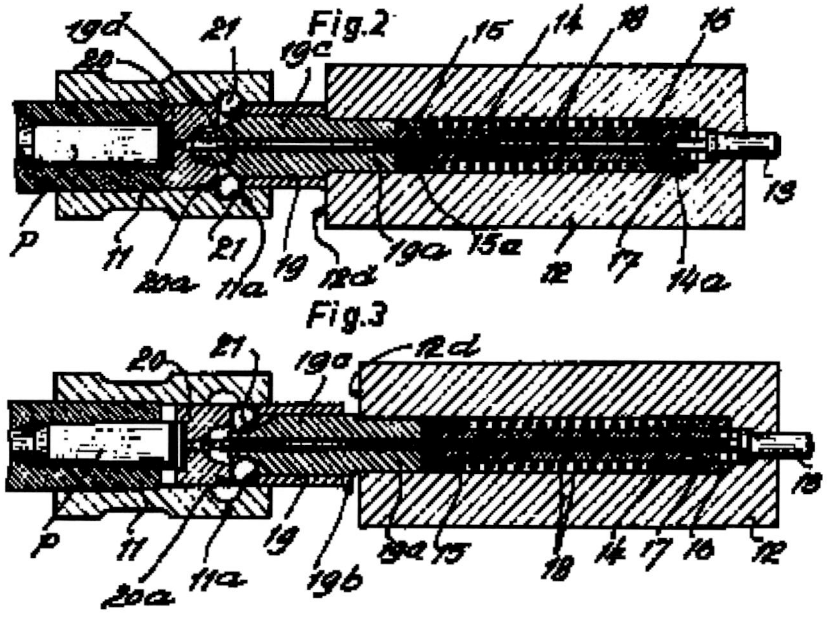

Diagram from the patent, issued to Ludwig Vorgrimler after WWII, which describes his roller-delayed action.

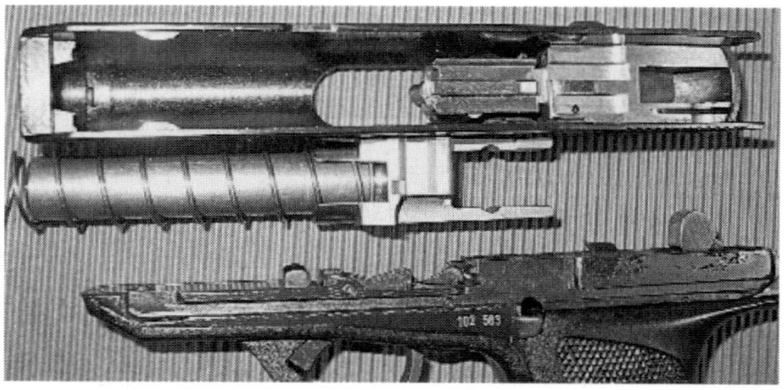

The German HK P9S, partially disassembled to show key parts of its roller-delayed blowback action. Note the barrel that is separated from the slide to show its side-mounted extensions with cuts made for retarding rollers; also note the separate two-part breechblock of complicated shape, mounted inside the stamped-steel slide. Rollers are slightly projecting from the sides of the breechblock.

pistol. Therefore, the major downsides of this system are its complicated design and disassembly procedures that require additional tools to remove the separate breechblock from the slide.

Locked-Breech Actions

The locked-breech action took its name from the simple fact that at the instant of firing the barrel breech is rigidly locked to the breechblock (bolt or slide), to avoid any blowback of the case when the pressure is still high. Therefore, this system needs some means for unlocking the action and cycling the slide or breech-block to reload the gun. There are several ways to provide the necessary energy for unlocking and reloading actions. The simplest is to use the muscular power of the shooter (as with bolt-action rifles), but as we are talking about semi-automatic weapons, the guns either bleed some powder gases into the separate part of the action, which then drives the mechanism (known as gas operation, as is used in most modern automatic rifles and many machine guns), or use the recoil force generated by the discharge against the rigidly locked-breechblock and barrel (known as recoil operation).

Gas-operated actions are almost unknown in the pistol world and are encountered only in a very few hunting-type pistols that fire Magnum class cartridges, such as the Israeli/American Desert Eagle or the American Wildey.

The much more common type is recoil action, which, in turn, can be further separated into two sub-classes, namely short recoil and long recoil. The short/long here refer to the length of the recoil movement by the locked barrel/breechblock group before their unlocking, as related to the length of the fired cartridge. In short-recoil weapons, this length is significantly less than the overall length of the cartridge and usually is about 3–8mm (0.1–0.3in) for most pistols.

The long recoil system, in which the barrel and breech-block are locked together for the full length of the recoil stroke, is very seldom used in pistols; probably the only long-recoil-operated pistol to ever see military service was the Hungarian Frommer Stop pistol of 1912, distinguished by an overly complicated design while firing relatively mild cartridges like the 7.65mm Browning and 9mm Browning Short.

This leaves us with the short-recoil system, which dominated in semi-automatic pistol designs right from the beginning. The very first practical self-loading pistols such as the Borchard C93 (1893), Mauser C96 (1896) and Colt-Browning (1900) used various forms of this system. There are a great many variations of locking systems but these are based on a few major ideas. These include the following:

- The toggle-lock systems of Borchard and Luger, which had fallen out of favour by the end of WWII. This system has not been used in any military-type firearm made since WWII and so it will not be discussed here in detail.
- Systems with a vertically tilting locking block originated in the Mauser C96 and then were extensively modified and perfected in Walther and Beretta pistols.
- The rotary barrel system, first patented by Browning in 1896–7 but actually made about ten years later in Austria by Karel Krnka at Roth and, later, Steyr factories. Today, this system has limited use in a number of pistols from China, Italy and Russia.
- Tilting barrel, swinging-link systems, designed and perfected by John Browning and used in a number of Colt pistols including the famous M1911, as well as in the Soviet Tokarev TT, the French Mle.1935 and Mle.1950, the American Ruger P85 and several other pistols.
- Tilting barrel, cam systems, also developed and perfected by John Browning. First made in the famous Belgian FN High Power/GP-35 pistol, this system is now made in a variety of versions and dominates the pistol world.

System with vertically tilting locking block (Walther type) This system uses a barrel that recoils straight back in the frame. Below its chamber is located a wedge-shaped locking piece, which is allowed to swing up and down through a short arc. This piece has two lugs, projecting upwards at either side of the barrel. When the barrel and slide are in battery (in the forward position), the locking wedge rests on a special shelf made in the frame and its lugs are engaged with the respective locking cuts made on the underside of the slide walls. Upon recoil, the wedge slides back in the

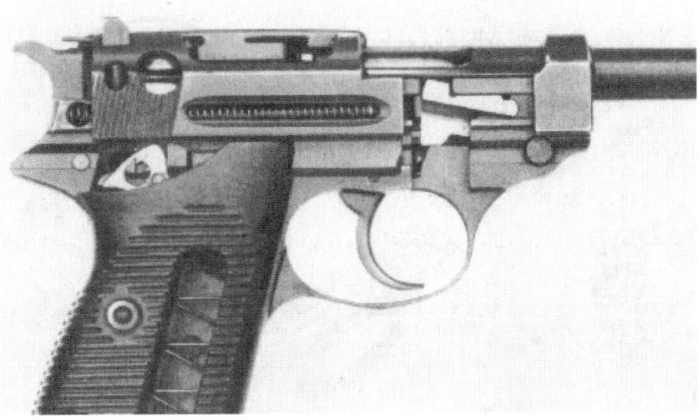

Walther P38 pistol factory cut to show its locking system. Note how the locking wedge (the white part above the front part of the trigger guard) engages the cuts in the slide and how it is supported by the frame from below.

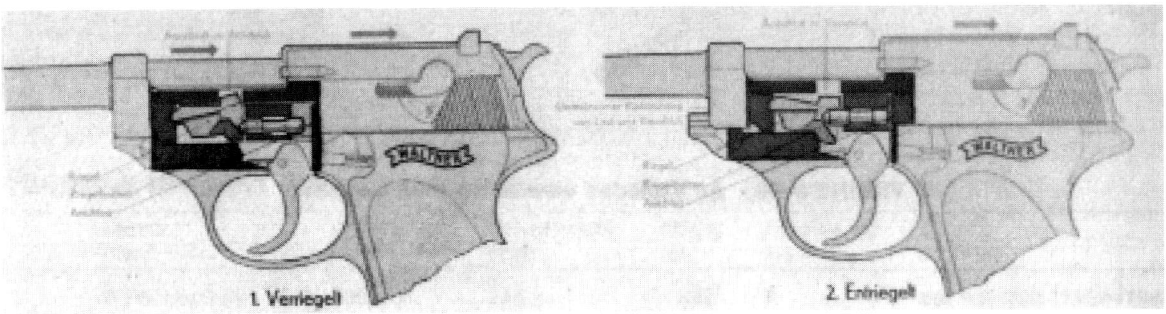

Illustration from the original German Walther P38 manual that shows the locking action.

frame along with the barrel and after a short amount of travel it falls down from the shelf into the recess cut in the frame, thus disengaging the slide from the barrel. Once the slide is unlocked, the barrel stops against the frame and the slide continues its rearward movement to complete the reloading cycle. Upon the forward movement of the slide it first strikes the barrel and then pushes it forwards into battery. During this movement the locking wedge slides up the sloped shelf in the frame to rise and engage the slide with its locking lugs.

This system has one advantage: the linear recoil of the barrel, which is good for accuracy. On the other hand, the locking wedge is subjected to substantial forces and thus must be very strong; this system also usually results in excessively thick slides, as these need to have enough space at either side of the barrel for the locking recesses.

Systems with rotary barrel locking action (Steyr-Roth type) The rotary barrel locking action first appeared in 1907 in the Roth-Steyr pistol made in Austria and next in the Steyr-Hahn pistol of 1911. Later on, it was tried in a number of prototypes and revived during the early 1990s in the Beretta Cougar 8000 series pistols. It was also used in the unsuccessful Colt All American 2000 pistol and in the more successful Russian GSh-18, which bears some resemblance to the unlucky Colt, although it is much more reliable.

This system relies on several lugs (from one to ten, depending on the design), which are formed on the barrel and fit into the slide to provide rigid locking at the moment of firing. To provide the necessary means of unlocking, the barrel is allowed to recoil and rotate in the frame, thanks to an additional lug, also formed on the underside of the barrel. This additional lug follows a slot, made in the frame or frame insert, which causes the barrel to rotate on recoil. This rotation (usually through a relatively short arc of 15 to 90 degrees) causes the locking lugs on the barrel to leave their mortises in the slide and thus unlock the action. Once the locking lugs are cleared from the locking recesses, the barrel is stopped and the slide is free to recoil; on its return movement, the slide hits the recoiled barrel and forces it forwards into the battery. This movement causes the barrel to rotate and lock into the slide ready for the next shot.

This system is used relatively seldomly in modern pistols, although it has the benefits of linear recoil of

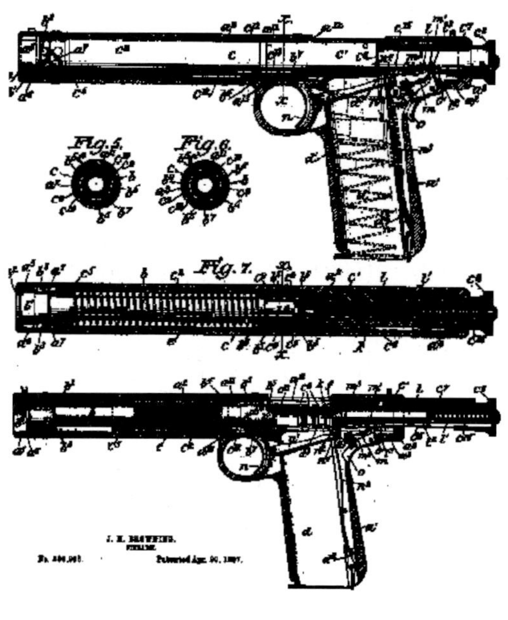

Diagram from the patent issued to John Browning in 1897 explaining his version of the rotary barrel locking system. Probably the first design of this type, it was never put forward by Browning and existed in few prototypes.

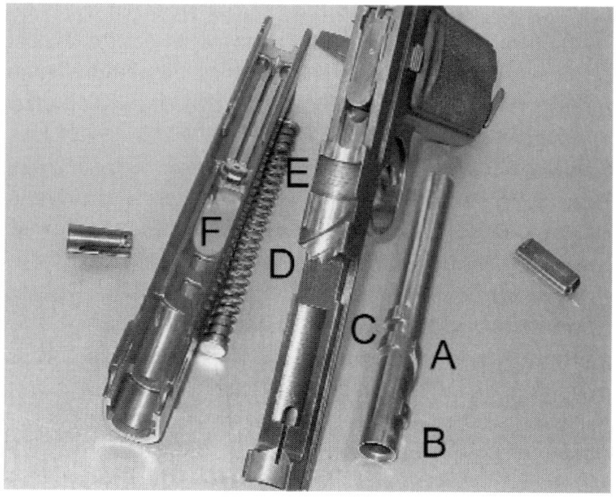

Rotary barrel locking system of the classic Steyr M1911 pistol.
***A** – helical lug on the underside of the barrel that is responsible for the barrel rotation upon recoil, when following through the slot **D** in the frame.*
***B** – second barrel lug that limits the short recoil of the barrel, by engaging slot **E** in the frame.*
***C** – twin locking lugs on the top of the barrel, which lock into recesses **F** in the slide.*

Diagram showing the rotary locking action of the modern Beretta Cougar 8000 pistol. (Courtesy of Beretta)

the barrel (good for accuracy) and, in the multiple lug systems, it also provides an extremely strong locking that allows the most powerful and high-pressure cartridges to be used in relatively lightweight pistols (as in the case of the Russian GSh-18, which has as many as ten radially spaced lugs on the barrel).

Tilting barrel systems (Browning type) John Browning, one of the world's most famous gun designers, patented his first pistol with tilting barrel locking in 1896–7; the first production pistol made to this design appeared in 1900, from the Colt Company of the USA. These early Colt-Browning pistols relied on two swinging links which connected the barrel to the frame. When in the battery, the barrel was rigidly locked to the slide with several lugs (two or three), made at the top of the barrel. These lugs fitted into slots cut into the underside of the slide. Upon recoil, the slide carried the barrel backwards, but the swinging links caused the barrel to go down simultaneously, so, after a short recoil, its lugs were pulled clear of the slide recesses. From this moment on, the slide was unlocked and free to recoil and the barrel was stopped. Upon its return into the battery, the slide struck the barrel and pushed it forwards, so the swinging links caused the barrel to go up and re-engage with the slide for the next discharge. This action, produced by Colt up until 1929, is often referred as 'Browning's parallel ruler' system, because the barrel stayed parallel to the slide at all times, as opposed to the later system that appeared in 1909.

This later system was a simplified version of the 'parallel ruler', as it got rid of the front swinging link. Instead, the barrel was held at the muzzle by the bushing set into the slide and upon recoil the rear link caused only the rear part of the barrel to tilt down and thus disengage with the slide. This system became extremely popular with Colt's M1911 pistol and several foreign pistols also employed this same action, such as the French M.1935 and M.1950 and the Soviet Tokarev TT.

In 1923, while working on a new 9mm pistol for the Belgian FN Company, Browning applied for a patent for an improved version of his tilting-barrel system, which replaced the swinging link with a cam-shaped slot made in a lug located below the barrel. Upon recoil, this open slot was followed by a transverse pin, set into the frame, so it caused the rear part of the barrel to tilt down and unlock from the slide. The patent for this system was issued in 1927, several months after

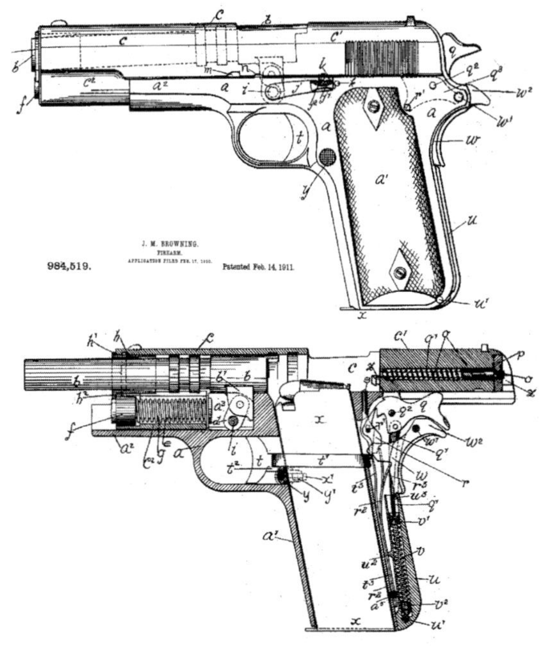

The first version of the Browning system patented in 1897 and usually referred to as the 'parallel ruler'. In this system, the barrel is connected to the frame with two swinging links and upon recoil it lowers itself while remaining parallel to the slide.

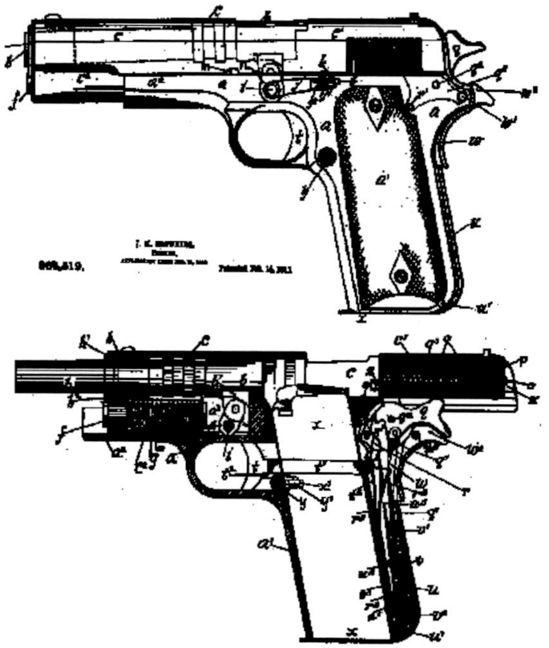

Browning's modified tilting-barrel system with a single swinging link at the rear, as used in the famous Colt M1911 pistol. In this system, only the rear part of the barrel is lowered upon recoil to unlock it from the slide.

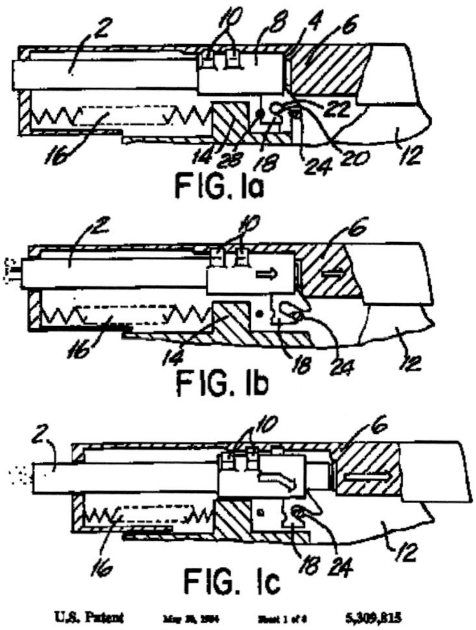

Illustration from the more recent patent that shows the last form of the Browning tilting barrel system. First used in the FN Browning High Power pistol, this system replaced the swinging link and its pins with single cam-shaped slot made in the lug below the barrel, which interacts with the transverse pin in the frame to lower the rear part of the barrel upon recoil.

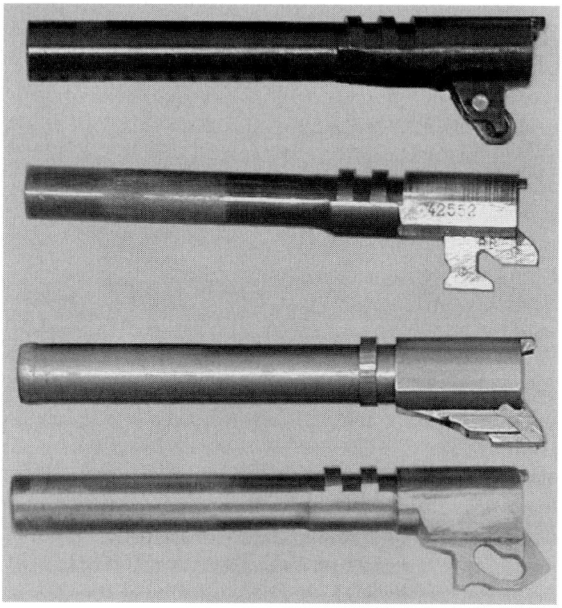

Various interpretations of the Browning tilting barrel system, top to bottom: Colt M1911 (barrel with link); FN Browning High Power; Smith & Wesson M4506; CZ 75.

the death of its inventor. It must be noted, however, that Browning had patented similar actions in around 1900, but had not made use of them. The reason for reviving this system for a weapon that later became the famous FN High Power pistol was, most probably, an attempt to avoid patent infringements with Browning's own patents that were owned by Colt.

Since the end of WWII, this Browning High Power-type action, in a number of modifications, has dominated the service-type pistol world. The open slot can be replaced with a closed slot in the lug below the barrel, or with a sloped projection; the number of the locking lugs could be anything from one to three and these can lock either into recesses made on the underside of the slide, or into the ejection port opening on the slide.

In any case, this system is known for reliability and simplicity and, while a tilting barrel system may promise less potential for accuracy, with proper implementation it can shoot at least as well as any other action, as proved by some American M1911-type or Swiss SIG P210 pistols.

Trigger Units

The trigger unit is responsible for the actual firing of the pistol. Its primary functioning starts with the pull of the trigger, which is done by the shooter, and ends when the firing pin hits the primer of a cartridge, thus causing the discharge. The trigger unit must therefore comprise at least the following parts:

- Trigger – a sliding or pivoted part, usually of crescent shape, which must be pressed by the shooter's finger to fire the weapon.
- Firing pin – a small pin which strikes the primer of a cartridge in the chamber, thus igniting the charge and causing the round to fire.
- Main spring – the source of energy for the firing pin.
- Sear – an internal latch which is used to hold the main spring in a compressed state and then release it after the pull on the trigger.
- Trigger bar – an internal or external link which connects the trigger and the sear.

There are two major ways to compress the main spring and then transfer its energy to the firing pin upon discharge. These are distinguished as hammer-fired designs and hammerless or striker-fired designs.

Hammer-fired design This is the older system; it involves a relatively heavy pivoting hammer, which first must be cocked against the pressure of the main spring to accumulate some energy in it; once cocked, the hammer is captured by the sear and held back until the sear is retracted by the pull of the trigger. Once the sear is taken off, the hammer pivots back to strike the firing pin and thus discharge the weapon.

Hammerless or striker-fired designs These have no hammer. Instead, they rely on a large firing pin (also known as a striker), which is directly connected to the main spring. In this case, the sear directly engages the striker when the weapon is cocked.

Each system has its own advantages and disadvantages. The advantage of the hammer system is that whether or not the gun is cocked is easily discernible by the position of the exposed hammer; also, it allows the user to cock and decock the hammer manually as needed. On the other hand, protruding hammers sometimes have a tendency to catch the clothing or webbing, especially during emergency draws from the holster or pocket. Also, if the gun is carried with the hammer cocked, dirt or other debris can enter the opening between the hammer and firing pin and jam or block the action.

Striker-fired designs usually need fewer parts and have no additional openings or projections that may catch on clothing. Their primary downside in pistols is that in most cases the user cannot manually operate the striker and in many cases has to pull the trigger to release the striker from the sear during disassembly. It is also harder to distinguish the state of the striker (whether it is cocked or not), although this can be solved to some extent.

Furthermore, trigger units, either hammer or hammerless, can be split into two major groups, known as single action and double action:

- **Single action (SA)** trigger systems are the oldest of all and the pull of the trigger has only one action – withdrawal of the sear and release of the hammer or striker.
- **Double action (DA)** triggers became popular in the nineteenth century, with the introduction of cap-and-ball revolvers; in pistols, they first appeared circa 1906 in a small pocket pistol – the 'Little Tom' made in Austria – and later became popular thanks to several Walther pistols, designed in Germany prior to WWII. In this system, the pull of the trigger performs two actions – first, it cocks the hammer via an internal linkage; then, once hammer is cocked, a further pull on the trigger withdraws the sear and causes the gun to discharge. By necessity, this system requires a longer and heavier trigger pull, compared with the single-action system. Most modern double-action pistols can fire in either mode, depending on the initial state of the hammer or striker, whether it was cocked previously or not. Therefore this system is sometimes referred to as double/single action.

Double-action only (DAO) Recent decades have seen the rise of so-called pistols, in which each and every shot is made in double-action mode and the hammer or striker cannot be held in a cocked state unless the trigger is being pulled. There are certain advantages and disadvantages to each of these systems, which are discussed below.

Single-action triggers are favoured for their short trigger travel and the relatively low force necessary to pull the trigger (usually around 1–2.5kg/2–5lb, although in some sport guns can be as low as 100g/4oz). Combined with the consistency of trigger pull from shot to shot, this system is the most favoured for accurate shooting. Disadvantages of this system include the necessity of some form of manual safety, if the gun is to be carried ready for immediate firing, that is, cocked. The few attempts to make single-action pistols without any positive manual safety turned out as predictable failures from the safety point of view. Because of the low force required to activate the trigger, this system also emphasizes a proper trigger discipline. However, with sufficient training of the users and properly designed safety arrangements, this system is as safe as it could be and is still favoured by many professionals who have to use their pistols in real-life scenarios.

Double-action systems allow users to carry firearms with a live cartridge in the chamber and the hammer or striker lowered. This minimizes the chances of an accidental discharge during (improper) handling of the pistol and in some cases allows the manual safety system to be discarded and replaced with safe decocking arrangements. This simplifies the handling of the weapon and minimizes the requirements for training; the penalty is a decrease in accuracy when the gun is fired in double-action mode, because of the long and relatively heavy trigger pull (usually 4–5kg/10–12lb, sometimes more). It should be noted that the double-action feature applies only to the first shot: after that, the hammer or striker is recocked by the recoiling slide and the gun functions in single-action mode until it is decocked. Another advantage of the double-action system is that it allows for a quick second strike on a failed primer in the case of a misfire. This might be a rare case for many civilian or police shooters, but for military users, whose ammunition may be stored in combat conditions for longer periods of time, this could be useful. One criticism of this system is that the trigger pull is inconsistent, since a long and heavy pull for the first shot is followed by shorter and lighter pulls for the subsequent shots. It is also necessary for the shooter to remember to decock the weapon or apply a safety catch after shooting.

Double-action-only systems further simplify the use of firearms, as the hammer or striker is always uncocked and, even with the cartridge in the chamber, the chances of an accidental discharge are minimized: every shot is fired in double-action mode. This system also allows manual safety controls to be dispensed with, converting the gun into a simple 'point and shoot' device. This is as good as it gets for defensive weapons which are to be shot very seldom and mostly under intense stress; however, the consistent but long and heavy trigger pull is less suitable for fast and accurate shooting.

There have also been several attempts to combine the positive features of all the aforementioned systems without most of their disadvantages, but, in every case, these new systems had their own downsides, sometimes serious.

The first such attempt was made by the Austrian Glock company which 'reinvented the wheel' (or, to be precise, the pre-cocked striker system) in around 1982. This system, originally designed by Karel Krnka in Austria for the Roth-Steyr pistol of 1908, is a clever combination of the single- and double-action system. During the reloading cycle, the main spring that operates the striker is only partially cocked. Full cock is achieved by the double-action-only trigger pull; therefore, when the gun is carried the mainspring is not compressed enough to provide ignition in the case of a sear failure. As a result, the trigger pull is shorter and somewhat lighter compared with true DAO designs, because the user has to compress the pre-loaded mainspring only partially. The first obvious downside to this design is the lack of a second-strike capability, although with modern-day ammunition quality, misfires occur quite seldom. Another downside is that,

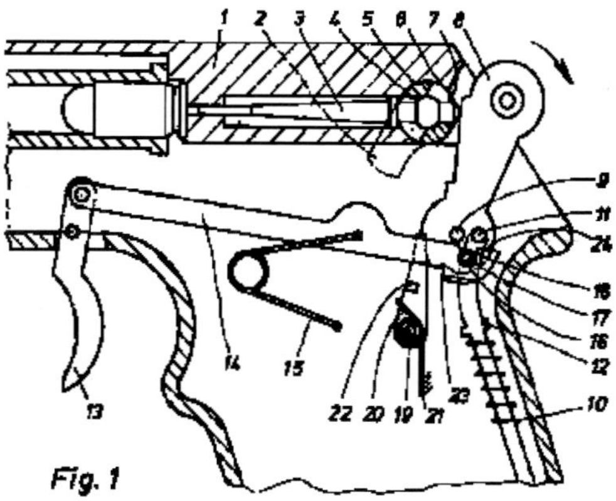

Fig. 1

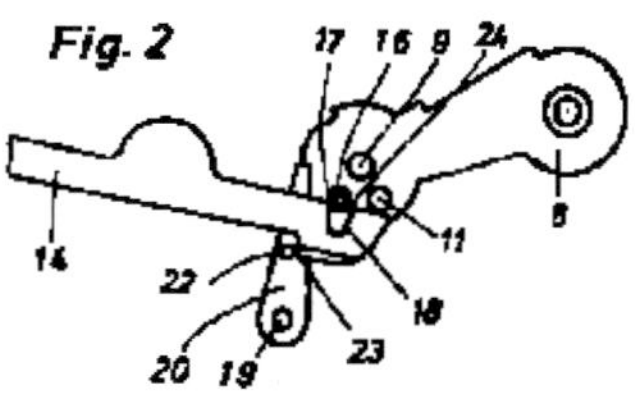

Fig. 2

Patent diagram explaining one of the simpler versions of a double-action trigger system. This patent has been issued to the Carl Walther Company of Germany.

like most true DAO designs, it is often made without any manual safety, but the relatively short and light trigger pull is less forgiving of careless handling by poorly trained users.

A second attempt was inspired by FN's Fast Action system (later renamed SFS – Safety Fast Shooting), in which the main spring is mechanically uncoupled from the hammer while the gun is being carried. This allows for safe carry with the hammer down on a loaded chamber (with minimal chances of accidental discharge), yet without a heavy double-action trigger pull for the first shot, as the main spring is already fully cocked. In the original Fast Action system the hammer was manually lowered after cocking and then brought back to full-cock position with a push on the 'safety' lever. In several later modifications, such as the Triple Action system from Daewoo, the LEM system from HK or the LDA system from Para-Ordnance, the hammer is cocked for the first shot with a double-action-like pull of the trigger, although with significantly less effort required for the compression of the relatively weak hammer spring.

The key problem with this system is that in most cases (with the exception of the Daewoo Triple Action), it does not allow for a fast second strike in the case of a misfire, as the only way to compress the mainspring is to cycle the slide or manually cock the hammer. Another issue is that even if the hammer is down, in many cases the mainspring will remain compressed and the only possible way to release it will be to dry-fire the pistol (true only for certain models like the FN GP-SFS or Para-Ordnance LDA). And as we learned from the Glock experience, there is a distinct possibility of a negligent discharge being caused by an irresponsible or poorly trained user who forgets to clear the chamber.

Safeties

The key function of any safety system in a firearm is to prevent accidental discharges during handling and storage. There are two major types of safeties – manual and automatic. This distinction is based on the key source of activation of the safeties. In the case of a manual safety, the shooter must make a deliberate action to engage (turn on) or disengage (turn off) the safety mechanism, that is, pull the safety lever or push the button. In the case of an automated safety, it is disengaged automatically when the user handles the firearm in a predetermined manner and is engaged at all other times. The two best examples of automated safeties are the grip safety (which is disengaged once the user properly grasps the pistol in the shooting position) and the trigger safety (which is disengaged only when the shooter pulls the trigger properly). There are some other types of automatic safeties that are described below.

Manual safeties appeared on the earliest pistols such as the Mauser C96 or FN Browning M1900. The most common type of manual safety control is a lever, located at the side of the pistol, usually within the reach of the shooting hand's thumb. These could be located on the frame, on the side of the slide, or even on the back of the slide. Most modern pistols have ambidextrous safeties, which means that manual safety levers are duplicated on either side of the weapon.

The less common type of manual safety is the cross-bolt button type, which was employed on several post-war Beretta pistols and one Walther (P1A1), but has now gone out of favour. This system was quite positive in functioning but awkward to operate, especially under the stress of combat.

Once engaged, a manual safety usually locks the sear to avoid accidental release of the hammer or striker. Additionally, it can block the hammer or striker itself and can also block the slide from cycling.

Automated safeties were invented to make weapons even safer in handling. Many modern pistols (especially those with DAO triggers) do not have any manual safeties and rely solely on several automated safeties to avoid accidental fire. These are outlined below.

Half-cock safety The oldest form of automated safety is the so-called half-cock safety. It was already known during the cap-and-ball era and was the simplest safety of all. In fact, it was just an additional notch in the hammer, located in such manner that once the sear engaged this notch, the hammer then rested in the 'half-cock' position. Its main goal was to capture the hammer if it slipped from the shooter's thumb during manual cocking, or in the case of primary (full-cock) notch failure.

Grip safety This is probably the second oldest automated safety. It works by blocking the trigger unit in one way or another unless the user grips the pistol in a proper manner. This avoids accidental discharge if the pistol is dropped or otherwise handled roughly. This safety is usually made in the form of a button or lever built into the backstrap or frontstrap of the pistol grip. This control is pushed outwards from the grip by its own spring and in this position blocks the trigger unit. Once the shooter grips the pistol, the hand pushes the grip safety lever or button inwards, thus unblocking the trigger unit. This system became famous with several

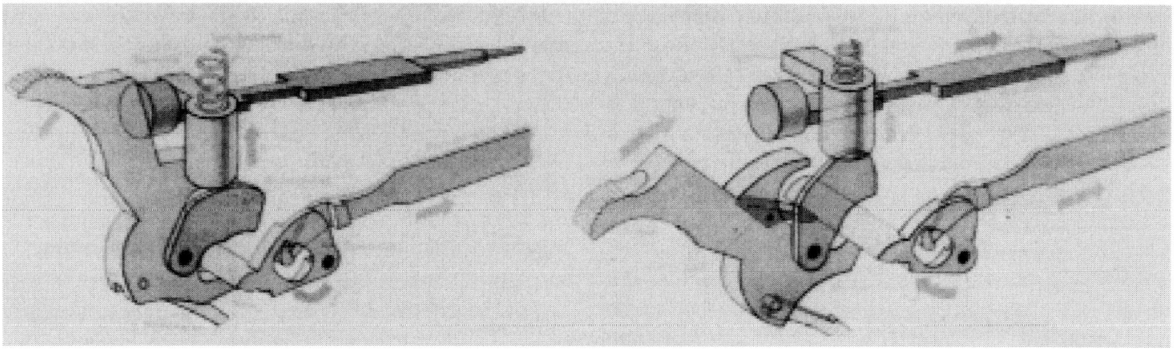

Illustration from the original manual for the German Walther P38 pistol that shows the working of a firing-pin block safety.

of John Browning's designs and is still used in several pistols.

Disconnect safety This is the other older type of automated safety. It works by blocking the trigger unit unless the slide is fully closed and is 'in battery' (that is, ready to fire). This is necessary to avoid firing with the cartridge not fully chambered and supported by the barrel and slide. Usually this safety mechanism is made in the form of an internal lever or projection on the trigger bar, which decouples the trigger bar and the sear once the slide leaves its in-battery position. Therefore, if the slide is not fully closed, a pull on the trigger will not cause the weapon to discharge.

Magazine safety or magazine disconnect This method blocks the firing of the pistol while the magazine is being removed. The key intent is to avoid negligent discharges during cleaning and disassembly procedures, if the user forgot to remove the cartridge from the chamber. This feature was rather popular for a time in European pistols, but on the US market it was always viewed with suspicion. There were two main reasons for that – first, in most designs the magazine safety somewhat affects the trigger pull, thus, in theory, decreasing the weapon's potential for accuracy. Second, many users liked the option of single-loading their pistols directly to the barrel, via the ejection window in the slide. However, this feature is still retained in several modern pistols, if only to avoid potential liability issues.

Firing-pin safety or firing-pin block First implemented in revolvers at about the turn of the nineteenth/twentieth centuries, it was tried on a self-loading pistol in the USA between the wars and rejected, before being (in a very different form) implemented in the Walther P38 pistol. The role of this safety was to block any movement of the firing pin towards the primer of the cartridge in the chamber at all times except when the trigger was pulled. This avoided accidental discharges caused by the inertia of the firing pin when the gun was dropped, or by the failure of the sear or hammer. A firing-pin bock became a necessity for most modern pistols, especially those DA or DAO weapons that do not have manual safeties. The most common form of firing-pin block is a small cross-bar that can slide up and down in the vertical channel formed within the slide. When the trigger is not pressed, this bar stays in lower position under the pressure of its own spring, and blocks the forward movement of the firing pin. When the trigger is being pulled, it raises this block out of the way of the firing pin, via some form of internal linkage. There have been other versions of this safety, such as one where the firing pin rested in a specially formed channel in the slide, which prevented it from moving forwards, and, furthermore, the firing pin was located out of the reach of the hammer. When the trigger was pulled, a special lever raised the firing pin out of its channel so that its rear part was accessible by the hammer and it was then allowed to fly forwards and hit the primer. Once the trigger was released, the firing pin was again put into its safety channel.

Sights

Sights are necessary to enable the weapon to be aimed properly towards the desired target. There are two major types of sights: so-called 'iron sights', which actually can be made not only from metal but also from plastics; and optical sights, which use optical lenses or translucent screens, or both, to provide the operator with an aiming reticule displayed over the target. Optical sights are rarely seen on combat pistols, mostly because of their size, weight and fragility, as well as because of the limited engagement ranges typical of pistol shooting.

The standard iron sight consists of two parts, namely the front sight and the rear sight. The standard aiming procedure is to align the target and front sight picture through the centre of the opening or aperture made in the rear sight. It also must be noted that aperture rear sights (that is, rear sights which have a circular hole instead of a U- or V-shaped notch) are almost unknown in the combat pistol world. The simplest and also most rugged are fixed sights, which are either permanently fixed to the barrel or slide, or made as an integral part of the respective pistol sub-assemblies. However, such sights have serious disadvantages. First, in the event of the point of aim (PoA) of the sights not being aligned with the point of impact (PoI) (which may happen at different ranges, or when using different types of ammunition), the only way to make corrections is to aim at a spot which is higher or lower than the desired point of impact, which may be difficult, especially in the stress of combat. The initial zeroing procedure (the process of bringing the PoI to the same spot as the PoA with selected ammunition and at a predetermined range) also may be complicated and could require use of a file or the replacement of some parts. The easiest way to improve and simplify the zeroing procedure is to use removable sights; this is usually done by cutting dovetail-shaped cross-cuts in the slide or barrel and then inserting the sights into these cuts. Horizontal adjustments are made by drifting the tightly fitting front or rear sight sideways with a suitable tool; vertical adjustments are made by the replacement of the front or rear sight with another one of different height.

During the first part of the twentieth century, a few military pistols that had been issued with detachable shoulder stocks to act as a kind of carbine were also fitted with adjustable tangent or flip-up sights that allowed for easy and fast adjustments for different ranges, by the simple movement of an adjustment slide on the rear sight. These sights were usually scaled from 25m or 50m, up to the rather optimistic 500m, or even 1,000m. To assess the usefulness of such a complicated and expensive sighting system, one must know that the relatively long-barrelled Mauser C96 pistol, fitted

Typical fixed-sight arrangement for a modern pistol (in this example, a SIG-Sauer P225). Both front and rear sights are inserted into dovetail-shaped slots made in the slide (that is, dovetailed). Note that the sights are fitted with contrasting white inserts to improve aiming against dark-coloured targets.

Quick-adjustable tangent-type rear sight of the Belgian FN Browning High Power pistol made during WWII. It is marked up to 500m, with the intention of using the pistol with a detachable shoulder stock as a carbine.

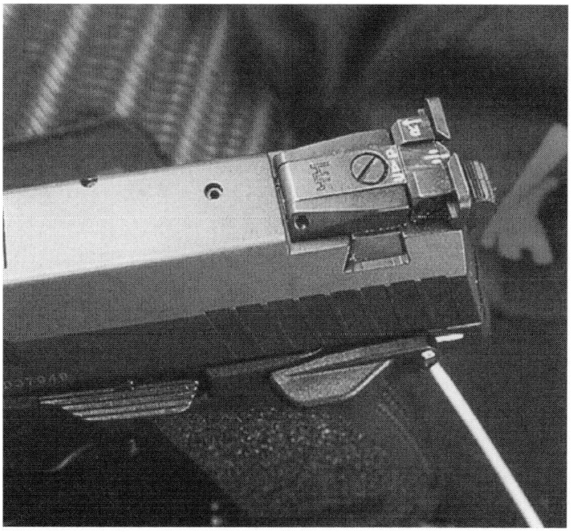

Adjustable sport-type rear sight of the German Heckler & Koch USP Tactical pistol; note that it requires a screwdriver to make adjustments.

Simple but practical three-dot sight arrangement on the Smith & Wesson Sigma pistol.

with the shoulder stock, in ideal conditions spread the hits about 4m (13ft) across the point of aim at a range of 1,000m (3,280ft). Real-life experience during both world wars and many smaller conflicts proved that most pistol shooting occurred at ranges well below 25m (82ft), and thus such sights became obsolete. The only military pistols made after WWII with sights instantly adjustable for range were the Belgian FN High Power pistol (this version was made in limited numbers) and the Soviet Stechkin APS.

More common are so-called target-type sights, which could be adjusted for range and windage using a screwdriver or other similar tool. These adjustments are made during the zeroing-in procedure at the range for a predetermined type of ammunition and at one specified range; in combat situations, these sights are used as standard fixed sights. The major advantage is that this set-up allows the zeroing of the pistol at a standard range with different types of ammunition, such as subsonic and standard, depending on the type and profile of the upcoming mission. Such sights are therefore most often used on sport-type pistols and on the few military-type pistols which are to be used with Special Forces personnel.

One useful and relatively simple modification that can be made to any type of open sight is the addition of contrast inserts to both the front and rear sights. This allows for a better sight picture against darkened backgrounds and generally makes for faster aiming. The simplest way is to paint a white dot on the front sight and two more white dots on the rear sight, at either side of the aiming notch. These inserts also can be made luminous to be used under poor light conditions. Such sights are usually marketed as 'night', but must not be confused with true night sight equipment, which uses electronic-image-intensifying or thermal vision technologies. Inserts also may have various shapes, such as simple dots, dot plus a U-shaped mark, two vertical stripes and so on – the general idea is the same. The colours also may be different – white, orange, bright (luminous) green or yellow. However, white is dominant.

Appendix Two

THE BASICS OF SOUND MODERATORS (SILENCERS)

The key source of the sound of a gunshot is the shock wave created by the hot powder gases that emerge from the muzzle of the firearm at a significant pressure. Therefore, the simplest way to achieve silent shooting, at least in theory, is to lock the powder gases inside the gun or cartridge. In practice, this is the most complicated way, as it requires special ammunition and special firearms. Very few specialized arms have been built using this principle; these include several Soviet-made pistols like the MSP, S-4M and PSS, as well as semi-experimental American 'tunnel' revolvers and the German HK P11 underwater pistol. In most of these cases, the hot gases are locked inside the cartridge case and thus ammunition is of a complicated design and very expensive.

The technically easier but significantly less effective way is to use some sort of device to cool down the powder gases before releasing them into the atmosphere. This is exactly similar to exhaust silencers used in automobile engines. The basic principle of most sound moderators ('silencer' is the more common, but less accurate term, as the gun in most cases in not silenced completely) is to allow the muzzle blast to occur within the internal volume of the moderator, more or less isolated from the atmosphere. The simplest sound moderators are just enclosed metallic cans attached to the muzzle of the firearm, with a hole drilled in the base, opposite the muzzle, for the bullet to exit. Being the most simple, such devices are also the least effective. Their effectiveness directly depends on the internal volume of the can – the bigger it is, the more effectively it moderates the sound of the gunshot. To increase further the effectiveness of a moderator of any given size, there are several ways, some of which may be used together.

The first way is to slow down the muzzle blast gases artificially, simultaneously taking off their energy by allowing them to heat up parts of the moderator. Usually, this is done by inserting a number of metallic baffles inside the moderator case. These baffles, obviously, must have holes for the passage of the bullet. Upon firing, the muzzle blast enters the inside of silencer and powder gases impact on the baffles and heat them up, releasing their energy, which would otherwise be spent on the generation of the sound wave. This design has the additional benefit of minimizing the recoil, as the baffles of the silencer work the same way as a reactive muzzle brake.

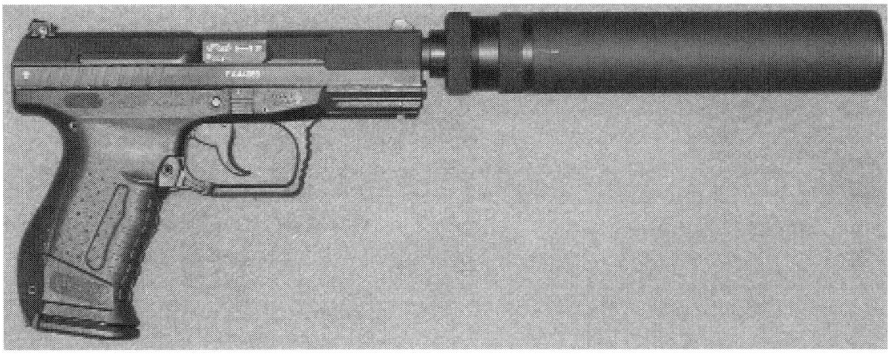

9mm Walther P99 pistol with the silencer attached.

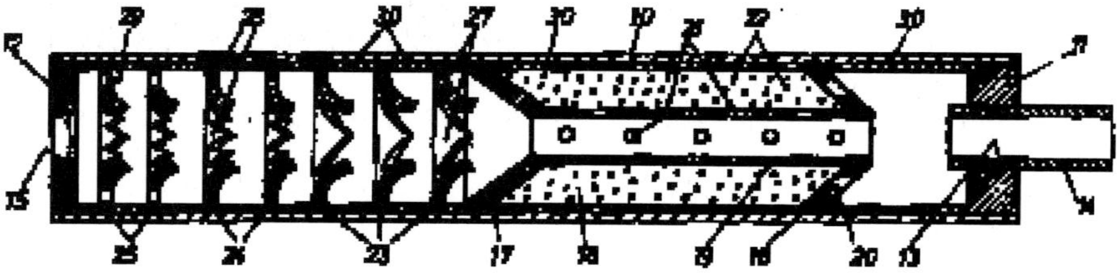

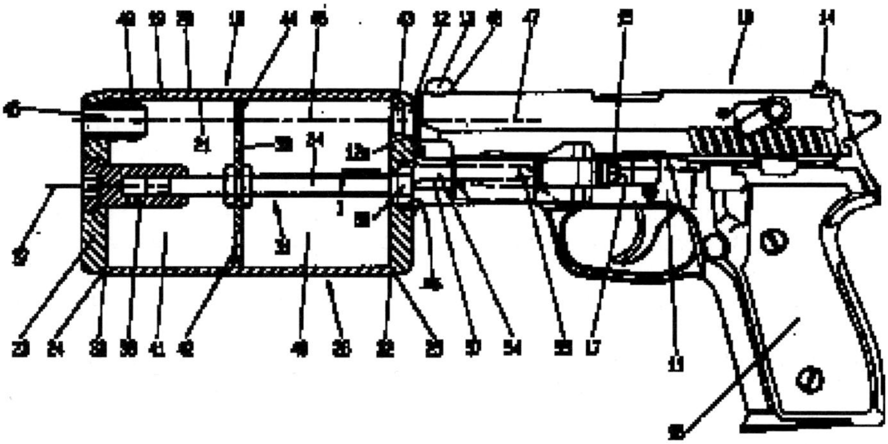

Just a few designs of add-on sound moderators (silencers) for firearms, patented in the USA during the twentieth century. *(From original patent publications)*

The second way is to isolate the powder gases completely within the moderator by closing down the bullet exit hole just after its passage. The easiest way to achieve such an effect is to use rubber or polymer baffles with no permanent bullet passage holes. Upon firing, the bullet will penetrate such baffles, which will self-seal afterwards, capturing most of the powder gases inside. The most obvious downside to this design is the life expectancy of the rubber baffles, which is usually about 100 shots, or even fewer. Another downside is that the bullet loses part of its energy upon penetration of the baffles, so accuracy may suffer somewhat if the baffles are not completely uniform. Lastly, such an arrangement will work only with relatively low-pressure and low-power ammunition, as the ability of an already pierced baffle to contain the pressure is very limited.

Sometimes these two principles are combined in a single design, so a moderator may have both fixed baffles with bullet passages and self-sealing rubber baffles.

The third way is to drain energy rapidly from the muzzle blast by using some physical process, such as vaporization of the water or some other liquid. Water vaporization is quite energy-consuming and therefore may cool down blast gases rather effectively. However, this method is never used alone, but often applied to the first or second method as an effectiveness multiplier. To do so, a small amount of water (about a teaspoonful or less) is poured into the sound moderator before shooting. Since the water is vaporized during the first shot or two, this method has limited combat usefulness.

One aspect that must be kept in mind when assessing overall effectiveness of the silenced weapon is the velocity of the bullet. Bullets that travel at supersonic velocities (roughly 320m/s (1,050fps) or faster) generate a 'sonic boom' on their travel, which sounds like a sharp 'crack' to observers. Therefore, it is reasonable to use ammunition with heavier bullets that travel at subsonic velocities along with the silencers.

Another aspect of add-on silencers is that they add significant weight to the barrel, which may impair the reliability of the functioning of most recoil-operated weapons. The weight of the silencer must therefore be kept as low as possible. This is not the case with pistols that have fixed barrels, such as blowback or retarded blowback designs.

9mm Heckler & Koch P7M13 pistol showing an extended threaded barrel, which is necessary to attach the silencer (visible below the pistol). *(Courtesy of Heckler and Koch)*

Another drawback of detachable silencers is that they add significant length to the pistol, which compromises their comfortable carry and speed of draw from the holster. One way to improve the handling of a dedicated silenced pistol is to adopt a so-called 'integral' silencer. Integrated designs are usually shorter because they move the silencer back towards the breech, so the rear expansion chamber of the silencer is placed around the front part of the barrel, rather than ahead of the muzzle. Another plus of an integral design is that it allows the drilling of pressure drain holes in the barrel, which are intended to drop working pressures upon discharge and slow down the muzzle velocity of the bullet. This option is very useful when using standard military ammunition such as 9 × 19 NATO that originally has supersonic velocities in standard weapons. The best examples of such integrally silenced weapons are Russian 9mm PB and Chinese 7.65mm Type 67 pistols, described in the respective sections above.

Appendix Three

Modern Pistol Ammunition

Calibres

The calibre of any firearm is the measurement of the bore of its barrel. It could be measured directly as the diameter of the bore, or some intermediate system could be used as in the case of shotguns, where the calibre or gauge equals the number of lead ball bullets of that diameter which could be molded from one pound of lead. In the case of rifled firearms, the calibre is the measured diameter between lands or grooves of the rifling (see the picture). However, for many reasons actual (measured) calibre may differ from the calibre designation. Most often this misnomer is based on historical or marketing issues. Another source of complications is that there are two measuring systems used worldwide – the metric system and the imperial or inch system. Metric calibres are measured in millimetres, i.e. '7.65mm' or '9mm'; Inch calibres are measured in hundredths or thousandths of an inch, with the omission of the leading zero, i.e. '.30' or '.300' (0.30 inch or 7.62mm) or '.45' (0.45 inch or 11.43mm). The direct relationship between metric and inch calibres is represented as 1 inch = 25.4 millimeter, or 1 millimeter = 0.039 inch. In some cases, the nominal inch calibre is the same as the bore diameter (between the lands), as in the case of many .30 calibre weapons that have bore diameters of 0.30 inch or 7.62mm. In other cases, the nominal calibre may match the bullet diameter (slightly wider than the bore) e.g. the .40 S&W. However, in a few cases, the nominal inch calibres have no direct relationship with actual bore or bullet diameter, as with .38 calibre rounds which have bullet diameters ranging from 0.357 to 0.401 inches; these cartridges retain their misleading designations from the age of black powder revolvers. Metric calibre designations tend to be more accurate, but may still vary between whether bore (e.g. 7.62mm) or bullet (e.g. 9mm) diameters are used.

Also, even if two firearms have exactly the same actual calibre, they may use cartridges of very different size and power, i.e. Soviet TT pistol and US M1 Garand rifle both have bores of 7.62mm diameter, but their cartridges are very different in size and power. Therefore, in most cases it is insufficient to know just the calibre of a firearm to procure suitable ammunition, and some additional information needs to be provided. The simplest way is to give any cartridge its own name,

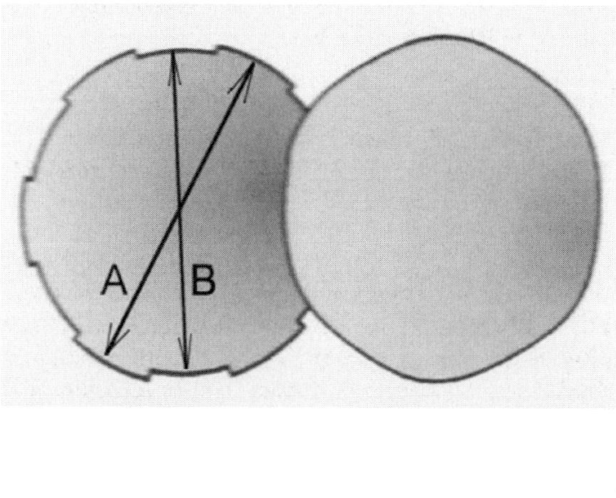

Calibre measurement and difference between standard and polygonal rifling. Left – cross-section of the bore with standard rifling, right cross-section of the polygonally rifled barrel.
A – calibre measured between grooves of the rifling
B – calibre measured between lands of the rifling

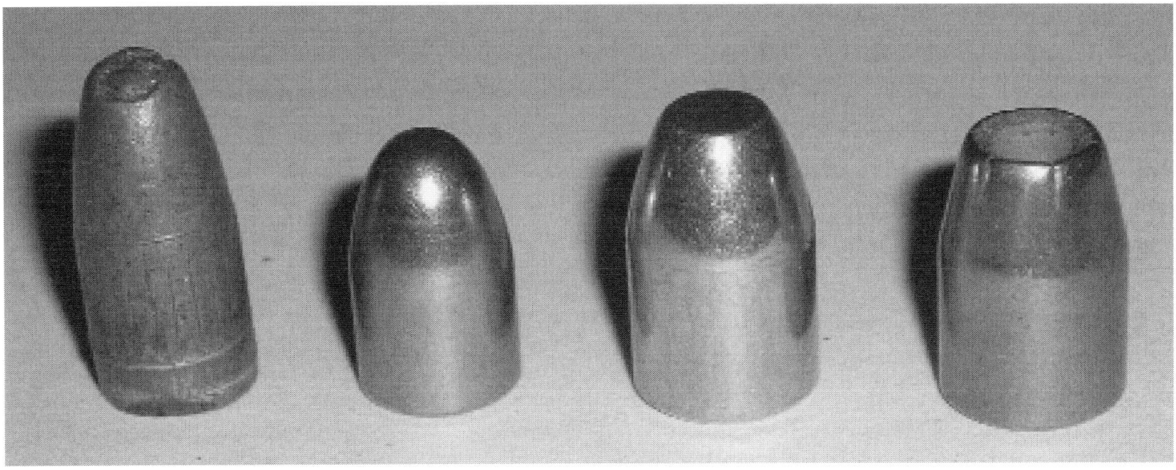

Most common types of handgun bullets, left to right: lead (unjacketed); jacketed (FMJ = full metal jacket); jacketed flat point (JFP); jacketed hollowpoint (JHP).

i.e. 9mm Steyr and 9mm Luger, or .357 Magnum and .357 SIG. In either case, the calibres (bullet diameters) are the same, but the cartridge shapes, dimensions and power are different, and they are NOT interchangeable. However, there are far too many cartridges to give them all names, so the most convenient (and most common) way with metric designations is to use the case length in conjunction with the calibre. The typical designation that follows this pattern is **9 × 19**, where '9' means the calibre and '19' is the cartridge case length, both measured in millimetres. If several cartridges of different properties have same calibre and case length, some additional information must be provided, usually in the form of a name, or a suffix which distinguishes the shape of case head. The sample of the 'name' use is 9 × 23 Largo/Bergmann and 9 × 23 Steyr cartridges, which were independent developments but are virtually indistinguishable in size and power. Another example is 9 × 23 Winchester, which, while having the same external dimensions as previous two 9 × 23 cartridges, has thicker case walls and thus can withstand heavier pressures; this cartridge can be easily loaded into firearm designed for either of former cartridges, but to do so would be extremely dangerous! Yet another example is a fourth cartridge with the same calibre and case length, the 9 × 23SR, more generally known as .38 Super Automatic or simply .38 Super. This cartridge has semi-rimmed case, that is, it has both the extraction groove and a diminutive rim, as it was designed in around 1898 to be used both in semi-automatic pistols and revolvers. Another example of similar designations but different actual dimensions are 9 × 18 PM and 9 × 18 Police cartridges. While these are identical in designations, actual calibres are different, as the 9mm PM bullet has an actual diameter of 9.2mm, and the 9mm Ultra has an actual bullet diameter of 9.02mm. Therefore, mismatching one such cartridge for another may be very dangerous for both gun and shooter. The source of this mismatch is that most western calibres are measured between the grooves of the rifling, and therefore are same as actual bullet diameter; in Russia and USSR, some calibres were measured between the lands of the rifling, therefore actual bullet diameter is bigger than measured calibre.

Considering all said above, great care must be exerted when selecting proper ammunition for any firearm.

The Most Widespread Cartridges for Combat Pistols

Bullet Types

Many and various bullet types have been developed for fighting, training and other applications; only the most common are mentioned below.

Lead bullets These are the oldest type and today are used mostly in revolver and small-bore rimfire ammunition. These are formed from lead, or, more often, an alloy of lead and antimony. Such bullets are inexpensive but usually cannot withstand higher velocities and

JHP bullets (Speer Gold Dot in this example), as loaded into the cartridge (right), and the bullet recovered from ballistic gelatin (left). Notice how significantly the bullet has expanded. (Courtesy of Oleg Volk)

produce significant lead fouling in a rifled bore during prolonged use.

Jacketed bullets These are the most common bullets and are used in military weapons because international treaties have banned expanding bullets. Such bullets are designed using a lead core that is enclosed by a gilding-metal jacket. These bullets are known for good penetration, although their stopping power is often less significant than that of expanding bullets. Jacketed bullets are sometimes referred as 'ball' bullets on historical grounds.

Hollow-point bullets These are currently the most popular choice for police and self-defence ammunition. Such bullets are designed with the hollow cavity in the nose (therefore the common name 'hollowpoint'). This cavity causes the bullet to expand once it hits the soft tissue of a human or animal body, which results in reduced penetration but a wider wound channel.

Armour-piercing pistol ammunition This is nowadays mainly intended for use against adversaries with body armour. The simplest AP bullets for handgun ammunition are usually made from solid brass or bronze; sometimes these bullets are made with pointed tips to improve penetration further. Since such bullets, because of their hardness, may cause excessive wear to the barrel, they may be covered with some a softer material, such as Teflon. In some cases, AP bullets are made with the traditional soft brass or other gilding-metal jacket and with a composite core made of a hardened steel penetrator together with some other filler. One example of such ammunition is the Belgian FN 5.7mm SS190 bullet, which has a core made partly of steel (front) and partly of aluminium (rear). Another example is the Russian 9mm 7N21 bullet, which has a hardened steel core that passes throughout the entire bullet and is exposed at the tip; the space between the jacket and the core is filled with polyethylene.

4.5 × 40R SPS

This unique ammunition was developed in the USSR in the late 1960s for combat divers. Development took place at the State Technical Institute of Precision Machine Building (TSNII TochMash), and this cartridge is used in only one weapon, unique to this cartridge, the SPP-1 underwater pistol. The SPS cartridge has a bottlenecked, rimmed brass case. The long projectile is made from mild steel, with a pointed tip. To protect the cartridge from salt water, the primer pocket and case mouth are protected with special sealing; the bullet is covered by a protective lacquer coating of dark green colour. When fired in water, the bullet is stabilized by drag forces. When fired in air, the bullet is not stabilized at all and tumbles in its trajectory.

4.5 × 40R SPS

Designation	Manufacturer	Bullet weight	Muzzle velocity	Muzzle energy	Comments
SPS	Tula Cartridge Works, Russia	13.2g (203 grains)	250m/s (820fps)	413J (306ft/lb)	Steel bullet

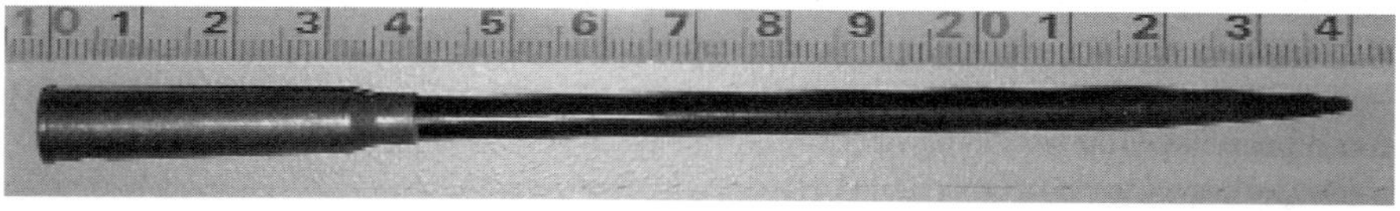

4.5mm SPS cartridge.

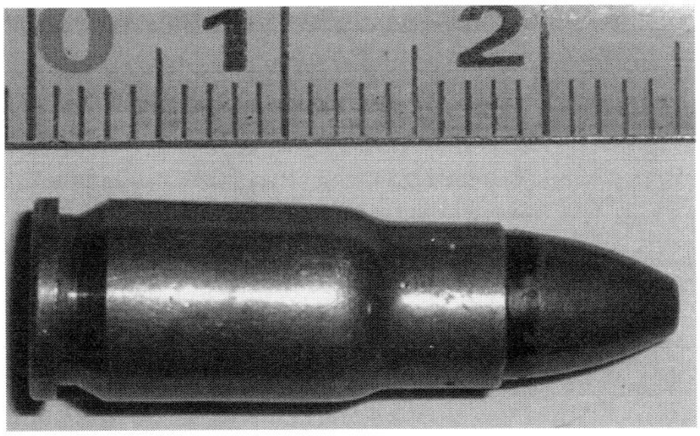

5.45mm MPts cartridge with ball bullet.

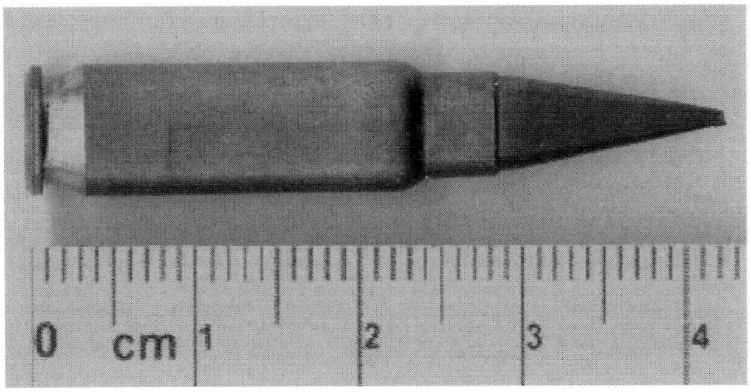

5.7mm SS90 cartridge. (Courtesy of Anthony Williams)

5.45 × 18 MPTs

The 5.45 × 18 MPTs (*Malokalibernyj Pistoletnyi Tsentralnogo boya* – small-bore pistol centrefire) cartridge was developed in 1971–2 in the State Technical Institute of Precision Machine Building (*TSNII TochMash*), for use in the easily concealable pistols then in development at the request of the KGB. The case is rimless and bottlenecked (that is, the diameter of the body of the case is greater than the calibre) and, unlike most other Soviet military cartridges, is made of brass rather than steel. The bullets are pointed and jacketed. Initially, only one bullet was produced, with a combination mild steel/lead core, but during the 1990s, when the PSM pistols were exported to Europe for sale in commercial markets, another type of bullet was produced, with a soft lead core. This cartridge is known for serious penetration (considering its tiny size), but the stopping power is marginal to say the least.

5.7 × 28 FN

The famous FN Herstal Company of Belgium started the development of this new ammunition in 1986. Originally intended for a sub-machine-gun-type personal defence weapon developed under the codename 'Project 9.0', the new cartridge represented a typical scaled-down rifle cartridge, with a relatively long, bottlenecked rimless case and a long, pointed bullet. Early bullets, known as the 5.7mm SS90, had metal jackets and lightweight plastic cores; however,

5.45 × 18 MPTs

Designation	Manufacturer	Bullet weight	Muzzle velocity	Muzzle energy	Comments
5.45×18 MPTs (7N7)	Tula Cartridge Works, Russia	2.5g (39 grains)	315m/s (1,033fps)	124J (92ft/lb)	Military issue
PSO 5.45 × 18.000-01	Tula Cartridge Works, Russia	2.68g (41 grains)	320m/s (1,050fps)	137J (101ft/lb)	Export version with lead core

5.7 × 28 FN

Designation	Manufacturer	Bullet weight	Muzzle velocity*	Muzzle energy	Comments
SS190	FN, Belgium	2.02g (31 grains)	650m/s (2,132fps)	427J (316ft/lb)	Armour piercing
L191	FN, Belgium	2.02g (31 grains)	650m/s (2,132fps)	427J (316ft/lb)	Tracer
SB193	FN, Belgium	3.6g (55 grains)	280m/s (918fps)	141J (104ft/lb)	Subsonic
SS192	FN, Belgium	1.8g (28 grains)	620m/s (2,034fps)	346J (256ft/lb)	Training

* When fired from the FN Five-seveN pistol.

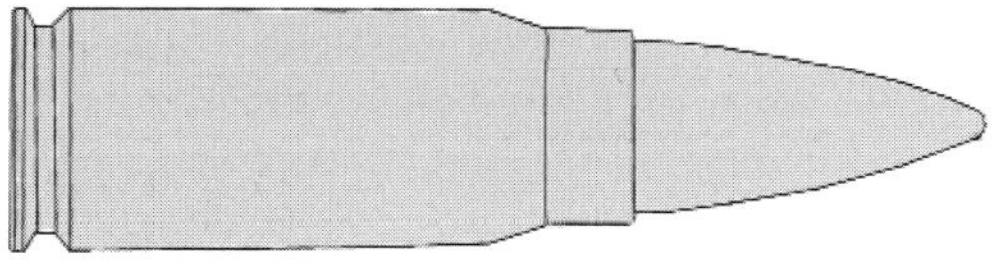

5.8mm DAP92-5.8 cartridge (drawing).

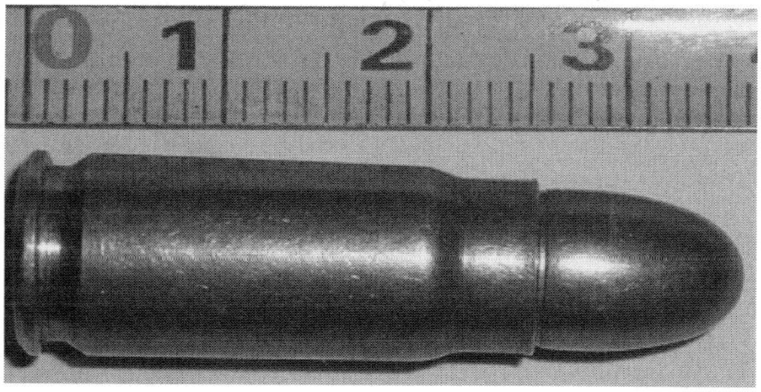

7.62mm TT cartridge with ball bullet.

with the intention of improving penetration against body armour, in around 1994 FN developed a heavier SS190 AP bullet with a composite core, consisting of hardened steel at the front and aluminum at the rear. Later on, several other types of bullets were introduced, including the heavier SB193 for subsonic loads, the tracer L191 and the 'practice' SS192 with a soft core instead of a steel penetrator. So far, the only handgun that fires this ammunition is the Belgian-made FN Five-seveN pistol; it is also used in FN's P90 sub-machine gun and is being considered for several more weapons in countries such as Sweden and Singapore.

5.8 × 21 DAP92-5.8

Chinese designers began development of a new small-bore pistol and sub-machine gun cartridge in around 1994. They copied the concept of the Belgian 5.7 × 28 cartridge/weapons system and so far this ammunition has been used in one military-type pistol and several prototype sub-machine guns; neither these weapons nor the ammunition has so far been offered for export from China. The main goal of this cartridge is to provide users with significant penetration against military helmets and body armour at ranges up to and beyond 50m (164ft). According to available information, the current versions of the DAP92-5.8 cartridge use composite core bullets, with steel front and lead rear parts of the core.

The DAP92-5.8 cartridge uses a lacquered steel case of rimless, bottlenecked design. The standard bullet type is FMJ with a composite core. Penetration is given as 'guaranteed penetration of PLA-issue steel helmet (1.3mm steel) at 100m plus 5cm of pine wood behind'. Apparently, a subsonic load is also available, with a muzzle velocity of about 290m/s (951fps); bullet weight for this round is unknown so far.

7.62 × 25 TT (7.63 × 25 Mauser, .30 Mauser)

This cartridge, initially known as the 7.63 × 25 Mauser, was developed in 1896 for the famous Mauser C96 pistol; it was developed from the 7.65mm Borchardt cartridge of 1893. During the early 1920s, Soviet Russia bought significant numbers of C96 Mauser pistols from Germany and, after close examination, in the late 1920s selected a slightly modified version of this cartridge as a future pistol and sub-machine gun round for the Red Army. At the same time, guns chambered for the original 7.63mm Mauser cartridge and made by Mauser and several Spanish companies (most notably Astra), were sold in significant numbers to China and various South American countries. The key differences between the original 7.63mm Mauser and the 7.62mm Soviet loadings were the size of the extraction groove and the diameter of the primer pocket (the Soviet cartridge used primers of larger diameter). Both cartridges were originally loaded to the same characteristics. It must be noted that several 'sub-machine

5.8 × 21 DAP92-5.8

Designation	Manufacturer	Bullet weight	Muzzle velocity	Muzzle energy	Comments
DAP92-5.8	State factories, China	2.99g (46 grains)	480m/s (1,574)*	344J (255ft/lb)	

* Muzzle velocity is given for the QSZ-92 pistol.

7.62 × 25 TT (7.63 × 25 Mauser, .30 Mauser)

Designation	Manufacturer	Bullet weight	Muzzle velocity	Muzzle energy	Comments
JHP	Reeds Custom Ammo, USA	3.9g (60 grains)	650m/s	824J (610ft/lb)	Modern commercial loading
DWM 403	DWM, Germany	5.5g (85 grains)	424m/s	494J (366ft/lb)	Original 7.63mm Mauser load
57-N-134S	State factories, USSR	5.5g (85 grains)	430m/s*	508J (376ft/lb)	Soviet military loading with steel-cored bullet (since 1951)
7.62-Pi	S&B, Czechoslovakia	5.5g (85 grains)	480m/s*	634J (469ft/lb)	Czechoslovak military load
7.62 M49	Prvi Partizan, former Yugoslavia	5.5g (85 grains)	518m/s*	738J (546ft/lb)	Yugoslavian 'sub-machine gun' military load; may be unsafe in most commercial pistols in this calibre

* Muzzle velocity as fired from the Tokarev TT-33 pistol.

gun' loadings, produced during and after WWII, produce pressures well beyond the levels of the original Mauser loading and are thus unsafe in older Mauser and Astra pistols. Also, certain loads produced since 1951 may have bullets with composite steel/lead cores, for both increased penetration and decreased cost.

These cartridges were used in pistols from Mauser, Astra and Star made during the interwar period, as well as in many sub-machine guns and Tokarev TT pistols and its clones, produced since 1933 in the USSR and several other countries.

This cartridge uses a rimless, bottlenecked case, usually made of brass. The standard loading is an FMJ bullet with a lead or composite lead/steel core. Special loads, such as tracers and AP, were also developed during WWII and afterwards for use in sub-machine guns. In recent times, some commercial loadings have become available with lightweight expanding bullets; these loads obviously are intended for civilian use in older, military surplus pistols.

7.62 × 41 SP-4

The SP-4 is a very specialized round, developed for the needs of the KGB and *Spetsnaz* elements of the Soviet armed forces. The key niche of this ammunition is noiseless and flashless firing during various special operations. The only weapons that use this cartridge are the PSS self-loading pistol and the NRS-2 scout knife with a single-shot weapon built into the handle. Both weapons are of Soviet origin. This cartridge uses a bottlenecked rimless case with a very short neck. The cylindrical bullet is made from mild steel, with a brass driving band at the front. The bullet is seated in the cartridge flush with the mouth and its base rests on the piston, which separates the bullet from the small load of special propellant. When the cartridge is fired, the

7.62 × 41 SP-4

Designation	Manufacturer	Bullet weight	Muzzle velocity	Muzzle energy	Comments
SP-4		9.3g (143 grains)	270m/s (886fps)	339J (251ft/lb)	Bullet is a steel unjacketed cylinder

7.62mm SP-4 cartridge.

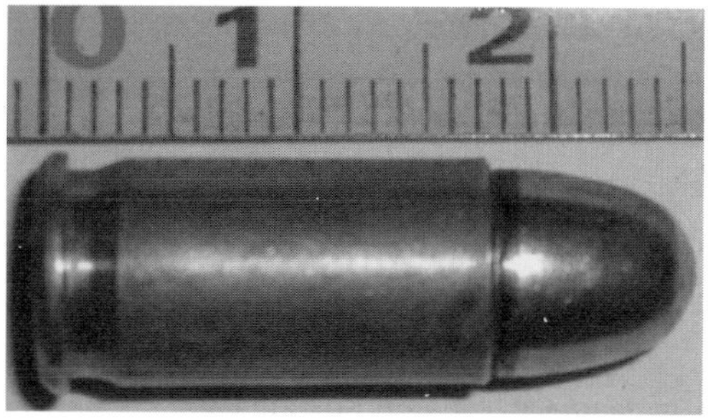

7.65mm Browning cartridge with ball bullet.

propellant gas pushes the piston forwards violently, driving the bullet forwards and out of the case; the piston is then jammed tightly at the neck of the case, locking the powder gases inside. Immediately after shooting, the fired cases are under strong pressure from inside and thus are unsafe; in time, the gases inside the case cool down and thus pressure decreases. However, Soviet manuals require special handling of the fired cases of SP-4 ammunition. Another interesting fact about this ammunition is that these cartridges are void of any headstamps, for obvious reasons.

7.65 × 17SR (7.65mm Browning, .32 ACP)

This cartridge, known in Europe as the 7.65mm Browning and in the USA as the .32 ACP, was developed by John Browning in around 1896–7, together with one of his pistol designs – a simple blowback pistol, which he sold to the Belgian FN company in 1899. In a very short time, the FN/Browning pistol of 1899/1900 pattern became very popular with European police and security forces. Just three years later, this cartridge was introduced into the USA along with the Colt Model 1903 pistol, another creation by John Browning. It must be noted that this cartridge never caught on in the USA and the only military use it saw there was in Colt's Model 1903 pistols, issued to general officers of the US armed forces until the mid-1970s. On the other hand, it found wide military acceptance in Europe, being issued to military personnel such as Naval officers and aviators, and even to infantry officers in countries such as France, Germany, Hungary and some others. In the police ranks, this cartridge served well into the 1970s, being issued along with smaller pocket-type pistols, of which the most notable were the famous Walthers – the PP and PPK.

This cartridge uses a brass cylindrical case with an extractor groove and a small rim (the so-called semi-rimmed design, hence the SR suffix in the metric designation). The presence of a rim means that this cartridge was intended for use in both semi-automatic pistols and revolvers (a great many such weapons being made in Europe before World War I, mostly in Belgium). The standard loading is a jacketed bullet, although today several good factory loadings with expanding bullets are available in this calibre.

7.62 × 17 Type 64

The 7.62mm Type 64 load was developed in China during the early 1960s for two pistols – the pocket-sized Type 64 pistol, intended for police use, and the specialized Type 64 silenced pistol. Basically, this round represents the 7.65mm Browning case with the rim removed and loaded to a lower velocity and muzzle energy. Later on, Chinese factories made more weapons in this calibre, such as the Type 67 silenced and Type 77 pocket-size pistols. During the early 1980s, China also produced a specialized version of the same cartridge for use by on-board security on aircraft. This loading, known as the 7.62mm Type 84, is a frangible semi-jacketed bullet, which is said not to penetrate a typical aircraft hull at ranges of 2m (6.6ft),

7.65 × 17SR (7.65mm Browning, .32 ACP)

Designation	Manufacturer	Bullet weight	Muzzle velocity	Muzzle energy	Comments
FMJ	FN, Belgium	4.6g (71 grains)	280m/s (918fps)	180J (133ft/lb)	Standard load

7.62 × 17 Type 64

Designation	Manufacturer	Bullet weight	Muzzle velocity	Muzzle energy	Comments
FMJ		4.8g (74 grains)	240m/s (787fps)	138J (102ft/lb)	Type 64 load

7.65 × 21 (7.65mm Luger/Parabellum, .30 Luger)

Designation	Manufacturer	Bullet weight	Muzzle velocity	Muzzle energy	Comments
FMJ	DWM, Germany	6.03g (93 grains)	370m/s (1,214fps)	412J (305ft/lb)	Original load

but is effective against human targets at ranges up to 15m (49ft).

The Type 64 cartridge has a straight, rimless brass case. The standard loading is a jacketed bullet with a lead core in Type 64 loading or a frangible bullet in Type 84 loading.

7.65 × 21 (7.65mm Luger/Parabellum, .30 Luger)
This cartridge was developed in around 1898–9 by Georg Luger from the earlier 7.65mm Borchard cartridge. Adopted in 1900 as a standard military cartridge by Switzerland, along with the famous Luger Parabellum pistol, this cartridge was also used as a military load in Brazil, Finland and Lithuania. However, since World War II this cartridge has been replaced by the more potent 9 × 19 Luger cartridge and by now is obsolete as a military load.

The 7.65mm Luger/Parabellum cartridge has a bottlenecked rimless brass case; the typical loading is a jacketed bullet.

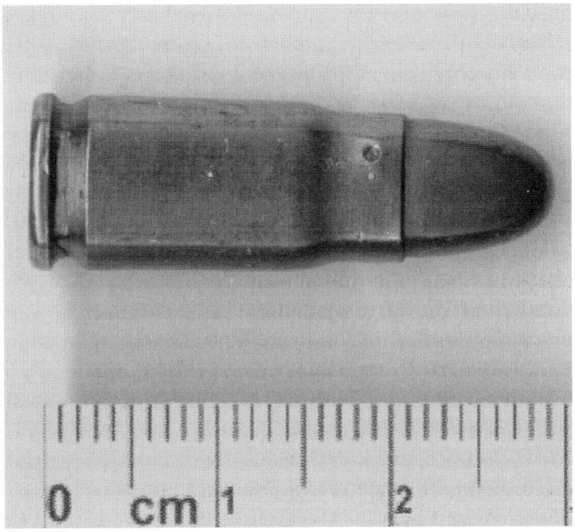

7.65mm Luger cartridge with ball bullet. (Courtesy of Anthony Williams)

9 × 17 Browning Short (9mm Short/Kurz/Corto, .380 ACP)
This cartridge, generally known in Europe as the 9mm Browning Short (to distinguish it from the earlier and now obsolete 9mm Browning Long, which had a case 20mm long), was developed by John Browning in the USA, to improve the effectiveness of the Colt M1903 pistol. Names like Kurz and Corto also mean 'short' in the German and Italian languages, respectively. In the USA this cartridge was introduced in 1908 along with the Colt M1908 pistol, as the .380 ACP. In Europe this cartridge was introduced in around 1912, in another Browning creation, the FN Model 1910 pistol. It must be noted that in the USA this cartridge was never popular as a service round; it was usually referred to as a 'pocket gun' round and regarded as the minimum adequate chambering for self-defence. On the other

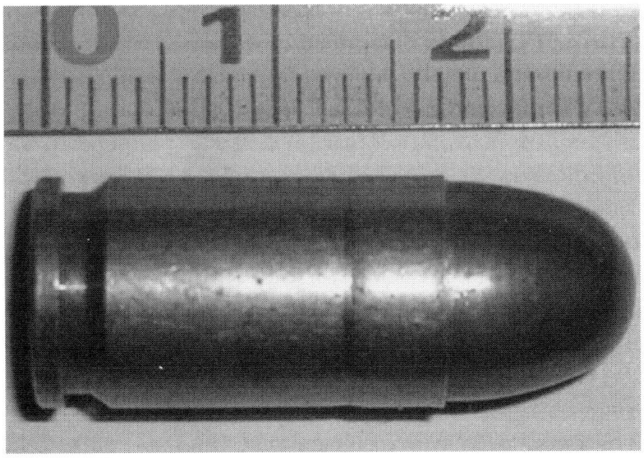

9mm Browning Short cartridge with ball bullet.

9 × 17 Browning Short (9mm Short/Kurz/Corto, .380 ACP)

Designation	Manufacturer	Bullet weight	Muzzle velocity	Muzzle energy	Comments
FMJ	Fiocci, Italy	4.66g (72 grains)	360m/s (1,181fps)	302J (224ft/lb)	Special service load issued by *Guardia di Finanza* (Italy)
SilverTip JHP	Winchester, USA	5.5g (85 grains)	305m/s (1,000fps)	255J (189ft/lb)	Modern commercial load
FMJ	FN, Belgium	6.6g (102 grains)	290m/s (951fps)	260J (192ft/lb)	Original load
JHP	Remington, USA	6.61g (102 grains)	285m/s (935fps)	268J (198ft/lb)	Modern commercial load

hand, in Europe this round quickly found favour among many military and police organizations and between the wars it was adopted as a standard military pistol loading in countries like Czechoslovakia, Italy and Yugoslavia. The military role of this cartridge mostly ended during the 1950s with the establishing of NATO and the adoption of the more powerful 9 × 19 round as the NATO standard. However, it served as a police round in many countries (such as Germany, Italy and Switzerland) up until the 1970s and is still used to a limited extent by law-enforcement services like the Italian *Guardia di Finanza*.

This cartridge is available in a wide spectrum of loadings, either with military-type jacketed bullets or more the popular self-defence and police loads with hollowpoint bullets. The case is straight, usually made of brass, but this ammunition also may be found with aluminum (made in the USA) or steel (made in Russia) cases.

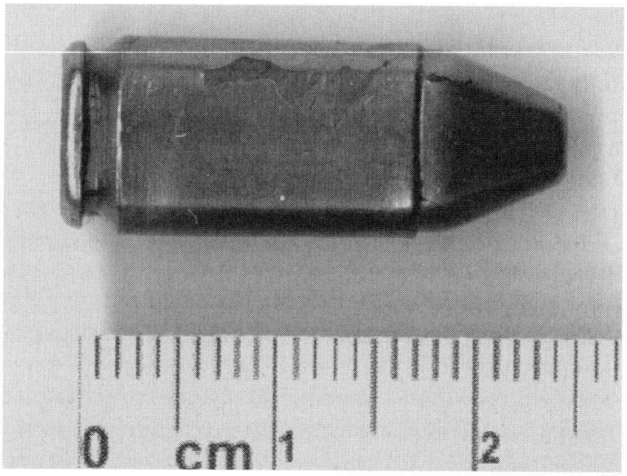

9mm Police cartridge with JFP bullet. (*Courtesy of Anthony Williams*)

9 × 18 Police/Ultra

This cartridge started its life under the name of 9mm Ultra during the mid-1930s as an experimental joint development of the GECO and Carl Walther companies, both from Germany. This development represented an attempt to create a cartridge powerful enough for military and police purposes, yet suitable for a relatively compact and simple blowback-operated pistol. GECO produced experimental ammunition and Walther made several prototype pistols, but at that

9 × 18 Police/Ultra

Designation	Manufacturer	Bullet weight	Muzzle velocity	Muzzle energy	Comments
JFP	GECO, Germany	6.09g (94 grains)	320m/s (1,050fps)	312J (231ft/lb)	Original Ultra loading
JFP	Hirtenberger, Austria	6.48g (100 grains)	323m/s (1,059fps)	338J (250ft/lb)	Police loading

time development never went any further. However, during the early 1970s this cartridge was revived as the 9mm Police in an attempt to provide German police with weapons more effective than the then-available pistols chambered for 7.65mm Browning and 9mm *Kurz*. Several police-type pistols were made for this cartridge in Germany and Italy and ammunition was produced by Hirtenberger in Austria. However, German police authorities finally decided to use the more potent 9 × 19 Luger ammunition, with the result that the 9 × 18 Police soon went into oblivion.

The 9 × 18 Police and Ultra rounds are similar except for headstamps; both have straight, rimless cases made of brass, and usually are loaded with a jacketed bullet of truncated cone shape.

Note: This cartridge must not be confused with the 9 × 18 PM described below.

9 × 18 Makarov PM

The development of a new cartridge, suitable for smaller, blowback-operated pistols, apparently began in the USSR just prior to the start of the Great Patriotic War. A few pistols were tested, but the outbreak of the war stopped the development. Soon after the war ended, the Soviet Army seriously reconsidered the role of the military pistol and selected the improved 9 × 18 round as a future military cartridge. While this round was only marginally powerful by Western standards, it proved to be adequate for personal defence weapons. Soviet tactical doctrine of the time had no use for submachine guns, which usually require more powerful pistol-type cartridges in order to be of any military use. Officially adopted in 1951, this cartridge was also imposed upon Warsaw Pact countries during the 1960s and 1970s in an attempt at standardization. So far, the 9 × 18 PM remains in active military and police service in most ex-USSR republics, including Russia. Pistols for this cartridge were manufactured in USSR and Russia, as well as in Bulgaria, China, Czechoslovakia, Hungary and Poland. Ammunition is still manufactured in Russia, as well as in the USA (for commercial use).

Just prior to the Moscow Olympic Games of 1980, Soviet ammunition factories developed several specialized loads at the request of the KGB; these included expanding and frangible bullets, as well as improved penetration loads. The Army and police continued to use standard issue jacketed ammunition. During the early 1990s, the Russian Army tried to extend the effectiveness of this round with the introduction of the 'Improved' or 'high impulse' PMM loading, with a more powerful propellant charge and a

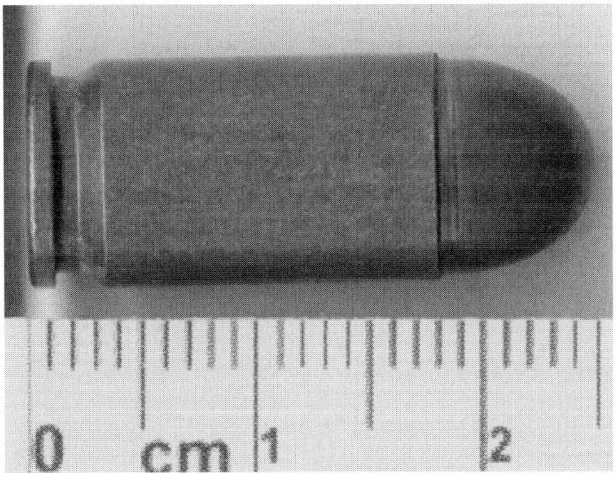

9mm PM cartridge with ball bullet. *(Courtesy of Anthony Williams)*

lighter bullet, but this was never adopted for safety reasons.

The 9 × 18 cartridge, generally known as the 9mm Makarov or 9mm PM after the most popular pistol that used it, was actually developed by the designer Syomin. It has a straight, rimless case usually made of lacquered steel (military issue) or brass (some commercial loads). The standard Soviet military/police load uses a round-nosed jacketed bullet with a composite core, made partly of lead and partly of mild steel (as a less expensive alternative to lead). The steel insert is of mushroom shape and does not help penetration much when compared with similar bullets with pure lead cores. Other loads include some specialized rounds, such as the frangible SP-8 (developed in 1980 to be used on board aircraft; many Soviet commercial aircraft crews of the time were armed with pistols, usually stored in the pilot's cabin in a special safe, to resist plane hijack attempts). Other specialized rounds include expanding and armour-piercing ammunition for law enforcement. Commercial ammunition, made in Russia for export as well as in other countries, is available with either jacketed bullets with lead cores, or with more or less traditional expanding JHP bullets.

9 × 19 Luger/Parabellum, 9mm NATO

This cartridge, generally known as *The* 9mm, is usually also referred to as the 9mm Luger (after the name of Georg Luger, who designed the pistol originally associated with this round), or the 9mm Parabellum (after the trademark of the German DWM company that originally made this round). It was developed at the request

9 × 18 Makarov PM

Designation	Manufacturer	Bullet weight	Muzzle velocity	Muzzle energy	Comments
57-N-181S	USSR, Russia	6.1g (94 grains)	315m/s	302J (224ft/lb)	Standard issue
57-N-181SM (7N16)	Russia	5.5g (85 grains)	410m/s (1,345fps)	462J (342ft/lb)	'Improved' PMM round
PBM	Russia	3.7g (57 grains)	519m/s (1,702fps)	498J (369ft/lb)	AP round, limited issue
Vz.82	Czechoslovakia	4.5g (69 grains)	400m/s (1,312fps)	360J (266ft/lb)	Czechoslovak military issue round for Vz.82 pistol and Vz.61 Scorpion sub-machine gun
SP-7	Russia	6.1g (94 grains)	325m/s (1,066fps)	322J (238ft/lb)	With expanding bullet; the hollow cavity in the bullet nose is covered by a polymer plug to improve feeding
SP-8	Russia	4.1g (63 grains)	255m/s (836fps)	133J (98ft/lb)	Frangible ammunition for onboard use in aircraft
FMJ	Winchester, USA	6.16g (95 grains)	310m/s (1,017fps)	296J (219ft/lb)	Commercial load
JHP	Russia	6.16g (95 grains)	321m/s (1,053fps)	317J (235ft/lb)	Commercial load
JHP +P	Cor-Bon, USA	6.16g (95 grains)	342m/s (1,122fps)	360J (266ft/lb)	Commercial load
JHP	Russia	7.45g (115 grains)	310m/s (1,017fps)	358J (265ft/lb)	Commercial load

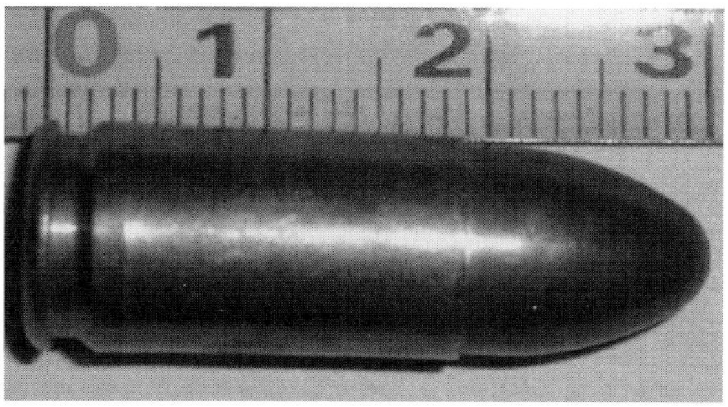

9mm Luger cartridge with ball bullet.

of the German Navy in 1902, which liked the 7.65mm Parabellum pistol developed by Luger and made by DWM, but wanted more stopping power. Following this requirement, DWM shortened the case of the original 7.65 × 21 Luger/Parabellum round to 19mm and loaded it with a 9mm jacketed bullet of truncated cone shape. The Parabellum pistol chambered for this cartridge was adopted by the German Navy in 1904 and (with a modified pistol) by the German Army in 1908. This round saw extensive use by German armed forces during World War I, in various pistols and in the first sub-machine guns. In around 1917 the original flat-point bullet was replaced by an ogive-shaped jacketed

9 x 19 Luger/Parabellum, 9mm NATO

Designation	Manufacturer	Bullet weight	Muzzle velocity	Muzzle energy	Comments
7N31	Russia	4.2g (65 grains)	600m/s (1,968fps)	756J (559ft/lb)	AP load with hardened steel core exposed at the tip of the bullet
EMB Police	Hirtenberger, Austria	5.0g (77 grains)	500m/s (1,640fps)	625J (463ft/lb)	Special police load with low-ricochet bullet
7N21	Russia	5.3g (82 grains)	460m/s (1,509fps)	560J (414ft/lb)	AP load with hardened steel core exposed at the tip of the bullet
M882	USA	7.26g (112 grains)	385m/s (1,263fps)	538J (398ft/lb)	Military-issue ammunition loaded to NATO standards
JHP	Remington, USA	7.45g (115 grains)	352m/s (1,155fps)	461J (341ft/lb)	Commercial round
SilverTip	Winchester, USA	7.45g (115 grains)	374m/s (1,227fps)	521J (386ft/lb)	Commercial round
JHP +P	Remington, USA	7.45g (115 grains)	381m/s (1,250fps)	540J (400ft/lb)	Commercial round
JHP +P+	Remington, USA	7.45g (115 grains)	400m/s (1,312fps)	596J (441ft/lb)	Commercial/ police round
2Z ball	Radway Green, UK	7.45g (115 grains)	427m/s* (1,401fps)	679J (503ft/lb)	Military issue ammunition for British Sterling sub-machine guns
JHP	Remington, USA	8.04g (124 grains)	341m/s (1,119fps)	467J (346ft/lb)	Commercial round
Balle 'O'	France	8.00g (123 grains)	385m/s** (1,263fps)	593J (439ft/lb)	Military issue ammunition for French MAT-49 sub-machine guns
L7A1 ball	Hirtenberger, Austria	8.04g (124 grains)	393m/s* (1,289fps)	620J (459ft/lb)	Military issue ammunition for British Sterling sub-machine guns
JHP Subsonic	Remington, USA	9.53g (147 grains)	302m/s (991fps)	435J (322ft/lb)	Subsonic round

Note: for sub-machine gun loadings muzzle velocities are given as fired from FN Browning High Power* or Mle.1950** pistols.

bullet that has remained standard for military loadings until now.

Between the wars, this cartridge continued its service with the German armed forces in a variety of pistols and sub-machine guns. It was also accepted as a military load by several other countries, starting with Belgium, which brought out one of the most successful and historically important 9mm pistols of all time, the FN Browning High Power. After World War II the 9mm Parabellum was quickly adopted as a standard pistol and sub-machine gun chambering throughout NATO, with the notable exception of the USA, which stuck to its .45 ACP cartridge for some forty more years. It was adopted as a military load by many other countries worldwide and since the 1970s it has found its way into the police weapons of many nations. A hundred years old, today this cartridge remains the most popular and long-lasting chambering for pistols and sub-machine guns in the world. It combines adequate combat effectiveness with relatively compact size that allows for a large magazine capacity. All in all, this is probably the best and most popular *compromise* combat pistol cartridge in the world.

The 9 × 19 cartridge uses a straight, rimless case with a slight taper. The cases are usually made of brass, but aluminum and lacquered steel cases may also be encountered. The standard military loading is an ogive-shaped jacketed bullet with a lead core, while police and commercial loadings may exhibit a wide variety of bullet types – expanding, frangible and so on. In recent years, several types of military armour-piercing loadings have been manufactured in countries like Austria, China and Russia.

One warning note must be made about military ammunition in this calibre. In many countries the 9 × 19 cartridge was considered primarily as a sub-machine gun loading and thus was loaded to higher velocities and pressure levels to provide the necessary effective range for infantry and other military personnel armed with SMGs. In many cases, the same ammunition was also issued to troops with pistols, which often resulted in a degraded lifespan or even damage to the guns and injuries to the shooters. Therefore, great care must be taken when shooting military issue or surplus ammunition in commercial or older military pistols, which may not stand up to the stronger recoil and higher barrel pressures of certain military 'sub-machine gun' loadings.

9 × 21 SP-10/SP-11/SP-12

The development of this relatively high-powered pistol ammunition, intended for both sub-machine guns and semi-automatic pistols, commenced in Russia during the early 1990s. The key goal was to provide military and law-enforcement personnel with pistol ammunition, which would be more effective against body armour than available rounds. Designers from *TSNII TochMash* first developed an armour-piercing bullet with a hardened steel core, exposed at the tip of the projectile, and then developed a new experimental round, initially known as RG052. Since 1993 this development has been funded by the FSB (Federal Security Service of Russia) and the first production rounds were issued under the codename SP-10 (SP stands for *Spetsialnyj Patron* – special cartridge).

Adopted in 1996, SP-10 ammunition is almost entirely used in one line of pistols (known under the codenames SR.1, *Vector* or SPS) and in one sub-machine gun, the SR.2 *Veresk*, both used by the FSB and special elements of other law-enforcement agencies in Russia. The original armour-piercing SP-10 cartridges were soon complemented by other types of

9 × 21 SP-10/SP-11/SP-12

Designation	Manufacturer	Bullet weight	Muzzle velocity	Muzzle energy	Comments
RG052	Russia	6.5–7g (100–108 grains)	415–30m/s (1,361–1,410fps)	560J (414ft/lb)	Original experimental loading, 1992
7N28	Russia	7.9g (122 grains)	390m/s (1,279fps)	600J (444ft/lb)	With low-ricochet FMJ bullet
7N29	Russia	6.7g (103 grains)	410m/s (1,345fps)	563J (417ft/lb)	With AP bullet
7BTZ	Russia	7.2g (111 grains)	395m/s (1,296fps)	562J (416ft/lb)	With AP-T (tracer) bullet

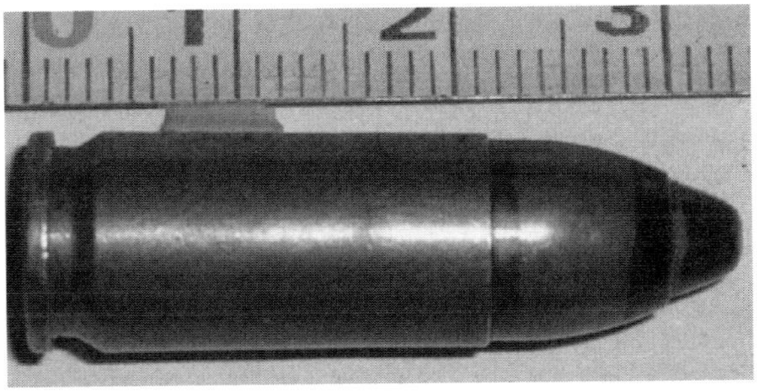

9mm SP-10 cartridge with AP bullet.

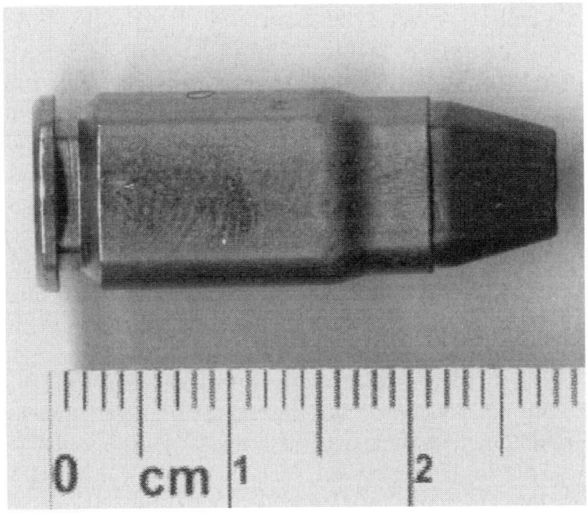

.357 SIG cartridge with JHP bullet. (Courtesy of Anthony Williams)

ammunition, including the SP-11 with a low-ricochet ball bullet (suitable for use against unprotected targets and for training), the SP-12 with an expanding bullet and the SP-13 with an armour-piercing tracer bullet. In about 2003 this nomenclature was changed to the Russian standard 7N*xx* designations, marking the official approval of this ammunition for all services.

The SP-10 cartridge, currently known as the 9 × 21 7N29, has a straight rimless case, usually made of steel and with a dark green lacquer coating. The armour-piercing bullet is of proprietary design; it has a hardened steel penetrator core that is exposed at the bullet tip, to ease the separation of the jacket upon the penetration of body-armour plates. The space between the penetrator and jacket is filled with polyethylene. The bullet is designed to stay intact upon impact with soft body tissue to conform to international treaties; in this case, it acts as a typical ball bullet of similar weight and energy. This cartridge is ballistically similar to some of the 'hotter' 9 × 19 loadings; the longer case is necessary because of the longer bullet, which has a relatively low-volume density because of its composite core that consists of polyethylene filler and a relatively long steel penetrator of 6mm diameter. Other types of bullets are fairly conventional.

These cartridges must never be used in pistols chambered for commercial 9 × 21 IMI ammunition because they generate pressures far beyond industry standards for the 9 × 21 IMI.

.357 SIG (9 × 22)

The .357 SIG cartridge is one of the most recent creations, developed as a joint effort between the US branch of the international SIGARMS Company (trademarks SIG, SIG-Sauer) and the US-based Federal Cartridge Company. This cartridge was obvi-

.357 SIG (9 × 22)

Designation	Manufacturer	Bullet weight	Muzzle velocity	Muzzle energy	Comments
JHP	Cor-Bon, USA	7.45g (115 grains)	457m/s (1,499fps)	778J (576ft/lb)	
JHP	Federal, USA	8.1g (125 grains)	411m/s (1,348fps)	684J (506ft/lb)	
JHP	Cor-Bon, USA	8.1g (125 grains)	434m/s (1,424fps)	763J (566ft/lb)	
JHP	Hornady, USA	9.5g (146 grains)	374m/s (1,227fps)	663J (491ft/lb)	

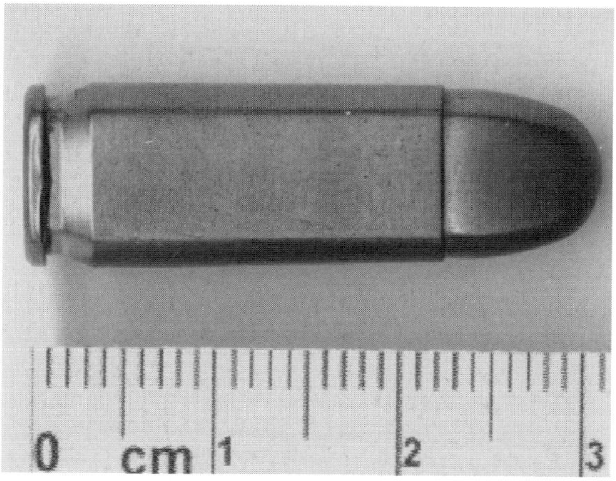

9mm Largo cartridge with ball bullet. (Courtesy of Anthony Williams)

ously intended primarily for the US market; its main goal was to provide American law-enforcement officers with a cartridge similar in ballistics and especially penetration to the famous .357 Magnum revolver round, yet suitable for use in conventional semi-automatic pistols. This cartridge has been very successful so far, being officially adopted by several major US law-enforcement agencies including the Texas DPS and the US Secret Service. Several companies make pistols for this cartridge, including SIGARMS, Beretta and Glock. It is worth noting that the latter company marks its pistols chambered for this cartridge as '.357', without any mention of the trademark of its major rival.

The .357 SIG cartridge case is based on the earlier .40 S&W cartridge, having a similar case head and body diameter, as well as a similar overall length, for obvious reasons of ease of conversion of available pistols from .40 S&W to .357 SIG by changing the barrel and return spring. The case design, however, is different, as the .357 SIG uses a rimless bottlenecked case with reinforced webbing (the base part of the cartridge), necessary because of the higher working pressures. The standard loading is a jacketed hollow-point bullet.

9 × 23 Largo (9mm Bergmann-Bayard)

This cartridge was developed in Germany in around 1903 for military-type semi-automatic pistols by the Theodor Bergmann Company. These pistols were later produced in Belgium by Pieper under Bergmann's licence and adopted by Spain and Denmark, hence the original Bergmann-Bayard name ('Bayard' was the trademark of the Pieper Company). This cartridge saw extensive use in Spain, being the standard pistol and sub-machine gun ammunition of the Spanish military and police between the start of the century and the late 1980s, hence the other common name, the 9mm Largo (or '9mm Big' when translated from Spanish, to distinguish it from other 9mm cartridges). Some of the Spanish ammunition may also be found labelled '9mm Camp Giro' after another pistol design, but it is the same cartridge.

The 9 × 23 Largo has a straight, rimless case usually made of brass. The standard loading has a full metal jacket (ball) bullet.

.40 S&W (.40 Smith & Wesson, 10 × 22)

The .40 Smith & Wesson cartridge was developed by the famous US arms-making company Smith & Wesson and another famous ammunition maker, the Winchester-Olin company, in around 1989–90. This cartridge was developed by shortening the 10mm Automatic (10 × 25) cartridge case by 3mm (0.1in), to match the overall length of the popular 9 × 19 cartridge, while providing users with more energy and impulse from a larger and heavier bullet. Compared with its parent 10mm Automatic cartridge, the .40 S&W allows the pistols to be made smaller and lighter, especially in the grip area. Although it is also somewhat less powerful, it still has enough potential to meet most police and self-defence requirements.

Today, the .40 S&W offers a good compromise between the two most popular service loads of the last century, the 9 × 19 Luger/Parabellum and the .45 ACP. It provides better terminal effectiveness than the 9 × 19, while permitting a bigger magazine capacity in the same space when compared with the .45 ACP. The one 'danger' of the .40 S&W is that its dimensions allow manufacturers to make shortcuts in design by simply rebarrelling existing 9mm pistols to the larger cartridge

9 × 23 Largo (9mm Bergmann-Bayard)

Designation	Manufacturer	Bullet weight	Muzzle velocity	Muzzle energy	Comments
FMJ		8.1g (125 grains)	341m/s (1,119fps)	470J (348ft/lb)	

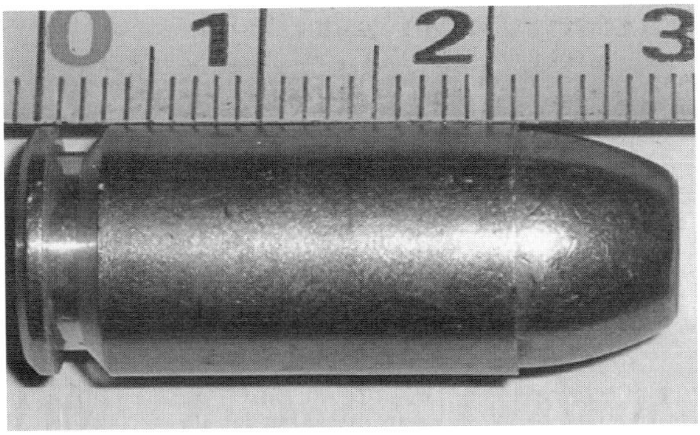

.40 S&W cartridge with JFP bullet.

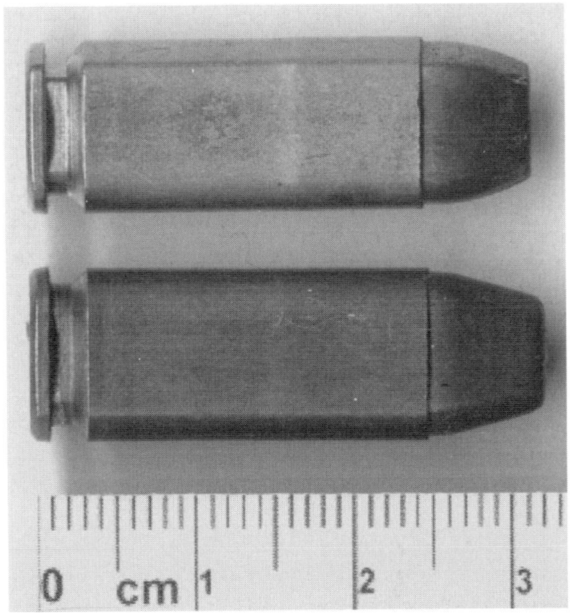

10mm Automatic cartridges with JHP (top) and JFP bullets. (Courtesy of Anthony Williams)

– due to the increased recoil impulse, this may cause excessive wear to frames and other parts of pistols initially designed for 9 × 19 ammunition. Not surprisingly, today many manufacturers prefer to design their new pistols for .40 S&W and then downgrade them to 9 × 19 if necessary, rather than vice versa. A great many companies make pistols for this round and several sub-machine guns chambered for .40 S&W also are known to exist, such as the MP5/40 and UMP-40, both made in Germany by Heckler & Koch, or the Taurus-FAMAE, made in Brazil.

The .40 S&W has a rimless, straight case, usually made of brass (steel-cased .40 S&W ammunition is made in Russia for export). The most common service loading uses an expanding jacketed or semi-jacketed bullet, but many other types of loading are also available.

10mm Automatic (10 × 25)

This cartridge was created in a search for the 'ultimate' combat round, which, while duplicating the terminal effectiveness of such proven 'man-stoppers' as the .45 ACP or .357 Magnum, should provide a bigger magazine capacity when compared with the .45 and be better suited for semi-automatics when compared with the .357. This cartridge retained the overall length of the .45 ACP, so it could be used in rebarrelled pistols originally designed for the .45; the case was entirely new. Originally loaded only by Norma of Sweden from 1983, this load was first used in the US-made Bren Ten pistol and later on in Colt's Delta Elite. In around 1988 this cartridge was picked up by the US Federal Bureau of Investigation (FBI) as a replacement for the 9 × 19 Luger/Parabellum, which had proved itself inadequate on several rather bloody occasions, such as the infamous Miami massacre of 1986. Smith & Wesson quickly produced several pistols in this chambering, while Heckler & Koch of Germany made a limited number of its MP5 sub-machine guns in 10mm for FBI agents. However, the FBI and several other US law-enforcement agencies soon discovered that this cartridge in its original loading was poorly suited for many officers because of its severe recoil, with the result that the FBI quickly adopted a reduced-charge

.40 S&W (.40 Smith & Wesson, 10 × 22)

Designation	Manufacturer	Bullet weight	Muzzle velocity	Muzzle energy	Comments
JHP	Remington, USA	10.69g (165 grains)	350m/s (1,148fps)	654J (484ft/lb)	
JHP	Winchester, USA	11.66 (180 grains)	308m/s (1,010fps)	553J (409ft/lb)	

10mm Automatic (10 × 25)

Designation	Manufacturer	Bullet weight	Muzzle velocity	Muzzle energy	Comments
JHP	Norma, Sweden	11.02g (170 grains)	408m/s (1,338fps)	917J (679ft/lb)	Original load, ca 1983
JHP	Winchester, USA	11.34g (175 grains)	393m/s (1,289fps)	875J (648ft/lb)	Modern commercial load
JHP	Norma, Sweden	11.66g (180 grains)	290m/s (951fps)	490J (363ft/lb)	Reduced power 'FBI' load
JHP	Norma, Sweden	12.96g (200 rains)	366m/s (1,200fps)	868J (642ft/lb)	Original load, ca 1983

.45 ACP cartridge with ball bullet.

loading, which then led to the shorter .40 S&W cartridge described above. Today, this round is most popular in the USA among civilian shooters, where it is used for self-defence and hunting. Several companies still make guns in this calibre, such as Tanfoglio of Italy and Glock of Austria.

The 10mm Automatic has a straight, rimless case, usually made of brass. Most common loads include various types of expanding or jacketed bullets.

.45 ACP (.45 Automatic Colt Pistol, 11.43 × 25)
This cartridge was developed as a result of extensive tests regarding the terminal effectiveness of various rounds that were conducted in 1904 by Colonel Thompson and Major La Garde on behalf of the US Army. As a result of these trials, replacement was recommended of the existing .38 calibre revolver

.45 ACP (.45 Automatic Colt Pistol, 11.43 × 25)

Designation	Manufacturer	Bullet weight	Muzzle velocity	Muzzle energy	Comments
M1911 ball	Various makers, USA	14.9g (230 grains)	275m/s (902fps)	563J (417ft/lb)	Original military load
JHP	Remington, USA	11.99g (185 grains)	305m/s (1,000fps)	557J (412ft/lb)	Commercial load
JHP +P	Remington, USA	11.99g (185 grains)	347m/s (1,138fps)	721J (534ft/lb)	Commercial load
JHP	Remington, USA	14.9g (230 grains)	267m/s (876fps)	531J (393ft/lb)	Commercial load
FMJ	Remington, USA	14.9g (230 grains)	255m/s (836fps)	484J (358ft/lb)	Commercial load
JHP +P	Cor-Bon, USA	14.9g (230 grains)	290m/s (951fps)	626J (463ft/lb)	Commercial load

cartridge with a cartridge that fired bullets of 0.45in calibre (11.43mm). The original recommendations were for a bullet of 200 grains (12.96g), but this was later changed to a 230-grain (14.9g) jacketed 'ball' bullet. The case was designed by cutting down the .30-03 rifle cartridge case to a length just below 1in (25mm). In 1911, this cartridge was officially adopted by the US military as the 'Cartridge, .45 calibre Ball, M1911', along with the famous Colt M1911 pistol. Since its introduction, this cartridge has been used in many semi-automatic pistols, as well as in a few revolvers (using half-moon or full-moon clips) and in several sub-machine guns, starting with the famous Thompson. This cartridge was never popular in Europe and was adopted only by Norway, in its M/1912 and M/1914 pistols, which were licensed Colt designs. On the other hand, this cartridge was and still is extremely popular in the USA, as well as in South and Central Americas; for many years this round was a standard military issue in countries like Argentina and Brazil.

Despite the fact that the US Army replaced the .45 ACP as its standard-issue pistol ammunition in 1985, this cartridge still fares quite well, thanks to its large bullet, which offers significant terminal effects even in the military 'ball' configuration. What is even more interesting is that in 2005, the US Army was again looking for a military pistol firing the same old .45 calibre round, although this has since been dropped.

The .45 ACP, also known as the .45 Automatic, has a rimless, straight case, usually made of brass, although it also may be encountered with aluminum (US-made) or lacquered steel (Russian-made) cases. The standard military bullet is of the jacketed type, weighting 230 grains (14.9g). Self-defence and police ammunition can be loaded with great variety of bullets, including frangible, expanding and other types of bullets. Most common police or defensive loadings use jacketed hollowpoint bullets of various makes, weighing from 185 to 230 grains (11.99 to 14.9 gram).

Appendix Four

Conversions Between the Metric and Imperial Systems

Please note that conversion factors and sample conversions below are rounded for more conventional use.

To Convert from the Metric System to the Imperial

Millimetres to inches	Multiply by 0.039; i.e. 20mm = 0.79in
Metres to feet	Multiply by 3.28; i.e. 50m = 164ft
Metres to yards	Multiply by 1.1; i.e. 100m = 110yd
Grams to grains	Multiply by 15.4; i.e. 10g = 154 grains
Grams to ounces	Multiply by 0.035; i.e. 700g = 24.7oz
Kilograms to pounds	Multiply by 2.2; i.e. 1kg = 2.2lb
Metres per second to feet per second	Multiply by 3.28; i.e. 500m/s = 1,640fps
Joules to foot/pounds	Multiply by 0.74; i.e. 500J = 368ft/lb

To Convert from the Imperial System to the Metric

Inches to millimetres	Multiply by 25.4; i.e. 7in = 177.8mm
Feet to metres	Multiply by 0.305; i.e. 50ft = 15.25m
Yards to metres	Multiply by 0.91; i.e. 100yd = 91m
Grains to grams	Multiply by 0.0648; i.e. 200 grains = 12.96g
Ounces to grams	Multiply by 28.4; i.e. 20oz = 568g
Pounds to kilograms	Multiply by 0.454; i.e. 3lb = 1.36kg
Feet per second to metres per second	Multiply by 0.305; i.e. 1,000fps = 305m/s
Foot/pounds to joules	Multiply by 1.36; i.e. 300ft/lbs = 408J

Appendix Five

The Elusive Stopping Power

Unfortunately, there is no single agreed definition of stopping power with regards to firearms ammunition. In different sources the term 'stopping power' may be used for some numerical characteristics calculated for specific rounds fired from specific weapons, or the same term may be used for some comparative purposes with no actual values. In general, any *power* has the *capability* to do something in a predefined amount of time. Referring to firearms ammunition, stopping power may be described as a capability to stop the live target (animal or human) in a predetermined amount of time. We therefore now have to define the desirable amount of time and, more importantly, what 'stop' actually means. The time part is relatively simple, as obviously the faster the better. Ideally, a bullet must have an effect on the target immediately (practically, within several tenths of a second from the impact), so that it will be incapable of *any* action after being hit with the bullet). This also more or less defines the 'stop' part – the target, ideally, must be incapable of any action once it has been hit. And the only sure way to do so is to kill the target, at least until someone invents the 'paralyzer' weapon, so well described in many sci-fi books.

In real life, however, we are unable to achieve this ideal. Even the 'best' bullets take some time to kill either an animal or a human being. What is worse, the bullet design and velocity is only a small part of the general equation that describes what happens when a bullet strikes a body. Two other major variables are the position of the strike (or, more precisely, the wound path within a particular body) and the actual properties, physical and psychological, of the target. The position of the strike is determined by many factors, of which the most important are accuracy of fire and the position of the target. In all available stopping-power theories the accuracy factor is eliminated, as only proper hits into the desired area of the target are counted. Other parts of the equation are far more complicated to deal with, as the same wound may produce quite different effects on various targets, even of the same class, be they game animals or *homo sapiens* of different fighting nations or on different sides of the law.

To provide at least some scientific basis to their theories, most authors either use only statistical analysis of actual shooting results, or substitute unknown variables with a more or less predictable alternative (for example, by replacing human flesh with ballistic gelatin or modeling clay), and then make assumptions by extrapolating the results to actual shooting situations. In either case, all calculations are suitable mostly for the purpose of comparison of some more or less imaginable properties and characteristics of specific loadings; however, the prediction value of most such values is questionable.

We will not discuss many of the theories that are available concerning stopping power, because in most cases modern combat ammunition that is actually adopted and issued to law-enforcement agencies produces results which closely accord with the theories, and military ammunition is limited to FMJ bullets only, of which most are of about the same effectiveness (small 'pocket pistol' rounds excluded). The only consideration that must be kept in mind at all times is that there is *no* magic bullet and *no* modern combat ammunition (with the possible exception of a direct hit with something like a 5in artillery shell) that can provide instant and certain incapacitation of the intended human target. Therefore, the most important factors in the effectiveness of any firearm are the accuracy of fire and the number of hits on the target. In most scenarios, it is advisable to keep firing at the target until it is visibly defeated. Furthermore, it is hardly useful to search for the most effective calibre/bullet combination, as in most cases a significant increase in effectiveness means an increase in the size, weight and recoil of the weapon. It is more useful to find an overall compromise between the calibre of the weapon, the number of rounds in the magazine and the handling characteristics, as even the most theoretically effective weapon will be useless in a fight if its owner is unable to bring it to the aim quickly enough and to then fire it accurately.

To sum it up in a few words, no 'man-stopping' bullet or groundbreaking calibre can substitute for training and practice, as well as for the availability of the weapon at the critical moment.

For those interested in a more thorough look at the problem of terminal effectiveness, the author recommends the works of authors such as Dr Martin Fackler or Evan Marshal and Ed Sanow, all from the USA. Furthermore, the author recommends studying several different theories before reaching any conclusions.

Appendix Six

Recommended Sources

Books

Edward C. Ezell, *Handguns of the World*, Stackpole Books, USA, 1973.
W.H.B. Smith and Joseph E. Smith, *Small Arms of the World*, 10th edition, A & W Visual library, USA, 1973.
Edward C. Ezell, *Small Arms Today*, Stackpole Books, USA, 1980.
George Nonte, Edward C. Ezell, *Combat Handguns*, Stackpole Books, USA, 1980.
G. Wollert, R Lidschun, W. Kopengagen, *Schutzenwaffen heute: illustriette Enzyclopadie der Schutzenwaffen aus aller Welt*, b. 1 und 2, Militarferlag der DDR, 1989.
Gene, Jr. Gangarosa, *Modern Beretta Firearms*, Stoeger Publishing Company, USA, 1994.
D. N.Bolotin, *Soviet Small Arms*, Polygon Press, Russia, 1995.
Gene, Jr Gangarosa, *FN Browning Armorer to the World*, Stoeger Publishing Company, USA, 1999.
Jan V. Hogg, *Small Arms, Pistols and Rifles*, Greenhill Books, UK, 2001.
Ian Hogg, John Weeks, *Military Small Arms of the 20th Century*, 7th edition, Krause publications, USA, 2002.
Leroy Thompson, *Combat Handguns*, Greenhill Books, UK, 2002.

Internet Resources

Modern Firearms website: www.world.guns.ru
V. Shilin gun club: www.club.guns.ru
Anthony Williams website: www.quarry.nildram.co.uk
The CRUFFLER: www.cruffler.com
Federation of American Scientists: www.fas.org
HKPro – unofficial Heckler & Koch resource: www.hkpro.com
ISAYERET.COM – Israeli Special Forces website: www.isayeret.com
North American Special Operations Group: www.nasog.net
China Defence: www.china-defence.com
The High Road internet forums: www.TheHighRoad.org

Internet Sites of Firearms Manufacturers

Beretta: www.beretta.com
Ceská Zbrojovka: www.czub.cz
Colt: www.colt.com
FEG Army: www.fegArmy.hu
FN Herstal: www.fnherstal.com
Heckler & Koch: www.Heckler-Koch.de
IMBEL: www.imbel.gov.br
Israel Military Industries: www.imi-israel.com
Izhevsk Mechanical Plant: www.baikalinc.ru
KB Priborostroenija: www.shipunov.com
Sturm, Ruger & Co: www.ruger.com
SIG Arms: www.sigarms.com
Springfield Armory: www.springfield-armoury.com
Smith & Wesson: www.Smith-Wesson.com
Tanfoglio: www.tanfoglio.it
Taurus: www.taurus.com.br
Tula Arms Factory: www.toz.vpk.ru
Zastava Arms: www.zastava-arms.co.yu

INDEX